Skeletal Anatomy

For Churchill Livingstone:

Editorial director: Mary Law
Project development editor: Dinah Thom
Project manager: Valerie Burgess
Project controller: Pat Miller
Copy editor: Mike Dean
Design direction: Judith Wright
Sales promotion executive: Maria O'Connor

Skeletal Anatomy

Glenda J. Bryan FCR SRR
District Superintendent Radiographer and Radiology Services Manager, Southmead Hospital, Bristol, UK

Foreword to the Third Edition by

E. Rhys Davies MA FRCR MRCP FFRRCSIre(Hon) FDSRCS
Emeritus Professor of Clinical Radiology,
University of Bristol, UK

Foreword to the Second Edition by

The late **Sir Howard Middlemiss** CMG MD
Former Emeritus Professor of Radiology and Past Dean of the Faculty of Medicine, University of Bristol, UK
Former President, the International Commission on Radiological Education

THIRD EDITION

EDINBURGH LONDON NEW YORK OXFORD PHILADELPHIA ST LOUIS SYDNEY TORONTO 1999

CHURCHILL LIVINGSTONE
An imprint of Elsevier Limited

First Edition 1961
Second Edition 1982
Third Edition 1996
Reprinted 1998, 2000, 2001, 2002, 2003, 2004, 2005, 2007, 2008 (twice)

ISBN: 978 0 443 05150 0

British Library Cataloguing in Publication Data
A catalogue record for this book is available from the British Library

Library of Congress Cataloguing in Publication Data
A catalogue record for this book is available from the Library of Congress

Printed and bound in the United Kingdom

Transferred to Digital Print 2010

Contents

Foreword to the Third Edition

Enthusiasm runs high in the expectation of a new edition from an author whose previous work has attracted uniform praise. The Third Edition of this volume is the second that Glenda Bryan has revised, and again she has done this thoroughly, while remaining true to the objectives of the First Edition.

It is arguable that much of the progress of modern medicine is derived from an understanding of biological events in individual living cells, and the application of these advances to holistic clinical methods still requires a fundamental knowledge of physiology and anatomy, particularly that of the skeleton. Thus the approach adopted in this book remains valid for the understanding of skeletal anatomy, which is so vital to radiographers, nurses, physiotherapists, and also occupational therapists and medical students seeking a straightforward, easily read introduction to the subject.

Glenda Bryan has again achieved a clear simple text which will appeal, not least, to those who do not have English as their first language, and indeed makes it likely that this volume will join her other work in being translated into many languages. These benefits are enhanced by the new layout and style, which include simple line diagrams of joints and muscle attachments, as well as radiographs showing the full range of possibilities to the most modern CT and MR.

The timelessness and the unsparing skills of a highly popular author in communicating technical details so effectively are a combination that is bound to appeal to students of medicine and radiology as well as to radiographers.

It is a privilege to be associated with this volume and I can only reiterate the commendation of my predecessor without hesitation.

1996 E. R. D.

Foreword to the Second Edition

A knowledge and understanding of skeletal anatomy are of vital importance to every radiographer and radiography student. The publishers were wise in inviting Miss Glenda Bryan to undertake the revision of this well established book for its new edition. Miss Bryan has already shown in her *Diagnostic Radiography*—now in its third edition and acknowledged throughout the English-speaking world as the most useful, concise and practical handbook on radiography available—that she possesses an ability to present and communicate detailed technical data in an acceptable and palatable, succinct form.

This present book has been extensively revised and in large part rewritten. The sequence in presentation of anatomical facts now comes in a logical order. Great attention has been paid to layout and the student in particular will find it an easy book from which to learn. However, it is not only the student who will benefit from it; in everyday life the practising radiographer will undoubtedly wish to refer to it. The trainee radiologist, too, who must become familiar with all normal anatomical landmarks and must learn how to demonstrate particular anatomical features, will find this book of inestimable value.

The illustrations, many of them from the first edition but now supplemented by well chosen new examples of standard views and special projections, and by many excellent line drawings, are of a high quality. The text is clear and uncluttered.

All in all, this is an excellent learning and reference book which should be in every School of Radiography library and every departmental library, and will undoubtedly be required by many student radiographers and trainee radiologists.

I commend it without hesitation.

Bristol, 1982 J. H. M.

Preface to the Third Edition

In preparing a new edition of *Skeletal Anatomy* my aim has been to provide a comprehensive yet simple textbook which, it is hoped, will fulfil the requirements of students of many disciplines needing to learn and understand skeletal anatomy. The title has been changed from the previous edition and the word 'Radiographic' omitted in recognition of the fact that it will be of value not only to those studying radiography but also to other groups of students.

The layout has been redesigned with the object of facilitating learning and revision. The index has been made as user-friendly as possible, with cross-references, to save the reader time in searching.

I gratefully acknowledge the help kindly extended by colleagues in many parts of the UK in providing possible illustrations, particularly CT scans and MRI. I am particularly grateful to Joanne Waring and John Dickinson, of G.E. Medical Systems, who contacted several hospitals for me and who provided me with many examples for possible use. I am also very grateful to Pamela Kimber, of Wessex Neurological Centre, Southampton, for providing me with Figures 2.19, 2.26, 3.17, 3.45, 6.18 and 7.30; to Hillingdon Hospital for Figures 2.47 and 5.11; to Neath Hospital for Figures 3.29 and 3.35; to Bournemouth Nuffield Hospital for Figure 8.19; to Merthyr Tydfil Hospital for Figures 3.23 and 7.9; to Frenchay Hospital, Bristol, for Figure 2.50; and to Bristol MR Centre for Figures 6.42, 8.20 and 8.23.

A number of other illustrations are from the First Edition of this work by Johnson and Kennedy (under the title *Radiographic Anatomy of the Human Skeleton*), of which some were adapted from illustrations in *Positioning in Radiography* by K. C. Clarke MBE HonFSR.

I would like to express my thanks also to Dr John Haworth, Clinical Director of Radiology, to Dr Susan Armstrong and Dr David Glew, Consultant Radiologists, all of Southmead Hospital, Bristol, for their advice, and to Nicholas Bowyer of the Department of Medical Illustration, also at Southmead, who made the prints for the new illustrations.

I am deeply grateful to Professor Rhys Davies for his advice and encouragement and for writing the Foreword to this edition.

Finally I should like to express my appreciation to the staff of Churchill Livingstone for their advice and guidance in the production of this edition.

Bristol 1996 G. J. B.

EXTRACT FROM THE PREFACE TO THE SECOND EDITION

For this edition, my sincere thanks are due to many colleagues for their advice and interest. In particular I am deeply grateful to Dr John Craig, Senior Registrar in Radiodiagnosis, Bristol Royal Infirmary, for all his help, particularly with the new sections on joints. I would like also to thank Professor Sir Howard Middlemiss for his encouragement and advice and for writing the Foreword, and Professor R. E. Coupland, Department of Human Morphology, University of Nottingham, for his helpful advice and comments. I am grateful also to Dr Jonathan Musgrave, Department of Anatomy, University of Bristol, Professor E. Rhys Davies, Dr F. G. M. Ross and Dr John Roylance,

Department of Radiodiagnosis, Bristol Royal Infirmary. My sincere thanks go also to Mr Edwin Turnbull who drew all the new diagrams and to Mr J. Hancock and to the Department of Medical Illustration, Bristol Royal Infirmary, who made the prints for the new radiographic illustrations. The new illustrations in the ossification series came from the museum of the Department of Radiodiagnosis, The Royal Hospital for Sick Children, Bristol.

Finally, I should like to express my appreciation to Mr Martin Davies for his help and encouragement, to the staff of my Department for their loyal co-operation and support during the writing of this book and to the staff of Churchill Livingstone for their professional expertise.

Bristol 1982 G. J. B.

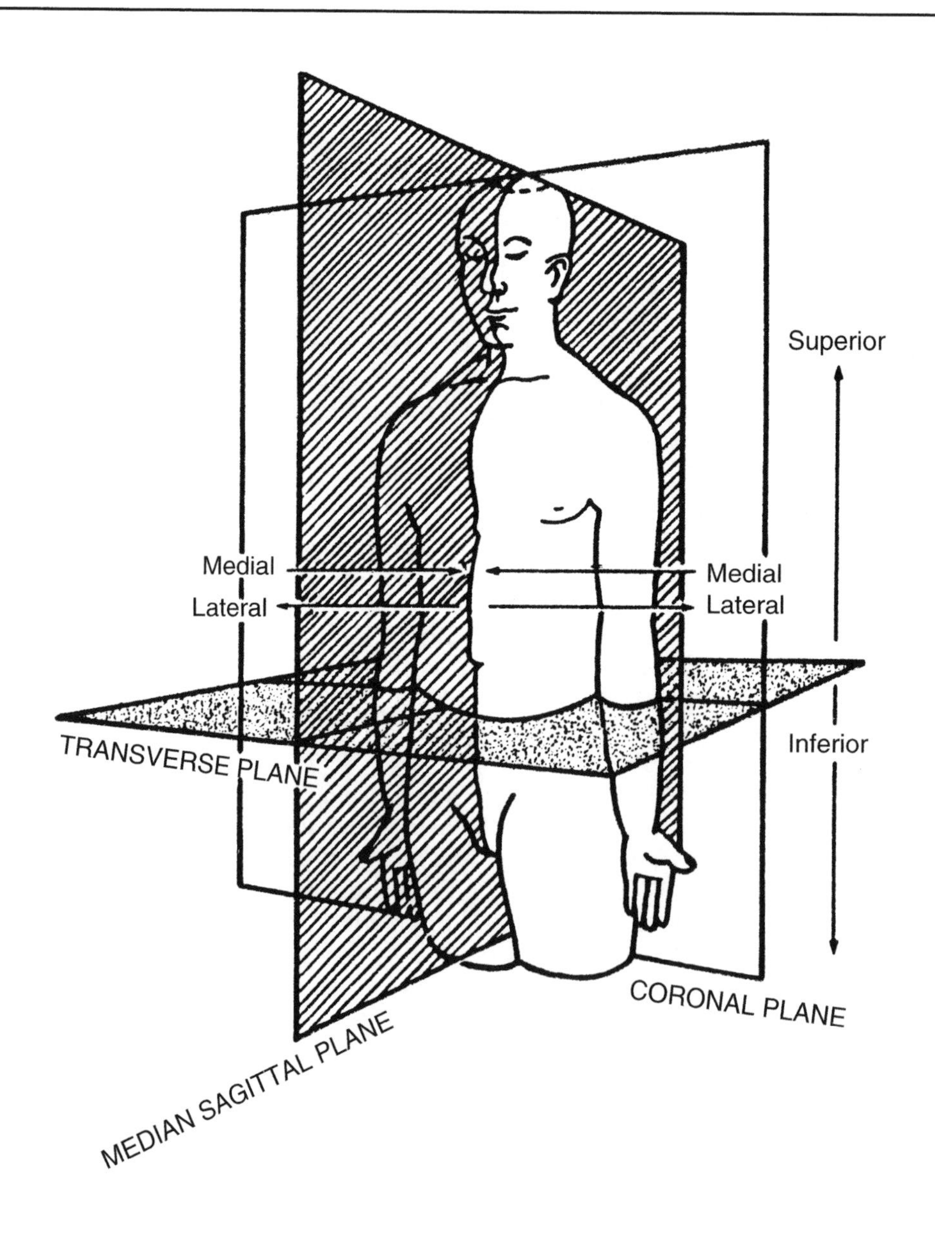

Glossary of Terms

Anatomical position
For the purposes of description, the human body is assumed to be in the upright position, the head facing forwards, the arms by the sides with the palms of the hands facing forwards and the feet together.

Median plane: a vertical plane passing through the centre of the body dividing it into a right and a left half.
Sagittal plane: any plane passing through the body parallel with, and including, the median plane.
Coronal plane: a plane passing through the body from side to side at right angles to the sagittal plane.
Transverse plane: any plane at right angles to the long axis of the body.

Medial: nearer the midline.
Lateral: farther from the midline.
Posterior (or dorsal): nearer the back of the body.
Anterior (or ventral): nearer the front of the body.
Superior: nearer the head.
Inferior: farther from the head.
Internal: inside.
External: outside.
Proximal: nearer to the axial skeleton.
Distal: further from the axial skeleton.

Origin: (used in describing attachment of a muscle): the end of a muscle that moves least, usually the distal end. (*Indicated in diagrams by* ▨ *areas.*)
Insertion: (used in describing attachment of a muscle): the end of the muscle that moves most, usually the distal end. (*Indicated in diagrams by* ☐ *areas.*)

Ala (pl. alae): a wing-like projection.
Border: a ridge or crest of bone separating two surfaces.
Canal: a tunnel.
Condyle: a rounded, usually articular, elevation.
Crest: a ridge.
Diaphysis: the part of a long bone formed from the primary centre of ossification, i.e. the shaft.
Epicondyle: an elevation or projection, usually adjacent to a condyle.
Epiphyseal plate: the layer of cartilage between the diaphysis and epiphysis.
Epiphysis: a secondary centre of ossification.
Facet: a small, smooth articular surface.
Fissure: a narrow space or cleft in, or between, structures.
Foramen (pl. foramina): a hole.
Fossa: a hollow or depression in a bone.
Hiatus: a gap.
Lamina: a thin plate of bone.
Meatus (pl. meatus or meatuses): a narrow passage.
Metaphysis: the end of the actively growing diaphysis of a long bone beneath the epiphyseal plate.
Septum: a partition dividing two cavities.
Sinus: a hollow or cavity.
Spine: a sharp spike of bone or an elongated projection.
Squama: a thin, flat plate.
Sulcus: an elongated groove.
Trochlea: a pulley-shaped articular surface.
Tubercle: a circumscribed bony projection.

Tuberosity
Trochanter } roughened bony projections for the attachment of ligaments or muscles.

Axial skeleton: the bones of the skull, vertebral column, ribs and sternum.

Appendicular skeleton: the bones of the shoulder girdle and upper limbs and of the pelvic girdle and lower limbs.

Flexion: bending, usually forward movement but can be backward, e.g. in the knee, or can be sideways, e.g. the vertebral column.

Extension: straightening, reverse of flexion.

Abduction: moving away from the midline of the body or agreed reference line of the hand or foot.

Adduction: moving towards the midline of the body or agreed reference line of the hand or foot.

Pronation: movement involving turning the palm of the hand posteriorly, rotating the sole of the foot outwards with flattening of longitudinal arch (= abduction and eversion together).

Supination: opposite of pronation; used by right hand in screwing in a screw with a right handed thread.

Circumduction: a circular movement involving a combination of flexion, extension, abduction and adduction.

Eversion: turning the sole of the foot away from the median plane.

Inversion: turning the sole of the foot towards the median plane.

1. Bones and joints

The skeleton forms the bony framework for the body, supporting the soft tissues and providing the necessary rigidity to the body as a whole. It supports the weight of the body and forms a number of levers which, when acted upon by muscles, bring about body movement. It also helps to form the walls which enclose and protect important structures such as the heart, the lungs and the pleura.

BONE

Bone is the hardest form of connective tissue in the body. It is made up of:

- 25% water
- 30% organic fibrous tissue
- 45% inorganic salts (calcium and phosphates).

Bone is a living structure and even when it is mature it is constantly being broken down and renewed throughout life, and its external and internal form changes in response to altered stresses and strains.

During the period of normal growth (up to about 18 years) bone synthesis and growth exceed reabsorption. In adult life the two processes are balanced and in old age reabsorption exceeds bone formation.

Bone disease is often reflected by the failure of bone synthesis and reabsorption to maintain a balance.

STRUCTURE OF BONE

There are two types of bone (Fig. 1.1):

- compact
- cancellous (spongy)

Compact bone

This is a dense, ivory-like substance. A section of compact bone, as seen microscopically, consists of units (**Haversian systems**, Fig. 1.2).

Each system comprises a central Haversian canal surrounded by concentric rings, known as **lamellae** (hence the term lamellar bone) which are composed of fibrous tissue impregnated with complex calcium salts. Small cavities (**lacunae**) lie between the lamellae and are interconnected by fine channels (**canaliculi**). The lacunae contain bone cells (**osteocytes**), which have been trapped in the lacunae by the formation of bone around their cells. Blood vessels and nerve fibres are present in Haversian canals and nutrient fluids pass out from them through the canaliculi to all parts of the system. The spaces between adjacent Haversian systems are occupied by irregular lamellae (**interstitial lamellae**). The circumference of compact bone is covered by **circumferential lamellae**.

Volkmann's canals, which are oblique or transverse channels running from the bone surface to Haversian systems, interconnect adjacent Haversian systems and by transmitting blood vessels help to maintain cell nutrition. The nutrition. The Haversian systems run approximately parallel with the surface of the bone and they branch and communicate with each other.

Because of the strength of compact bone, its presence in the cortex of the diaphysis of a long bone prevents the bone collapsing when stresses are applied.

Compact bone covers cancellous bone in all situations.

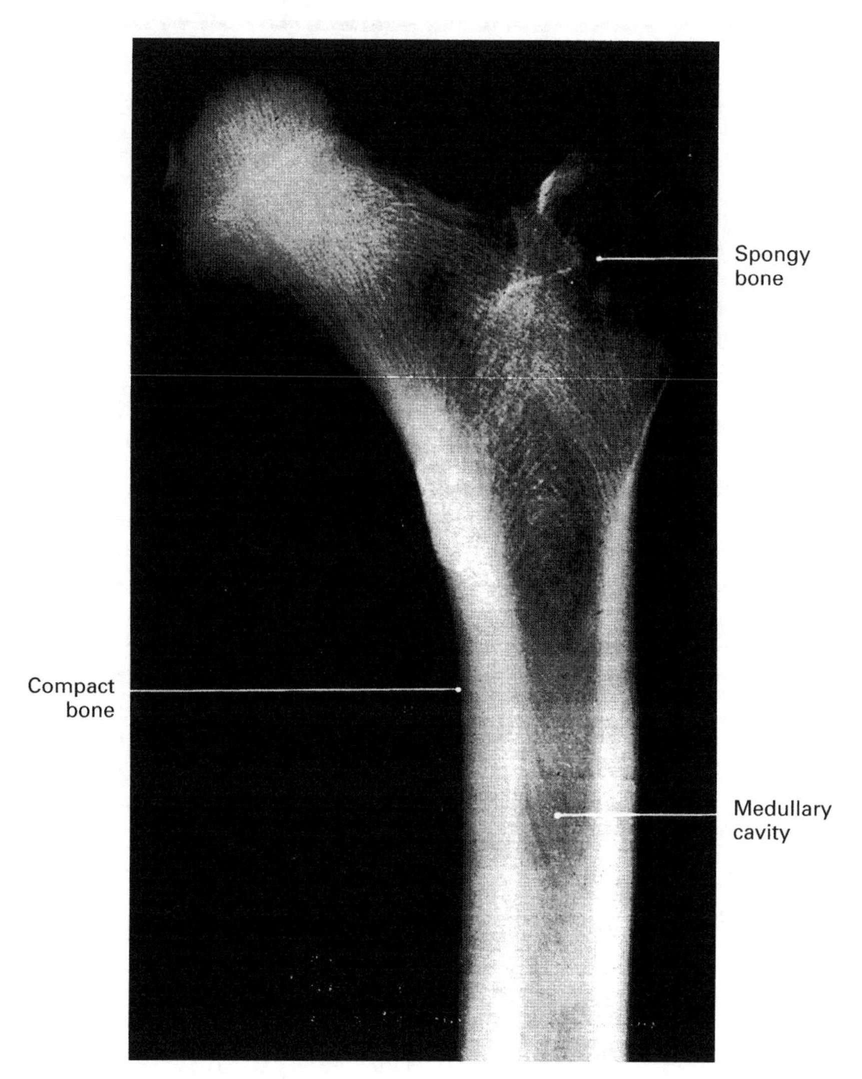

Fig. 1.1 Radiograph of the upper end of femur

Cancellous (spongy) bone

Cancellous bone has a spongy texture and is composed of spikules of bone making up a meshwork of trabeculae between which are interconnecting spaces.

In the axial skeleton, cancellous bone contains red bone marrow. In the limbs of the adult, cancellous bone normally contains yellow bone marrow. The irregular, spongy nature of cancellous bone means that it is lighter and less strong than compact bone but it provides a large internal surface area for the blood-forming cells to occupy. Where red bone marrow exists the medullary spaces continue to produce blood cells throughout life whereas at other sites yellow bone marrow is haemopoietically dormant. These dormant areas retain the ability to produce blood cells when appropriately stimulated, e.g. as of consequence of anaemia.

Periosteum

With the exception of articular surfaces, the external surface of bone is covered with a thin vascular membrane called periosteum which adheres closely to the bone.

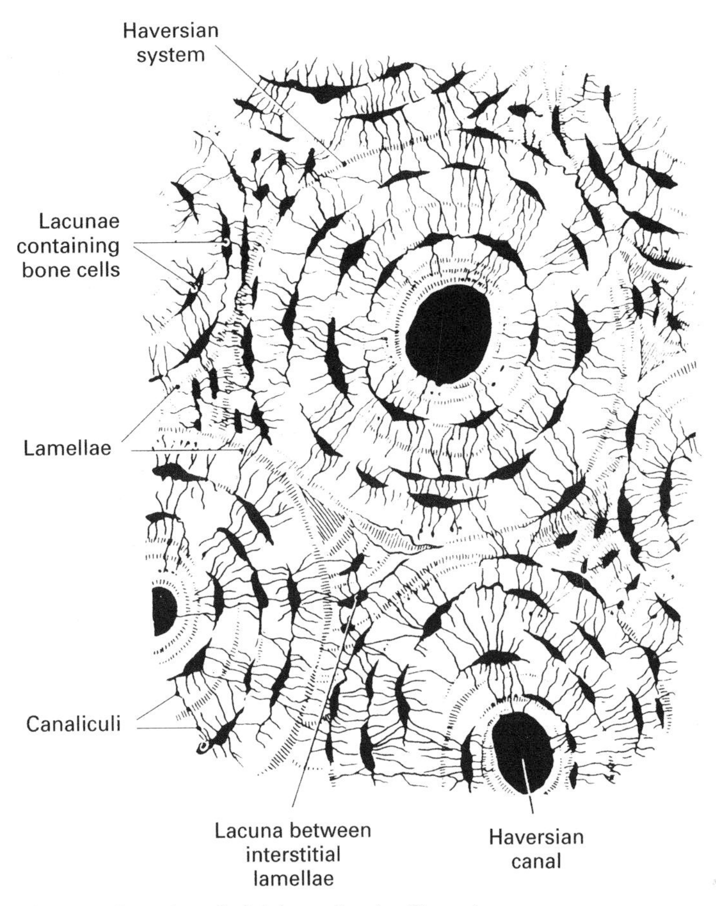

Fig. 1.2 Secretion of adult bone showing Haversian systems

Periosteum consists of two layers:

- an outer fibrous layer containing collagen fibres
- an inner vascular layer containing loose connective tissue and osteocytes.

Functions of periosteum

- forms a protective outer covering to the bone
- provides attachment for tendons, muscles and ligaments
- contains blood vessels which provide a substantial proportion of the blood supply to the bone
- forms new bone tissue (by means of osteoblasts found in the inner layer of periosteum).

BONE MARROW

Bone marrow is a soft pulpy substance, composed of connective tissue and cells. It is found in the marrow cavities of long bones and in the spaces between trabeculae of all bones.

There are two kinds:

- red bone marrow
- yellow bone marrow.

Red bone marrow is the centre for haemopoiesis (production of blood cells). It is responsible for the formation of red blood cells, white blood cells and blood platelets which, when mature, enter the blood circulation. (It should be noted that haemopoiesis occurs also in the liver and spleen of the foetus).

At birth, red bone marrow occupies the medullary cavities of all long bones and the cancellous bone of all bones. By about 7 years of age, yellow bone marrow appears in the centre of the medullary cavity and in some bones gradually replaces the red bone marrow. However, red bone marrow is present in the sternum, pelvis, ribs and vertebrae throughout life.

Yellow bone marrow is contained in the medullary cavity of all long bones and in the cancellous bone of long bones except the heads of the femora and of the humeri.

Yellow bone marrow is composed mainly of cells containing fat which accounts for its colour.

OSSIFICATION (OSTEOGENESIS)

Ossification is the process of bone formation. It can take place either within hyaline cartilage precursor (intracartilaginous or endochondral ossification) or within membrane (intramembranous ossification). Most bones ossify within cartilage but a few, e.g. the vault of the skull, the facial bones and the shaft of the clavicles, ossify within membrane. The process is almost identical in each case.

Ossification involves the laying down of complexes of calcium and phosphate salts (calcification) within a matrix (osteoid) which is secreted to replace either cartilage or membrane. The normal synthesis of the matrix and its calcification with the formation of calcium phosphate complexes are essential parts of the process of ossification. These changes necessitate adequate general nutrition with a normal intake and utilization of vitamins (especially vitamins C and D) and a competent endocrine system.

Initially, the perichondrium around a cartilage shaft forms bone from osteoblasts in the deepest layer, the perichondrium then being termed periosteum. This process is followed by the appearance of a bone-forming centre in the middle of the cartilage shaft. This is called the **primary ossification centre**. These centres appear during intrauterine life, or soon after birth, at various times which are fairly constant for a particular bone (Fig. 1.3). In the case of a long bone, the primary centre is in the middle of the shaft and from it the diaphysis is formed.

Secondary ossification centres (epiphyses) appear, some just before birth and the remainder during childhood, in an ordered sequence, continuing until late adolescence. In a long bone, the epiphyses are usually at each end. Bone formation in the epiphysis keeps pace, relatively, with the increase in bone length. The epiphysis is always separated from the diaphysis by the **epiphyseal cartilage (epiphyseal plate)**.

The area of diaphyseal bone immediately adjacent to the epiphyseal plate is the metaphysis and is the site of advancing diaphyseal bone formation. It is responsible for increase in bone length. It should be noted that the term metaphysis is sometimes inaccurately applied to the expanded end of a long bone whether it is growing of fused, rather than to the site of advancing bone formation during ossification.

When full growth has been achieved, the epiphyseal plate is invaded from both the epiphyseal and diaphyseal sides and it ossifies so that the epiphysis fuses with the diaphysis. In the case of long bones this occurs during late adolescence and the early twenties. The time of the appearance of the centres of ossification and the fusion of the epiphyses and diaphysis varies considerably from one person to another and the ages given for any bone represent an average figure. In general the largest epiphyses begin to ossify first and are the last to fuse with the diaphysis so that one end of a bone contributes more to growth in length than does the other.

The periosteum, in growing bone, sends fibres (Sharpey's fibres) deep into the bone it has formed and further still into the cartilage of the diaphysis. Passing along these fibres are small blood vessels and also **osteoclasts** (bone/cartilage eroding cells) and **osteoblasts** (bone forming cells). The osteoclasts erode the cartilage, thus allowing passage of osteoblasts which lay down bone in these channels. The channels unite with each other to form medullary spaces in an irregular network. The wafers of bone making up the margins of these spaces are the trabeculae which form spongy bone.

The expanding medullary spongy bone brings about the erosion, from within, of the diaphyseal compact bone which proceeds to lay down new compact bone under the periosteum. In this way the diaphysis increases in width. Increase in length of a bone occurs at the metaphysis where a sequence of recognized changes in the cartilage cells and the matrix takes place. The formation of bone in the epiphysis itself is an identical process

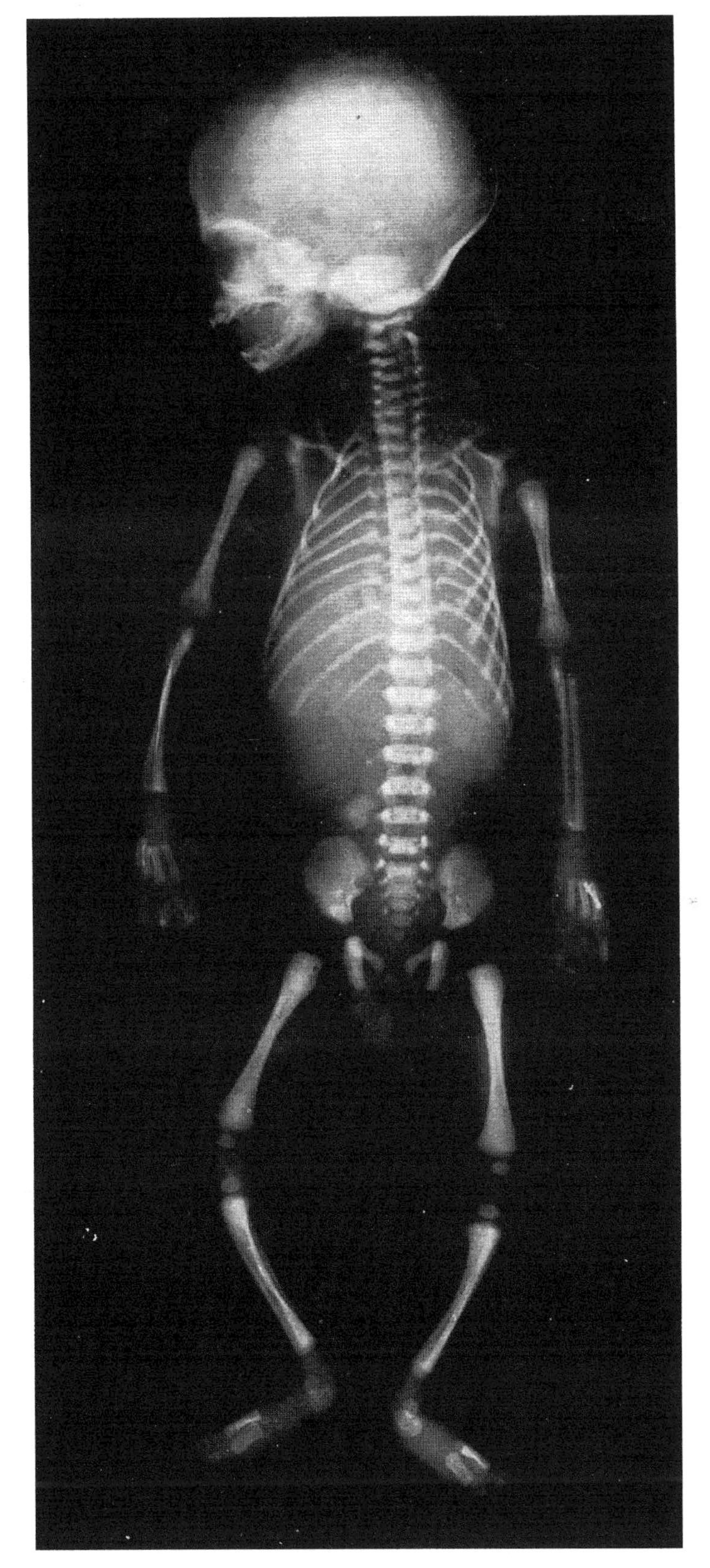

Fig. 1.3 Ossification of the skeleton at birth

except that it occurs in a concentric manner, always being separated from the diaphysis by the epiphyseal plate until maturity is reached.

The growing bone is stressed by movement and also by ligament and muscle attachments which cause characteristic alterations in shape, giving rise to tubercles and depressions at their origins and insertions. Once the trabeculae ossify they reflect the direction and magnitude of the stresses within the bone by their thickness and alignment. The overall shape of the bone is maintained while it increases in length by the process of remodelling whereby small areas are absorbed and the bone is redistributed by new bone formation elsewhere in the bone.

CLASSIFICATION OF BONES

Bones are divided into types according to their shape:

- long
- short
- flat
- irregular
- sesamoid.

Long bones consist of a shaft and two expanded ends or extremities. The shaft is a cylinder of compact bone with a central medullary cavity. In the adult the cavity contains yellow bone marrow. The ends of the bone consist of cancellous bone covered with a thin layer of compact bone. They provide a large surface for articulation and muscle attachment.

Examples: humerus, femur.

Short bones consist of cancellous bone covered with a thin layer of compact bone. They are found at sites where strength but limited movement are required.

Examples: carpal bones, tarsal bones.

Flat bones consist of a thin later of compact bone enclosing a later of cancellous bone. They are found where protection of underlying organs or the need for space for muscle attachment is the greatest need.

Examples: vault of skull, scapula.

Irregular bones consist of cancellous bone surrounded by a layer of compact bone. They vary in size and shape.

Examples: vertebrae, facial bones.

Sesamoid bones are nodules of bone which ossify within a tendon at a point of friction. They protect muscle tendons from wear and provide a channel for their movement as they glide over bony surfaces.

Examples: sesamoid bones of thumb and great toe.

JOINTS

A joint is the junction between two or more bones, whether or not movement occurs between them. There are three main types of joint:

- fibrous
- cartilaginous
- synovial.

FIBROUS

In these joints the bones are joined together by a fibrous ligament or membrane and no movement occurs between the bones.

There are three main types of fibrous joints:

- **Suture:** this type is found between the bones of the skull. A thin ligament (sutural ligament) separates the jigsaw-like bones.
- **Gomphosis:** peg-and-socket joint. This type is found between the teeth and jaws.
- **Syndesmosis:** the bones are joined together by a fibrous inter-osseous membrane. A small amount of movement is possible. This type is found between the tibial shaft and the fibular shaft at the inferior tibiofibular joint.

CARTILAGINOUS

In these joints the opposing bones are joined by a layer of cartilage and are bound together by ligaments. In general, no movement occurs. There are two main types of cartilaginous joints:

- **Primary:** synchondrosis. The two bones are joined by hyaline cartilage which is later replaced by bone. Examples are the joints between the diaphysis and epiphysis of a growing bone (where no movement occurs and the periosteum is continuous between the two bones), the joints between the sternum and the first ribs (where slight movement can take place during respiration) and the joints between the sternal segments.

- **Secondary:** symphysis or amphiarthrosis. The ends of the articular surfaces are covered by hyaline cartilage and joined together by a fibrocartilaginous disc. This type of joint occurs in the midline of the body, e.g. the symphysis pubis and the joints between the vertebral bodies.

SYNOVIAL—diarthroses

In these joints the ends of the bones are covered with a thin layer of hyaline articular cartilage and are separated by a joint cavity. The whole joint is surrounded by a fibrous capsular ligament which is thickened in the lines of stress to form the ligaments of the joints. The capsule is lined with synovial membrane which secretes synovial fluid for the lubrication of the joint. There are several different types of synovial joint:

- **Ball and socket** (spheroidal). These are multiaxial, e.g. hip, shoulder
  Movements: abduction, adduction, extension, flexion, rotation and circumduction
- **Hinge** (ginglyni). Uniaxial, e.g. elbow, interphalangeal joints
  Movements: flexion, extension
- **Gliding** (plane) e.g. intercarpal, intermetatarsal
  Movements: gliding one on another
- **Pivot** (trochoid). Uniaxial, e.g. superior radioulnar joint
  Movements: rolling, e.g. of radius and hand around the fixed ulna
- **Saddle** (sellar). Biaxial, e.g. 1st carpometacarpal joint
  Movements: flexion, extension, abduction, adduction, circumduction
- **Condylar.** Mainly uniaxial, e.g. temporomandibular joint
  Movements: extension, plus a limited amount of rotation
- **Ellipsoid.** Biaxial, e.g. wrist joint.
  Movements: flexion, extension, abduction, adduction.

STABILIZING FACTORS OF JOINTS

- Shape of the articular surfaces of the opposing bones
- Strength of the capsular and intracapsular ligaments
- Strength of the muscles surrounding the joint
- Strength provided by rings of cartilage (labra)
- Presence of fibrocartilaginous discs.

2. The skull

The skull is the skeleton of the head (Figs 2.1 to 2.3). It consists of 22 bones locked together at irregular fibrous joints called sutures. The only freely movable bone is the mandible. The bones of the skull can be divided into two groups:

- the cranium
- the facial bones.

The cranium. The cranium forms the upper part of the skull and it surrounds and protects the brain. It consists of eight bones in the adult:

- 1 frontal
- 2 parietal
- 2 temporal
- 1 sphenoid
- 1 ethmoid.

It should be noted that at birth the frontal bone consists of two parts separated by the frontal suture. Usually the two halves fuse by about eight years but occasionally they fail to do so.

The facial bones. The facial bones are situated below and anterior to the cranium. The principal bones are those of the upper and lower jaws. The facial bones enclose the cavity of the

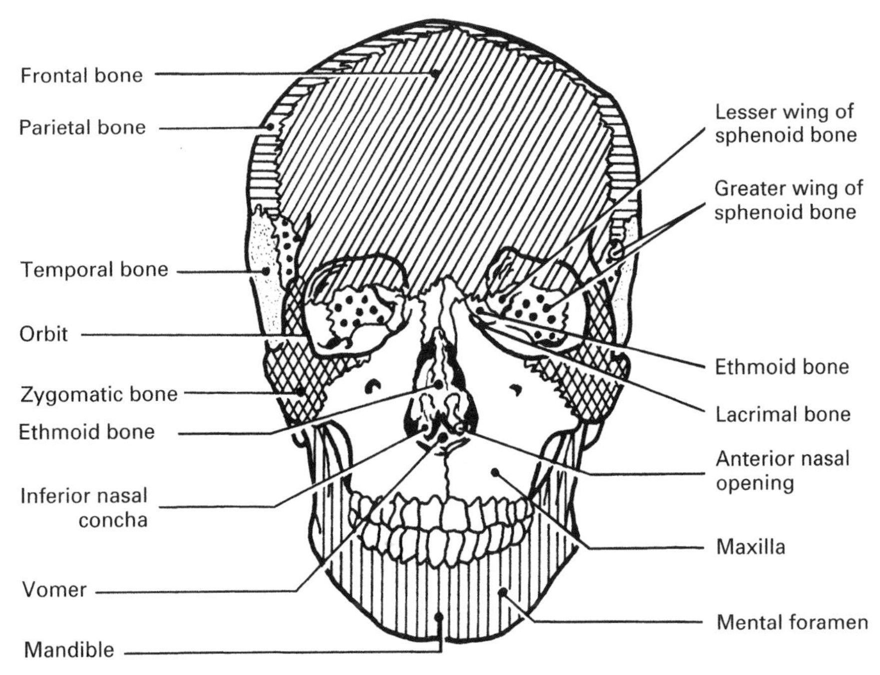

Fig. 2.1 Diagram of the skull: anterior aspect

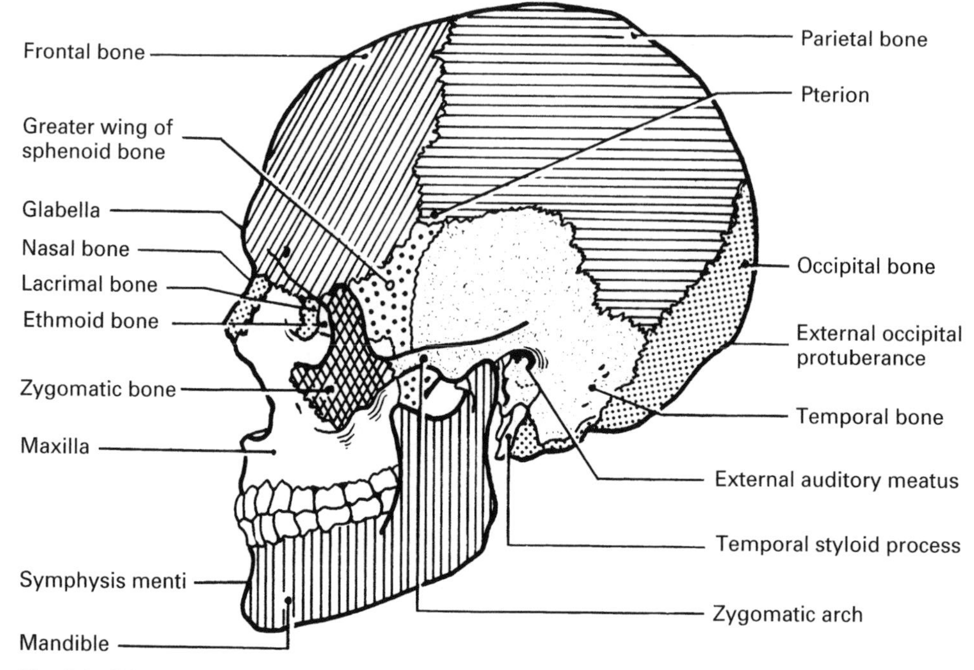

Fig. 2.2 Diagram of the skull: lateral aspect

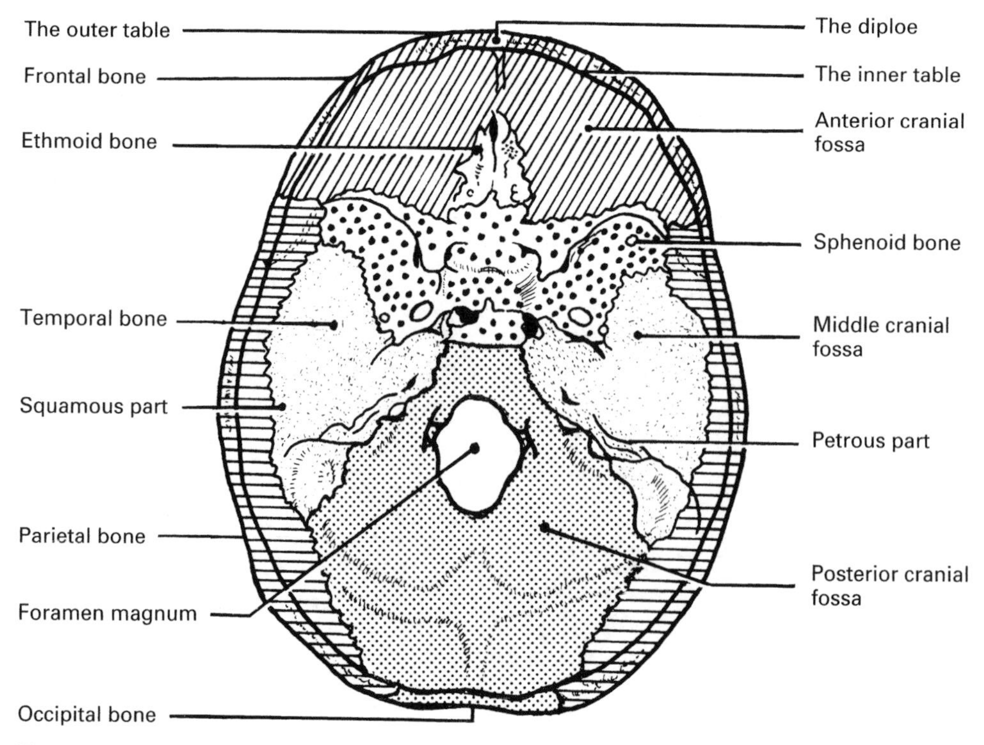

Fig. 2.3 Diagram of floor of cranial cavity

mouth and between them and the lower surface of the cranium are the orbits and the nasal cavity. There are 14 facial bones:

- 2 maxillae
- 2 zygomatic
- 2 nasal
- 2 lacrimal
- 2 palatine
- 2 inferior turbinates (conchae)
- 1 vomer
- 1 mandible.

THE SKULL AS A WHOLE

FRONTAL ASPECT (Fig. 2.4)

As seen from the front (Fig. 2.4), the skull appears slightly wider above than below. The frontal bone has a convex anterior surface and it forms the dome of the forehead and the upper margins of the orbits. The zygomatic bone forms the prominence of the cheek and completes the lower margin of the orbit. The two maxillae together form the upper jaw, the remainder of the lower and medial margins of the orbits, and enclose the

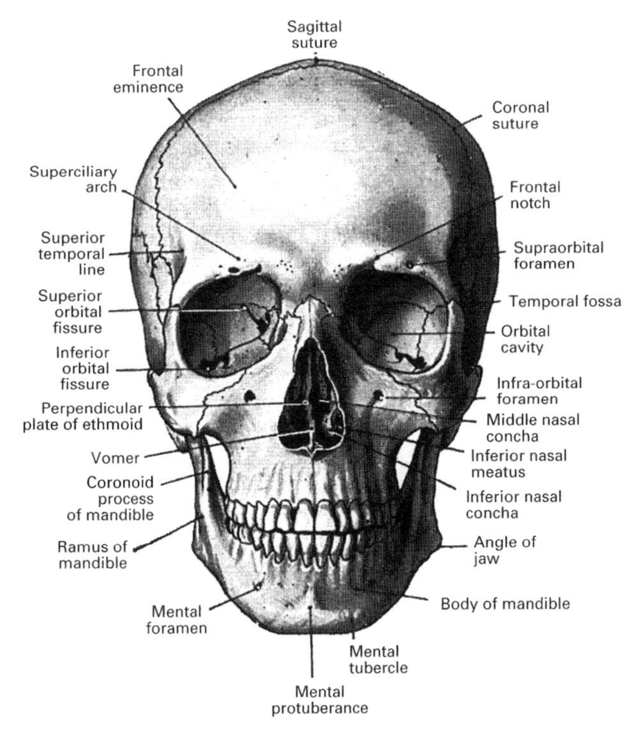

Fig. 2.4 Skull: frontal aspect

anterior nasal opening. A number of smaller bones take part also in the formation of the anterior nasal opening and of the orbits (Fig. 2.23 and Fig. 2.33). The supercilliary arch is the rounded ridge above the inner part of each supraorbital margin. The region above the bridge of the nose and between these arches is the glabella. A frontal sinus lies on each side of the frontal bone behind the supercilliary arch and communicates with the nasal cavity. The frontal sinuses are variable in size and may extend posteriorly into the roof of the orbit. The supraorbital margin ends laterally in a prominent downward projection—the zygomatic process of the frontal bone—which articulates with the zygomatic bone.

The nasal portion of the frontal bone projects downwards between the orbits and articulates with the nasal bones at the nasion and with the frontal processes of the maxillae.

LATERAL ASPECT (Fig. 2.5)

The lateral aspect of the skull may be divided into two parts by an imaginary line drawn through the supraorbital margin and the external auditory meatus. This line represents the general plane of the base of the cranium and is inclined at a slight angle to the horizontal. On and above this line lies the cranium and below and in front of it are the facial bones.

Two important sutures cross the vault of the cranium. The **coronal suture** separates the frontal from the parietal bones and the **lambdoid suture** separates the parietal bones from the occipital bone. The coronal and lambdoid sutures are connected in the midline by the sagittal suture which therefore lies between the two parietal bones. The meeting point of the coronal and sagittal sutures is called the **bregma** and is occupied in infancy by a diamond-shaped membranous defect called the anterior fontanelle. This normally closes by the end of the second year. The junction of the sagittal and lambdoid sutures is called the **lambda**. A small gap—the posterior fontanelle—is present here at birth but closes soon after. The junction of the parietal, frontal, greater wing of sphenoid and squamous temporal bones form the H-shaped **pterion,** medial to which lies the anterior branch of the middle meningeal artery.

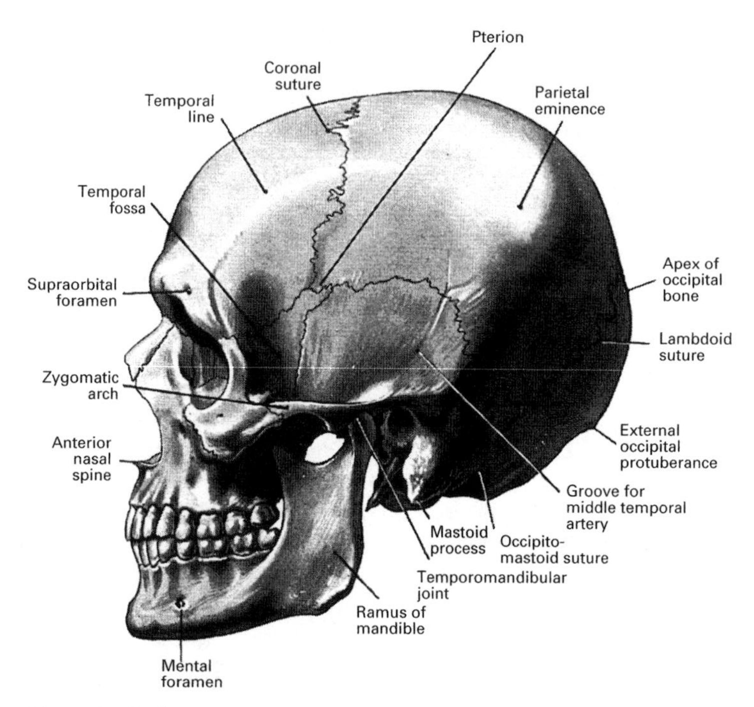

Fig. 2.5 Skull: lateral aspect

The main features on the lateral aspect of the skull are:

The zygomatic arch, a thin bony arch formed anteriorly by the zygomatic process (zygoma) of the temporal bone. On the under surface of the arch, in front of the mandibular fossa, is an eminence—the articular tubercle. The posterior part of the arch extends above the external auditory meatus and becomes continuous with the supramastoid crest.

The external auditory meatus consists of a short, outer, cartilaginous part and a longer, inner, bony part. Only the bony part is visible in the dry skull. It is directed anteriorly and medially and ends at the tympanic membrane which separates the external ear from the middle ear or tympanic cavity. The roof and upper half of the posterior wall of the meatus are formed by the squamous part of the temporal bone. The remainder of the circumference of the wall is formed by the tympanic plate of the temporal bone.

The temporal lines—superior and inferior—commence at the zygomatic process of the frontal bone and arch upwards and backwards across the parietal bone. The temporal fossa is the area bounded above by the temporal lines and below by the zygomatic arch. The floor of the fossa is formed by the frontal bone and greater wing of the sphenoid anteriorly, and by the parietal bone and squamous part of the temporal bone posteriorly. The temporal fossa gives attachment to the temporalis muscle—one of the principal muscles of mastication—which is inserted into the coronoid process of the mandible. The infratemporal fossa is the area below the zygomatic arch, bounded anteriorly by the posterior surface of the maxilla. The two fossae communicate via the wide gap that separates the zygomatic arch from the lateral wall of the cranium.

The mastoid process is the triangular projection behind the external auditory meatus. It gives attachment to some of the muscles of the neck and it contains the mastoid air cells.

THE UNDER SURFACE OF THE SKULL (Fig. 2.6)

The under surface of the skull is very irregular and presents a large number of foramina (p. 52)

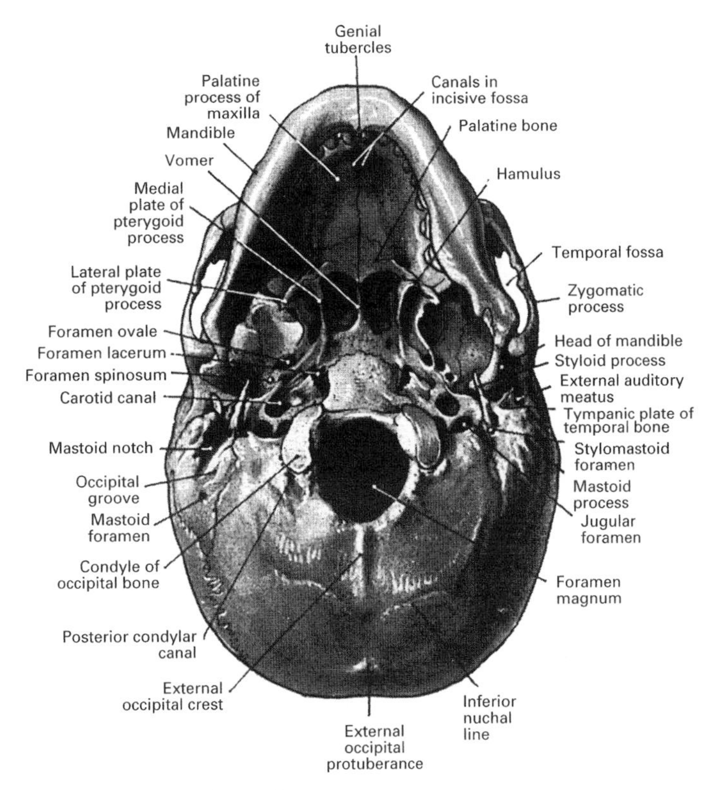

Fig. 2.6 Skull: under surface

and bony projections. As there are no natural boundary lines it is convenient to divide the whole area into three parts:

- anterior
- middle
- posterior.

by two transverse lines, one running through the posterior margin of the hard palate and the other running through the anterior margin of the foramen magnum.

The anterior part is formed by the hard palate which constitutes the roof of the mouth and the floor of the nasal cavities. It is bounded anteriorly and laterally by the alveolar processes of the maxillae. The anterior two-thirds is formed by the palatine processes of the maxillae. The posterior third is formed by the horizontal plates of the palatine bones. Midline and transverse sutures can therefore be seen on the hard palate. At the anterior end of the midline suture behind the incisor teeth is a small fossa, the incisive fossa, into which the lateral incisive canals enter to transmit the nasopalatine nerves from the nasal cavities to the mouth. The posterior margin of the hard palate is thin and sharp and gives origin to the soft palate. In the midline there is a small projection, directed backwards, called the posterior nasal spine.

The alveolar processes of the maxillae form the thick, horseshoe-shaped ridges on the anterior and lateral margins of the hard palate. They provide the sockets for the roots of the teeth of the upper jaw. The posterior end of the alveolar process is rounded and prominent and is known as the maxillary tuberosity.

The middle part extends from the posterior margin of the hard palate to the anterior margin of the foramen magnum and it includes the sphenoid bone anteriorly, the basi-occiput posteriorly and the temporal bones laterally.

The posterior nasal openings (choanae) lie between the posterior margins of the hard palate and the base of the cranium. They are separated in the midline by the vomer, a thin flat bone which forms the posterior margin of the nasal septum.

The pterygoid process of the sphenoid bone projects downwards from the base of the cranium, behind the maxillae and on each side of the posterior nasal openings. Behind the posterior nasal openings, the body of the sphenoid and the basilar part of the occipital bone form a broad bar of bone which extends to the foramen magnum. These two bones are separated by a cartilaginous plate in early life and bony fusion usually takes place at about the 25th year.

The area lateral to the basi-sphenoid is formed anteriorly by the greater wing of sphenoid and laterally by the squamous part of the temporal bone which bears the mandibular fossa and articular tubercle. The petrous temporal bone projects inwards and medially in the form of a wedge between the sphenoid and occipital bones.

Two important foramina are present in the greater wing of the sphenoid, between the lateral pterygoid plate and the mandibular fossa: (a) **the foramen ovale**, a large, oval, anterior opening, which transmits the mandibular branch of the trigeminal nerve, and (b) **the foramen spinosum**, through which the middle meningeal artery enters the middle cranial fossa. **The foramen lacerum** lies medial to the foramen ovale between the apex of the petrous-temporal and sphenoid bones. It is an irregular opening, partially closed in life by cartilage. A short distance posterolateral to it lies **the carotid canal** through which the internal carotid artery passes into the middle cranial fossa.

The cartilaginous part of the auditory tube (pharyngotympanic tube) lies in a groove between the carotid canal and the foramen ovale and it opens into a small bony canal which leads into the tympanic cavity.

Between the petrous-temporal and occipital bones, and behind the carotid canal, is **the jugular foramen** which transmits the sigmoid sinus to become the jugular vein. At the base of the styloid process, lateral to the jugular foramen, is **the stylomastoid foramen** through which the facial nerve emerges to supply the muscles of expression of the face. On the medial aspect of the mastoid process, posterior to the stylomastoid foramen, is a deep groove—**the mastoid notch (digastric notch)**—which gives attachment to the posterior belly of the digastric muscle.

The posterior part is formed mainly by the occipital bone. There are few prominent features on the occipital bone behind the foramen magnum except for the prominent external occipital protuberance. The external occipital crest and the nuchal lines are bony ridges which give attachment to the ligamentum nuchae and muscles of the neck (Fig. 3.43, p. 89).

SAGITTAL SECTION OF THE SKULL (Fig. 2.7)

Points to notice are:

- The base of the cranium is inclined at an angle to the horizontal and the facial bones are suspended from the anterior part of the base.
- The frontal sinuses are situated between the inner and outer tables of the frontal bone behind the supercilliary arches.
- The sphenoidal sinuses lie beneath the pituitary fossa and occupy the body of the sphenoid bone.
- Behind the pituitary fossa, the base of the cranium slopes downwards from the dorsum sellae to the anterior margin of the foramen magnum. This area is known as the clivus. The upper part of the clivus is formed by the body of the sphenoid bone and the lower part by the basilar part of the occipital bone. The two parts are separated in early life by a cartilaginous plate, bony fusion taking place at about 25 years.

THE CRANIAL CAVITY

The cranial cavity is occupied by the brain. It is lined by **dura mater** which is the outer of the three membranes or meninges surrounding the brain. (The **pia mater** is the innermost layer and closely covers the brain convolutions; between the pia mater and the middle layer (the **arachnoid**) lies the **subarachnoid space** in which the cerebrospinal fluid circulates.)

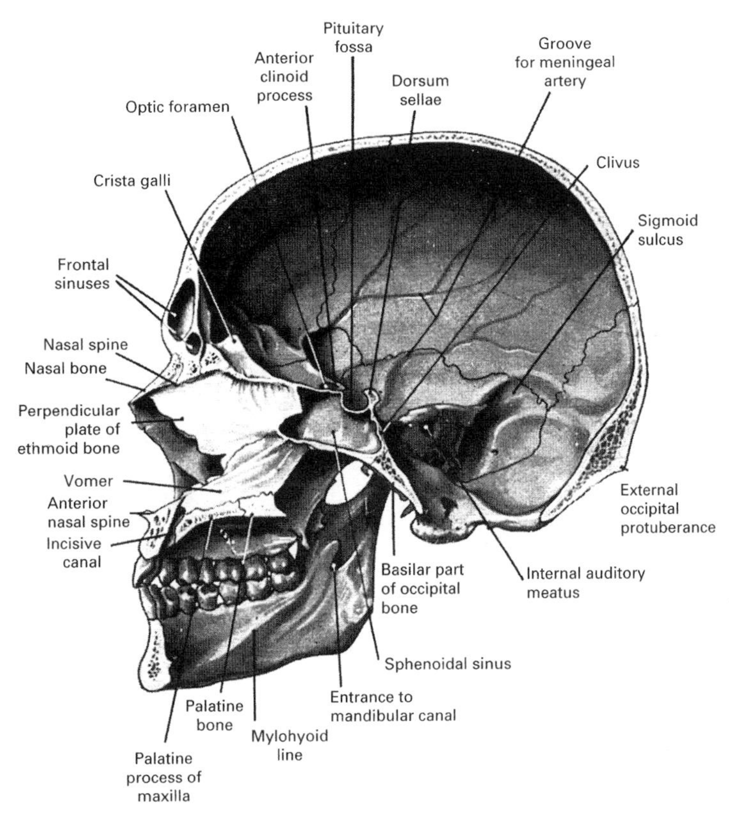

Fig. 2.7 Sagittal section of the skull

Internal surface of the roof of the cranial cavity

If the roof of the cranium is removed, the internal (or cerebral) surface of the cavity can be seen. Examination of the cut surface shows that the thickness of the bone varies. It is always thickest in most exposed areas, e.g. the frontal and occipital regions, and thinnest in areas well covered by muscles, e.g. the temporal region.

Most cranial bones consist of an inner and an outer table of compact bone, separated by cancellous bone containing red bone marrow (diploë) and diploic veins. The diploic veins are very variable in size and extent but are usually most prominent in the parietal bones. The sagittal groove extends in the midline from the frontal crest to the internal occipital protuberance. A large venous channel (the **sagittal sinus**) lies in this groove enclosed in a fold of dura mater (the **falx cerebri**) which projects downwards between the cerebral hemispheres. On either side of the sagittal groove there are numerous small depressions in the bone for protrusions of the subarachnoid space—**arachnoid granulations**. These granulations are found in the vicinity of large blood sinuses and through them there is an interchange of fluid between the blood and cerebrospinal fluid.

The inner surface of the bone is grooved also by branches of the middle meningeal artery. The anterior branch initially runs parallel with the coronal suture immediately medial to the pterion (see p. 11) and the posterior branch extends obliquely upward and posterior to the parietal bones. These internal grooves are variable in size and in some skulls may be almost absent.

Floor of the cranial cavity (internal base of skull (Fig. 2.8)

The floor of the cranial cavity is divided naturally by prominent bony ridges into anterior, middle and posterior fossae. The anterior fossa lies at the highest and the posterior at the lowest level.

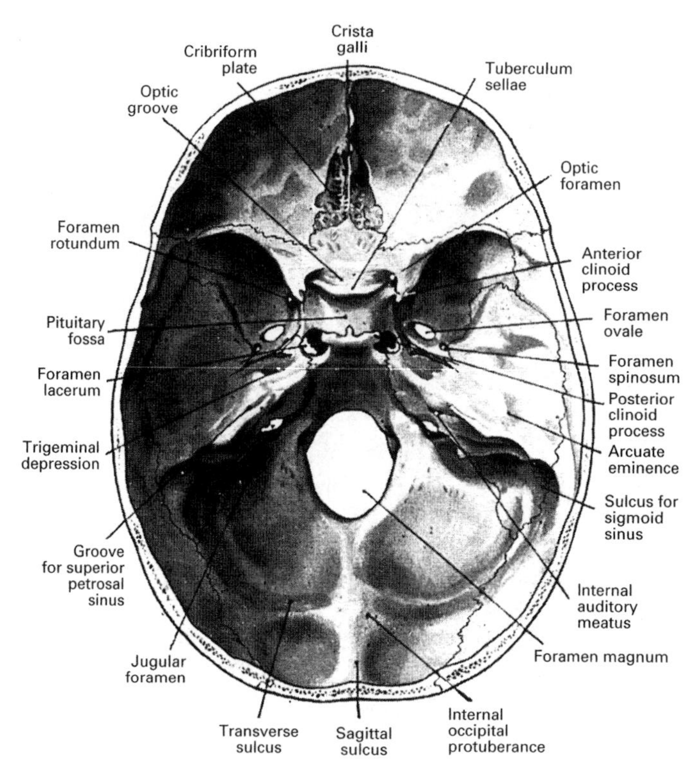

Fig. 2.8 Floor of cranial cavity

The anterior cranial fossa

The anterior fossa contains part of the frontal lobes of the brain. The frontal bone forms the anterior and lateral walls; the orbital plate of the frontal bone forms the anterior two-thirds of the floor and separates the fossa above from the orbits below. The posterior margins are formed by the body of the sphenoid bone centrally and by the lesser wings of the sphenoid on either side. The medial end of each lesser wing projects posteriorly as a prominent process—the anterior clinoid process.

The cribriform plate of the ethmoid bone lies in the midline between the orbital plates of the frontal bone, and is perforated by numerous small holes for passage of the olfactory nerves from the nasal cavity. A prominent crest of bone on the cribriform plate—the **crista galli** ('cock's comb')—gives attachment for the falx cerebri. The frontal sinuses lie in the bone of the anterior wall of the fossa but may sometimes extend backwards into the orbital plate.

The middle cranial fossa

The middle cranial fossa is bounded anteriorly by the posterior borders of the lesser wings of the sphenoid, posteriorly by the well-defined ridges on the upper surface of the petrous temporal bones and by the dorsum sellae of the sphenoid. The middle cranial fossa has a narrow middle part and two wider lateral parts.

The middle part is formed by the upper surface of the body of the sphenoid and is raised above the level of the lateral parts. The **optic foramina** (through which pass the optic nerves and ophthalmic arteries) lie between the lesser wings and body of the sphenoid and are directed forwards and slightly outwards to enter the apex of the orbit. A transverse groove—the **optic groove**—is present on the upper surface of the body of the sphenoid between the optic foramina and is bounded posteriorly by a ridge, the tuberculum sellae.

The upper surface of the body of the sphenoid is shaped like a Turkish saddle; hence its name, sella turcica. It is bounded in front by the tuberculum sellae and behind by an upward projecting plate of bone, the dorsum sellae. The concavity between the tuberculum sellae and the dorsum sellae, the pituitary fossa, is occupied by the pituitary gland (hypophysis cerebri). The upper corners of the dorsum sellae are prominent and are called the posterior clinoid processes. The anterior clinoid processes do not overhang the pituitary fossa but are situated a short distance lateral to it.

The lateral parts of the middle cranial fossa are wide and deep. Each contains one of the temporal lobes of the brain and lies in relationship with the orbit anteriorly, the temporal fossa laterally and the temporomandibular joint inferiorly. The greater wing of the sphenoid forms the anterior part of the floor of the fossa and a little of the lateral wall. The squamous part of the temporal bone forms the lateral part of the floor and most of the lateral wall. The petrous part of the temporal bone forms the posterior part of the floor. The overhanging margin of the lesser wing of the sphenoid forms the anterior boundary. The upper border of the petrous temporal bone forms the posterior boundary. The lateral part of the fossa communicates with the orbit through the superior orbital fissure, between the greater and lesser wings of the sphenoid. The lesser wing overhangs this fissure. Several important foramina are present:

- the foramen rotundum lies below the superior orbital fissure and transmits the maxillary nerve
- the foramen ovale lies behind the foramen rotundum and transmits the mandibular division of the trigeminal nerve
- foramen spinosum, lies close to the posterior margin of the foramen ovale and transmits the middle meningeal artery
- the foramen lacerum lies between the apex of the petrous temporal bone and the body of the sphenoid bone. It transmits the carotid artery.

The upper surface of the petrous part of the temporal bone forms the posterior part of the middle fossa and near its apex is a small hollow—the trigeminal impression—in which lies the ganglion of the trigeminal nerve. Halfway between the apex of the petrous temporal bone and the lateral wall of the cranium is a small elevation—the arcuate eminence—below which lies the superior semicircular canal of the inner ear. Between the arcuate eminence and the lateral wall of the cranium the bone surface is smooth and is called the **tegmen tympani**. It forms the roof of the middle ear (tympanic cavity).

The posterior cranial fossa

The posterior fossa is the largest and deepest of the fossae and it contains the hind brain (cerebellum, pons and medulla oblongata). The large oval-shaped **foramen magnum** lies in the anterior part of the fossa and transmits the lower part of the brain stem and its membranes.

The anterior boundary of the posterior fossa is formed in the midline by the posterior surface of the dorsum sellae and on either side by the petrous parts of the temporal bones. The floor of the fossa and its lateral and posterior walls are formed by the occipital bone, apart from a small but important region at the junction of the anterior and lateral walls where it is formed by the mastoid portion of the temporal bone and is grooved by the sigmoid sinus.

Three foramina lie one above the other on the medial part of the anterior wall above the foramen magnum:

- the anterior condylar canal (the lowest foramen) transmits the hypoglossal nerve which supplies the muscles of the tongue
- the jugular foramen (the middle foramen) is large and irregular in shape and lies in a gap between the occipital and petrous temporal bones. It transmits the sigmoid sinus, the inferior petrosal sinus and the 9th, 10th and 11th cranial nerves
- the internal auditory meatus (the highest foramen) enters the posterior surface of the petrous temporal bone and transmits the auditory and facial nerves.

A bony ridge, the internal occipital crest, extends along the occipital bone behind the foramen magnum to end at the internal occipital protuberance. A shallow, wide groove, the transverse sulcus, sweeps round the posterior and lateral walls of the fossa from the internal occipital protuberance. The transverse sulcus becomes the sigmoid sulcus on the posterior surface of the temporal bone and ends at the jugular foramen. These sulci contain the transverse and sigmoid sinuses respectively, which form the main venous drainage from the brain. The upper part of the sigmoid sinus is only separated from the tympanic (mastoid) antrum by a thin layer of bone.

The posterior fossa is roofed over by a fold of dura mater called the **tentorium cerebelli**. This is attached posteriorly to the edges of the transverse sulci (where it encloses the transverse sinuses) and anteriorly to the petrous temporal and sphenoid bones. The tentorium is joined in the midline above to the falx cerebri and separates the cerebellum below from the posterior part of the cerebrum above.

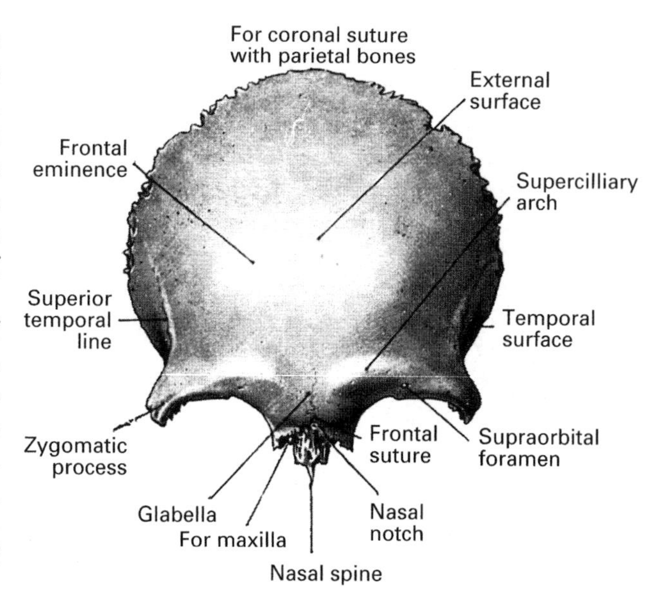

Fig. 2.9 Frontal bone: anterior aspect

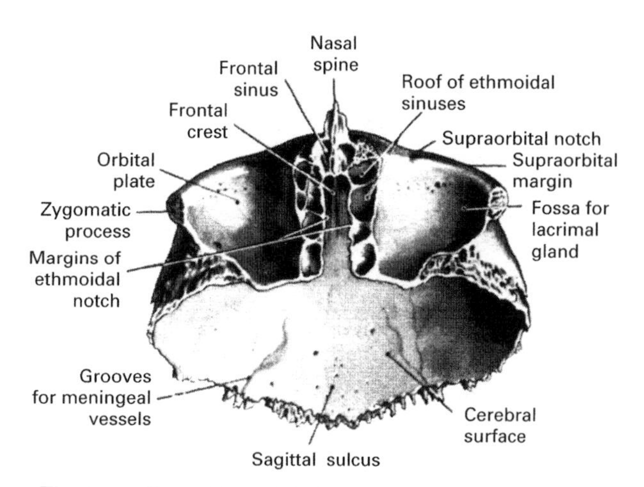

Fig. 2.10 Frontal bone: inferior aspect

THE INDIVIDUAL BONES OF THE SKULL 1. CRANIAL BONES

FRONTAL BONE (Figs 2.9 and 2.10)

The frontal bone consists of two parts:

- frontal (squamous)
- horizontal.

The frontal part forms the dome of the forehead. It is continuous with the horizontal part at the supraorbital margin which terminates laterally at the zygomatic process. Above each supraorbital margin is a rounded ridge—the supercilliary arch. The glabella lies between the supercilliary arches. The nasal part of the frontal bone projects downwards between the supraorbital margins. It articulates with the nasal bone at the nasion and with the frontal processes of the maxillae. The nasal spine projects downwards from the nasal part and forms a small part of the nasal septum.

The frontal sinuses lie between the inner and outer tables of the frontal bone posterior to the supercilliary ridges.

The horizontal part consists of two orbital plates which are separated in the midline by the ethmoid notch. The orbital plates form the roof of the orbits and part of the floor of the anterior cranial fossa. The inferior surface of each orbital plate is smooth and concave. Laterally there is a shallow depression for the lacrimal gland. The openings of the frontal sinuses lie in the front of the ethmoid notch on either side of the nasal spine.

PARIETAL BONES (Figs 2.11 and 2.12)

The two parietal bones are situated one on either side of the midline and form a large part of the roof and sides of the cranium. Each parietal bone is quadrilateral in shape. They overlie the main

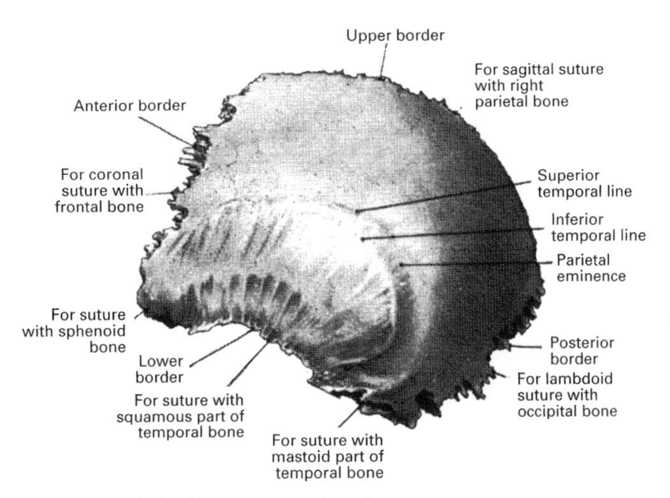

Fig. 2.11 Parietal bone: external surface

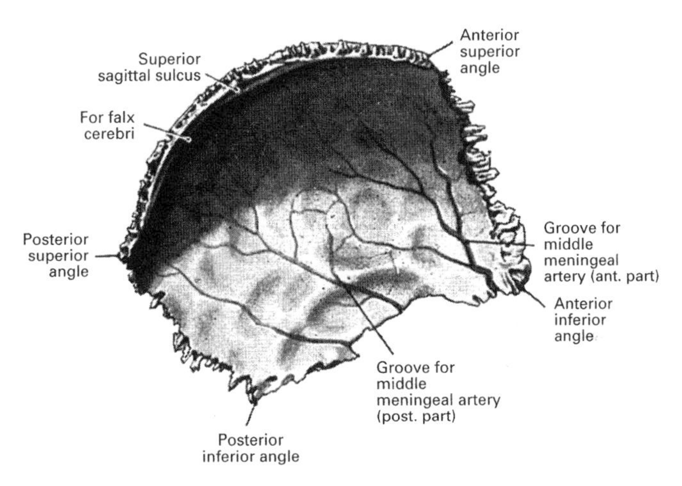

Fig. 2.12 Parietal bone: internal surface

motor and sensory areas of the brain and they contain grooves for the anterior and posterior branches of the middle meningeal artery.

Superiorly the two parietal bones articulate with each other at the sagittal suture. Posteriorly they articulate with the occipital bone at the lambdoid suture and anteriorly they articulate with the frontal bone at the coronal suture. Inferiorly the parietal bones articulate with the greater wing of the sphenoid anteriorly and the temporal bones posteriorly.

The external surface of the parietal bone is convex and has two well-marked ridges—temporal lines—which give attachment to the temporalis muscle. A prominent convexity on the outer surface is called the parietal eminence and is the site at which ossification commences.

The internal surface is marked by the convolutions of the brain and shows well-marked grooves for the two main branches of the middle meningeal artery. On the upper border is a groove for the superior sagittal sinus. The fold of the dura mater—the **falx cerebri**—which separates the hemispheres is attached to the margins of this groove.

OCCIPITAL BONE (Figs 2.13 and 2.14)

The occipital bone forms the posterior part of the base and vault of the cranium. It encloses the

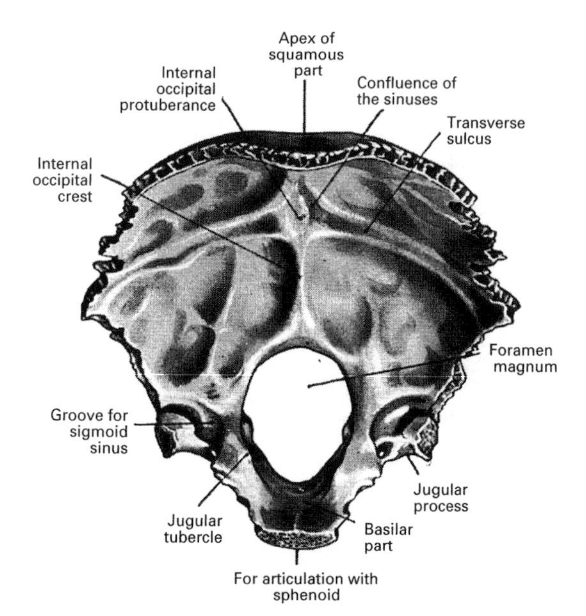

Fig. 2.13 Occipital bone: internal surface

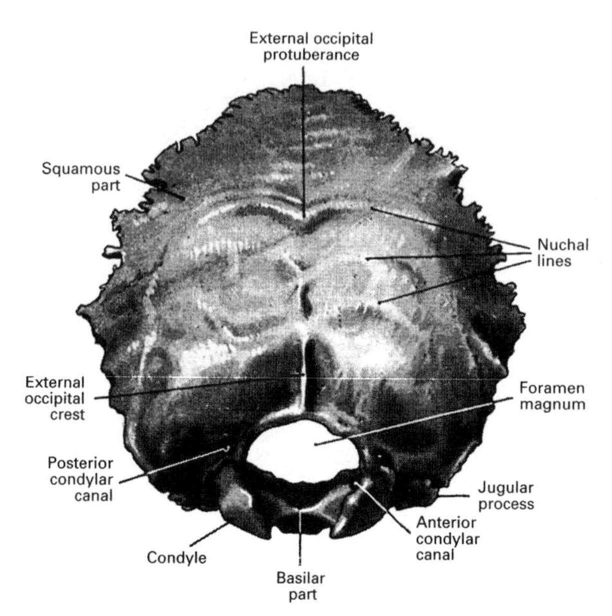

Fig. 2.14 Occipital bone: external surface

foramen magnum through which is transmitted the lower part of the brain stem. The occipital bone consists of three parts:

- the squamous part
- the lateral (condylar) part
- the basilar part.

The squamous part is a curved plate of bone posterior to the foramen magnum. On the internal surface are a vertical bony ridge and a horizontal groove. These meet at the internal occipital protuberance. The transverse venous sinuses lie in the horizontal groove. The lower part of the vertical ridge is known as the internal occipital crest and gives attachment to the fold of dura mater which separates the cerebellar hemispheres. On the external surface of the occipital bone the external occipital crest extends upwards in the midline from the foramen magnum to the external occipital protuberance. Extending outward on each side from the external protuberance is a curved line—the superior nuchal line—which gives attachment to the superficial muscles of the neck.

The lateral (condylar) parts of the occipital bone lie on either side of the foramen magnum. On the under surface of each part is an occipital condyle which articulates with the atlas vertebra at the atlanto-occipital joints (p. 89). The condyles lie obliquely beside the foramen magnum so that their anterior ends are closer together than their posterior ends. Above the condyles are the anterior condylar canals which transmit the hypoglossal nerves. The bone lateral to the condyle is the jugular process. Its cerebral surface is grooved by the end of the sigmoid sinus and forms the posterior margin of the jugular foramen. On the cerebral surface of the condylar part there is a bony eminence—the jugular tubercle—which overlies the condylar canal.

The basilar part is a thick oblong bar of bone extending forwards from the foramen magnum. It articulates anteriorly with the body of the sphenoid. Its cerebral surface forms the posterior part of the clivus and on its inferior surface there is a small elevation—the pharyngeal tubercle—which forms the posterior attachment of the pharynx in the midline.

TEMPORAL BONE (Figs 2.15 and 2.16)

The temporal bone develops from three separate parts:

- the squamous part
- the tympanic part
- the petromastoid part.

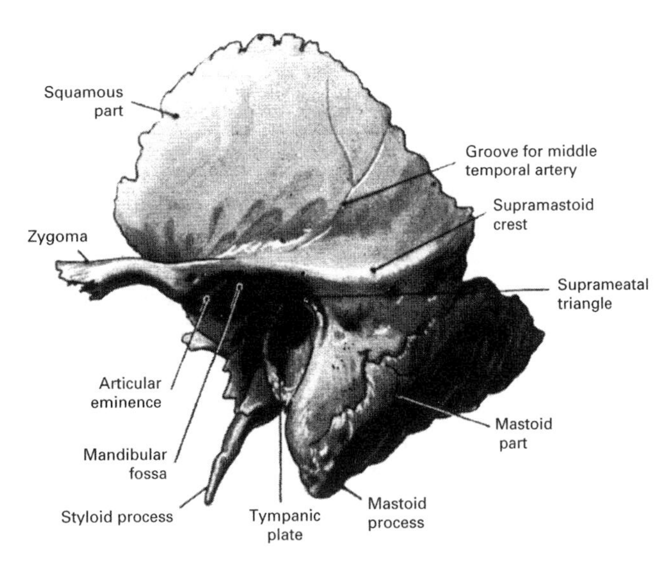

Fig. 2.15 Temporal bone: lateral aspect

The three parts fuse during the first year.

The squamous part forms part of the lateral wall of the cranium and the posterosuperior part of the external auditory meatus. The zygomatic process (zygoma) extends from the lower lateral surface and on the under surface is the articular mandibular fossa which receives the head of the mandible to form the temporomandibular joint (p. 48). The squamous part articulates with the greater wing of the sphenoid anteriorly and with the parietal bone superiorly.

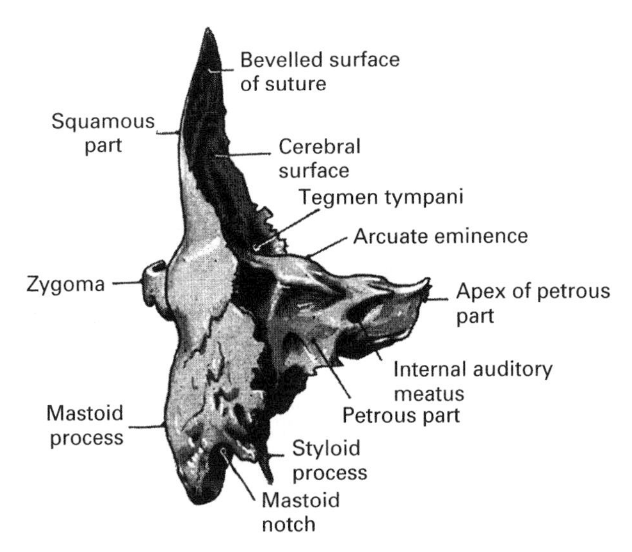

Fig. 2.16 Temporal bone: posterior aspect

The tympanic plate is a thin curved plate of bone which forms the floor, anterior wall and lower half of the posterior wall of the external auditory meatus (Fig. 2.15). It lies between the squamous and mastoid parts and envelops the root of the styloid process. The mandibular fossa lies immediately anterior to the tympanic plate. The styloid process emerges from the inferior surface of the tympanic plate and gives attachment to the stylohyoid ligament.

The petromastoid part is pyramidal in shape and forms part of the middle and posterior cranial fossae and part of the under surface of the base of the skull. It is subdivided into petrous and mastoid parts.

The petrous part is a wedge of dense bone directed forward and medially between the sphenoid and occipital bones. Its upper surface forms part of the middle cranial fossa and the posterior surface forms the anterior wall of the posterior cranial fossa. It contains the inner ear structures (p. 24).

On the upper surface is the arcuate eminence under which lies the superior semicircular canal of the inner ear. Between this eminence and the lateral wall of the cranium are the roof of the tympanic cavity (tegmen tympani) and the mastoid antrum. On the posterior surface is the opening of the internal auditory meatus through which pass the auditory and facial nerves.

On the under surface of the bone is the stylomastoid foramen which transmits the facial nerve. On the under surface also are the carotid canal and the jugular foramen (p. 52). At the junction of the mastoid and petrous parts is a groove for the sigmoid sinus. The carotid canal commences on the under surface of the bone.

The mastoid part is the posterior part of the temporal bone and it articulates posteriorly with the occipital bone. It consists of an upper part which contains the mastoid antrum and a lower part which is a conical projection called the mastoid process.

THE EAR (Figs 2.17 to 2.19)

The ear is divided into three parts:

- the external ear
- the middle ear (tympanic cavity)
- the internal ear (labyrinth).

The external ear

The external ear consists of the auricle and the external auditory meatus. The auricle and the lateral third of the external auditory meatus are cartilaginous. The bony part of the external auditory meatus is about 24 mm long and is directed medially, forward and slightly downward. It ends at the tympanic membrane (ear drum) which lies obliquely and which separates the external ear from the middle ear. The outer surface of the membrane is slightly concave.

The middle ear

The middle ear or tympanic cavity is a small narrow space in the petrous part of the temporal bone between the tympanic membrane and the lateral wall of the internal ear. Its long axis lies parallel with the tympanic membrane and it contains the three auditory ossicles. These very small bones form a chain which transmits vibrations of the tympanic membrane across the tympanic cavity to the inner ear.

The lateral wall of the tympanic cavity is formed mainly by the tympanic membrane. A

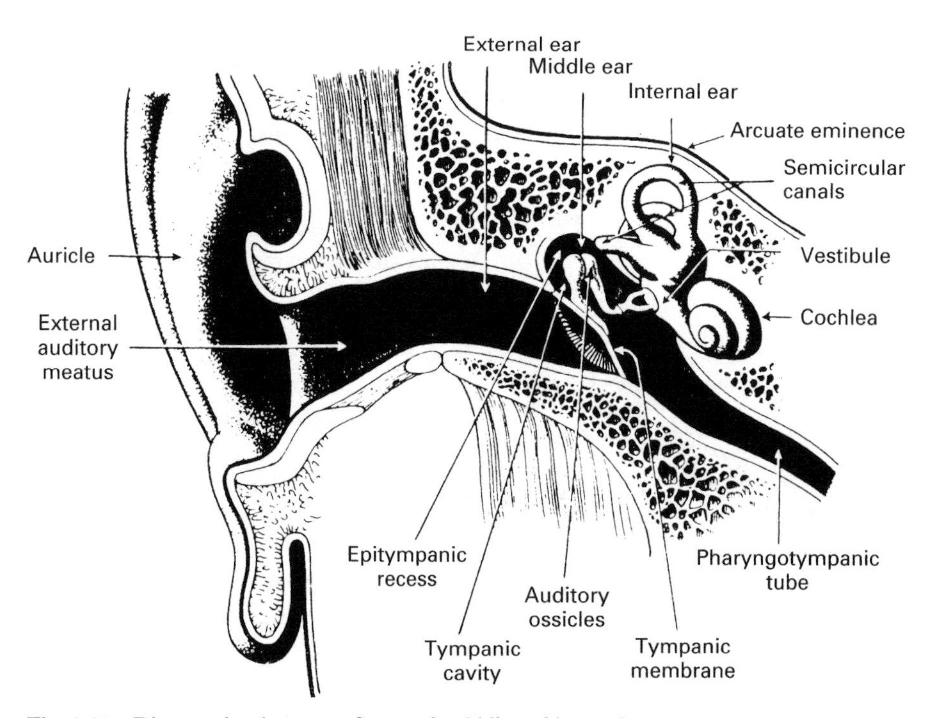

Fig. 2.17 Diagram showing parts of external, middle and internal ear

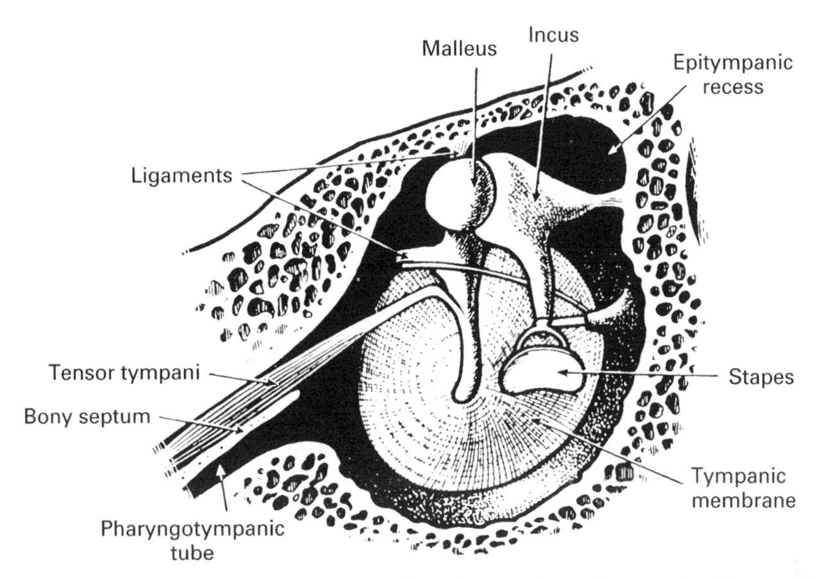

Fig. 2.18 Diagram showing internal surface of tympanic membrane and auditory ossicles

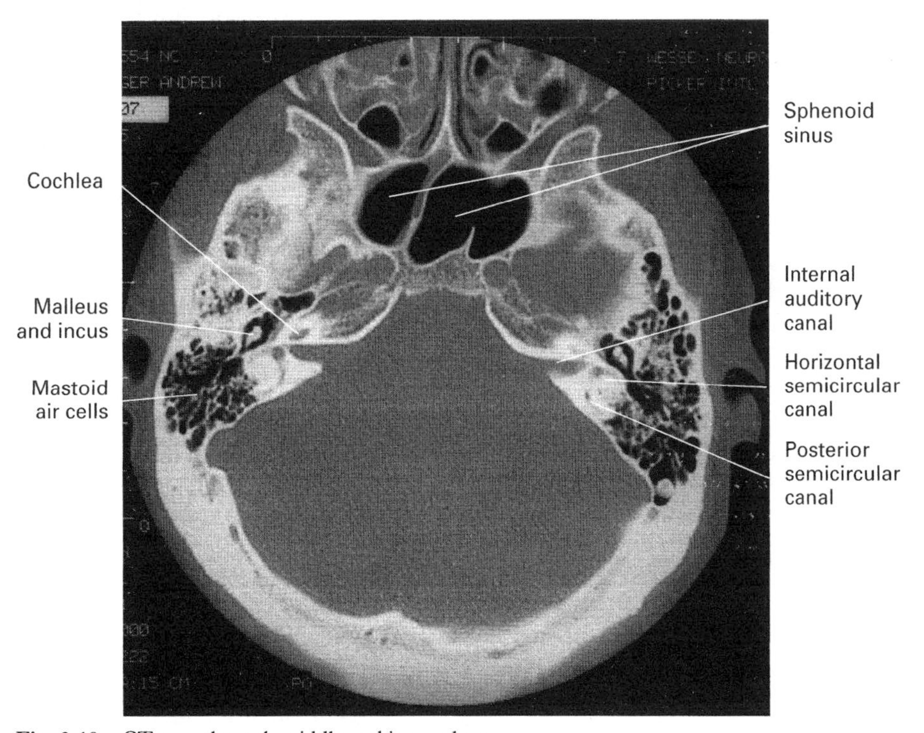

Fig. 2.19 CT scan through middle and internal ear

small part of the cavity lies above the level of the tympanic membrane and is called the epitympanic recess.

On the medial wall, near its centre, is a well-marked prominence, called the promontory, which lies over the basal turn of the cochlea of the inner ear. Above and posterior to the promontory is the fenestra vestibuli—an oval window which is closed by the footplate of the stapes (p. 24).

The roof is formed by a bony plate called the tegmen tympani which separates the tympanic cavity from the temporal lobe of the brain.

The floor lies immediately above the jugular fossa. Anteriorly the roof and floor converge to become the auditory (pharyngotympanic) tube which connects the pharynx with the middle ear cavity and allows air pressure on both sides of the tympanic membrane to remain equal.

On the upper part of the posterior wall is an opening—the aditus—through which the epitympanic recess communicates with the antrum and air cells of the mastoid process.

The auditory ossicles

There are three ossicles:

- the malleus
- the incus
- the stapes.

The malleus is hammer-shaped. It is the largest of the ossicles and is about 9 mm long. It lies in the lateral part of the middle ear. It has a head, a long process called the handle and two smaller processes. The handle is attached to the upper internal surface of the tympanic membrane and the head articulates with the body of the incus.

The incus is anvil-shaped. It is the middle of the three bones and consists of a body and two processes—one long and one short. The body articulates with the head of the malleus and the long process articulates with the stapes.

The stapes, the most medial of the bones, is stirrup-shaped. It consists of a small head, a neck, and two limbs which hold between their ends the oval base or footplate. The footplate is attached to the margins of the fenestra vestibuli (oval window) by an anular ligament and transmits the vibrations of the tympanic membrane to the fluid (perilymph) of the internal ear.

The internal ear

The internal ear is situated in the petrous part of the temporal bone, medial to the tympanic cavity and it contains the essential organs of hearing and balance. It consists of a communicating series of membranous sacs and ducts (the membranous labyrinth) enclosed in bony cavities of similar shape (the bony labyrinth). Both the membranous and the bony labyrinths are filled with a fluid similar to cerebrospinal fluid.

The bony labyrinth

This consists of three parts:

- the bony cochlea
- the vestibule
- the semicircular canals.

The bony cochlea which is shaped like a snail's shell and consists of a central column—the modiolus—around which a spiral tube winds for $2\frac{3}{4}$ turns. A spiral ledge of bone projects into the cochlea from the modiolus and supports the membranous cochlea.

The vestibule is situated between the cochlea anteriorly and the semicircular canals posteriorly. In its lateral wall lies the fenestra vestibuli which is closed by the footplate of the stapes.

The three semicircular canals—superior, lateral and posterior—lie at right angles to each other and surround the membranous semicircular ducts. The upper part of the superior semicircular canal lies under the arcuate eminence of the petrous bone.

The organ of hearing (organ of Corti) lies in the membranous cochlea. It picks up the vibrations in the fluid of the internal ear produced by movement of the base of the stapes and transmits them along the cochlear part of the auditory nerve to the brain, where they are interpreted as sound. In like manner there are special receptor areas in the walls of the semicircular ducts which are stimulated by the movement of the fluid. They transmit these impulses along the vestibular part of the auditory nerve to be interpreted by the brain as balance.

The mastoid antrum is an irregularly shaped air space posterior to the tympanic cavity. The air cells of the mastoid process develop as buds from it. At birth the middle ear and mastoid antrum are fully formed. Development of the mastoid cells usually begins when the mastoid process forms about the second year and is complete at puberty. The mastoid air cells vary greatly in size and distribution.

SPHENOID BONE (Figs 2.20 and 2.21)

The sphenoid bone is butterfly-shaped. It consists of:

- the body
- two lesser wings
- two greater wings
- two pterygoid processes.

It is situated in the centre of the base of the cranium and thus it articulates with most of the cranial bones and also with the maxilla, vomer, zygomatic and palatine bones.

The body is cube-shaped and contains the sphenoid sinuses. On the anterior part of the superior surface (Fig. 2.20) is the ethmoid spine which articulates with the cribriform plate of the ethmoid. On the superior surface, between the ethmoid spine and the optic groove, the bone is smooth and is called the jugum sphenoidale. The posterior edge of the jugum forms the anterior part of the optic groove (optic chiasma) which connects the optic canals.

Posterior to the optic groove is the tuberculum sellae—the anterior part of the sella turcica (p. 17). Posterior to the dorsum sellae the sphenoid articulates with the occipital bone.

On the anterior surface (Fig. 2.21) there is a midline ridge—the sphenoidal crest—which forms part of the nasal septum and which continues on the under surface of the body as the rostrum. The crest articulates with the perpendicular plate of the ethmoid and the rostrum articulates with the vomer.

The lesser wings project laterally from the upper anterior part of the body and form the posterior edge of the anterior cranial fossa. They are thin and triangular in shape. Each lesser wing is attached to the body by two roots and between these roots lies the optic foramen. The medial end of each lesser wing projects backwards to form the anterior clinoid process.

The greater wings extend laterally from the body. The concave superior surface forms a large part of the middle cranial fossa. The temporal (external) surface forms part of the temporal fossa. The orbital surface forms a large part of the lateral wall of the orbit. The upper border of the orbital surface is separated from the lesser wing

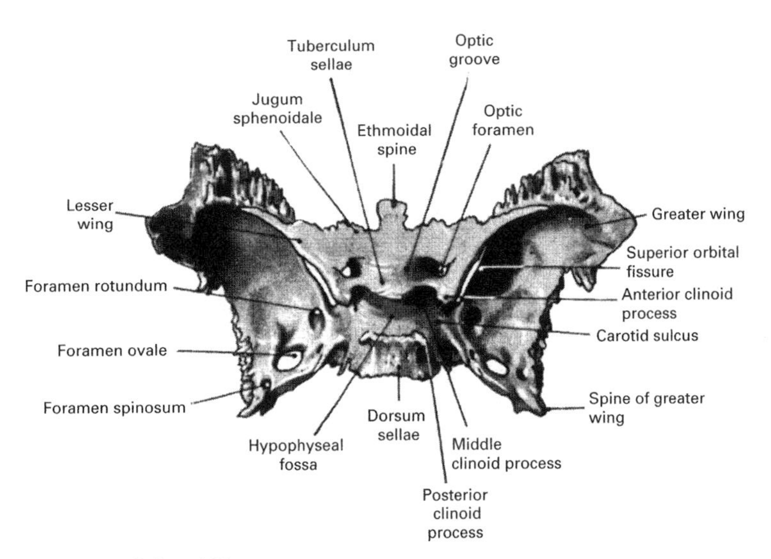

Fig. 2.20 Sphenoid bone: upper aspect

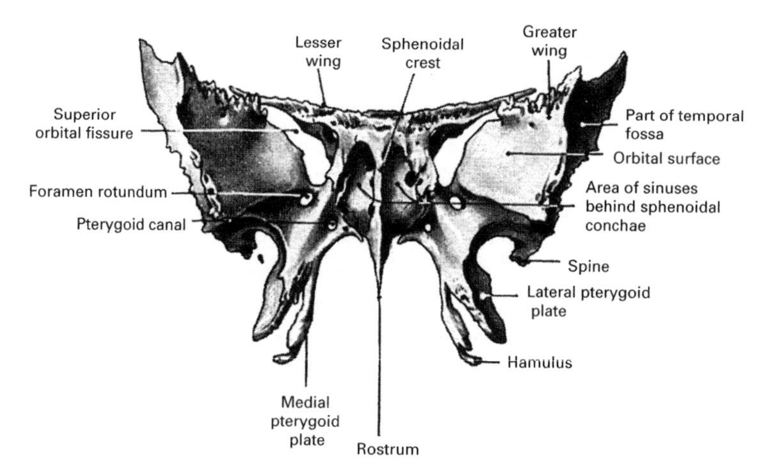

Fig. 2.21 Sphenoid bone: anterior aspect

by the superior orbital fissure, and the lower border from the maxilla by the inferior orbital fissure.

The pterygoid processes project downwards from the junction of the body and the greater wings. They form posterior buttresses to the facial bones. Each process consists of two plates, the medial and lateral laminae, which are fused together superiorly and separated behind by a fossa—the pterygoid fossa. At the lower end of the medial lamina is a delicate hook, the hamulus. The medial lamina forms part of the posterior nasal opening (choana). The lateral lamina forms the medial wall of the infratemporal fossa.

Three foramina are present on the upper surface of the sphenoid bone:

- the foramen rotundum
- the foramen ovale
- the foramen spinosum.

The foramen lacerum is a gap between the apex of the petrous part of the temporal bone and the greater wing of sphenoid (see p. 17).

ETHMOID BONE (Figs 2.22 and 2.23)

The ethmoid bone is a delicately constructed bone which occupies the superior part of the

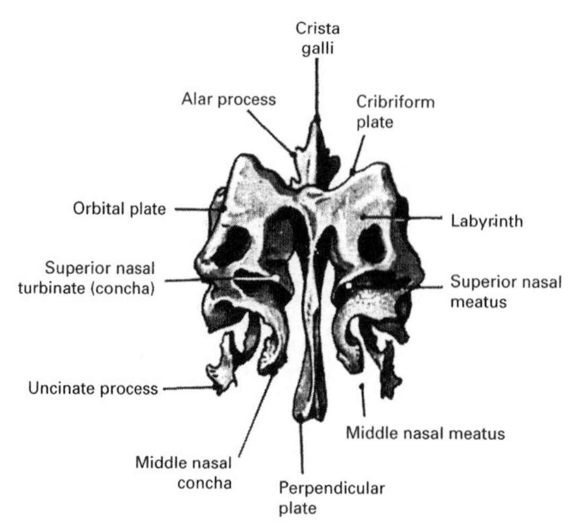

Fig. 2.22 Ethmoid bone: posterior aspect

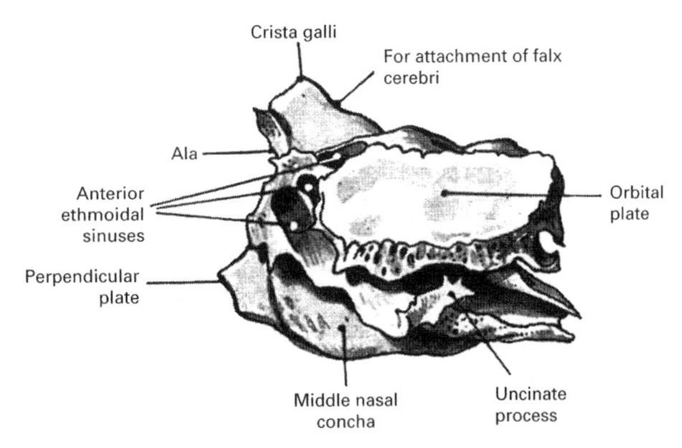

Fig. 2.23 Ethmoid bone: lateral aspect

nasal cavity, lying between the orbits and anterior to the sphenoid. It consists of:

- the perpendicular plate
- the horizontal plate.

The perpendicular plate forms the upper part of the nasal septum and is continued above the horizontal plate as a bony projection called the **crista galli** ('cock's comb') to which is attached the falx cerebri.

The horizontal plate consists of two parts—medial and lateral. The medial part is the cribriform plate which contains many perforations for the passage of the olfactory nerves. The cribriform plate lies between the orbital plates of the frontal bone and forms part of the floor of the anterior cranial fossa and the roof of the nasal cavity.

The lateral part of the horizontal plate suspends the labyrinth of ethmoidal air cells which open into the nasal cavity. The medial wall is irregular because of curved bony projections—the **superior** and **middle nasal turbinates** (conchae)—which project from it (see also p. 30). From the under surface of the labyrinth there is a downward pointing projection called the **uncinate process**, which forms part of the medial wall of the maxillary antrum.

THE ORBITS (Figs 2.24 to 2.26)

The orbital cavities are pyramidal-shaped cavities which contain the eyeballs, their associated muscles, nerves and blood vessels and most of the lacrimal system. The cavities are wide anteriorly and they narrow to an apex which is directed backwards and slightly medially.

Boundaries of the orbit

The supraorbital margin is formed by the frontal bone. The infraorbital margin is formed by the maxilla and the zygoma. The lateral margin is formed by the frontal process of the zygomatic bone and the zygomatic process of the frontal bone. The medial margin (less well-defined) is formed by the frontal bone and by the lacrimal crest of the frontal process of the maxilla.

Structure of the orbit

Each orbit consists of:

- the roof
- the floor
- the medial wall
- the lateral wall.

The roof of the orbit is formed mainly by the thin orbital plate of the frontal bone which separates the orbit from the anterior cranial fossa. The **optic canal** (optic foramen) (Figs 2.24 and 2.25) lies at the apex of the orbit, between the body of the sphenoid and the roots of the lesser wing. It transmits the optic nerve and ophthalmic artery.

The floor of the orbit is formed by the orbital plate of the maxilla and the zygomatic bone. Anteriorly it is continuous with the lateral wall but posteriorly the two walls are separated by the

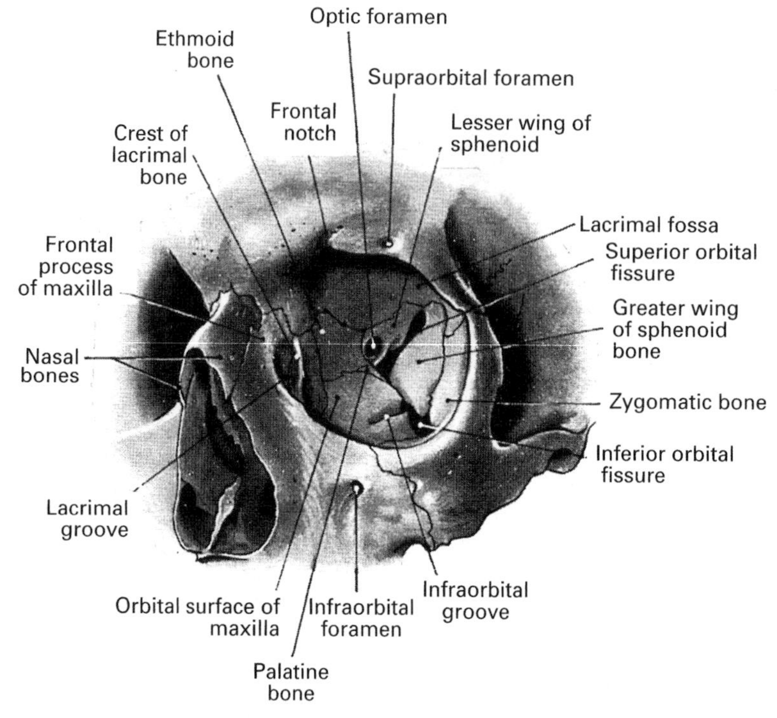

Fig. 2.24 The left orbit

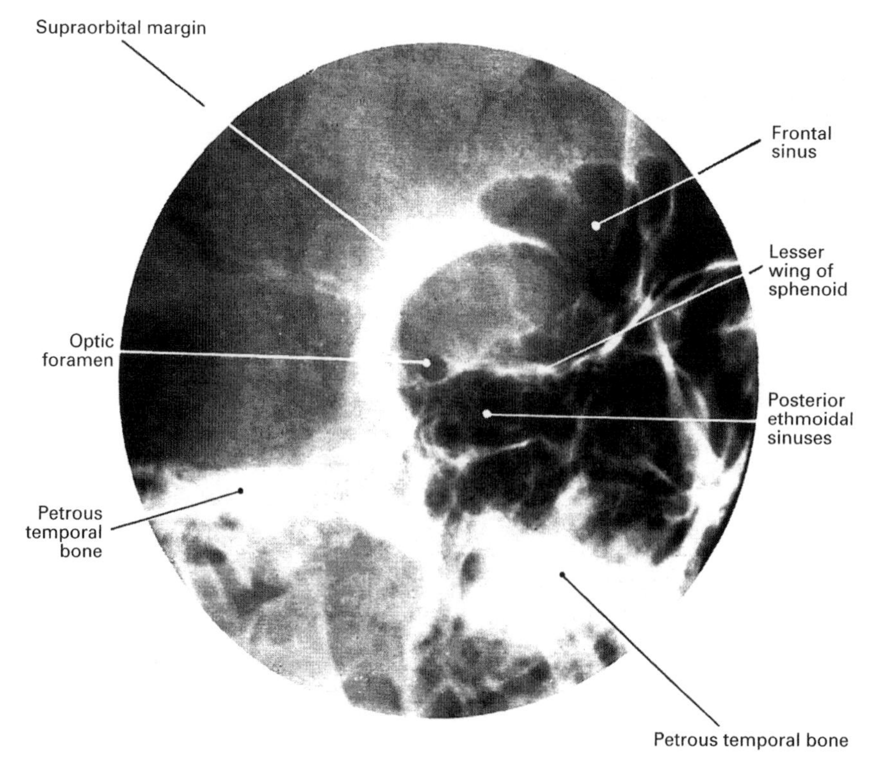

Fig. 2.25 Optic foramen: right oblique view

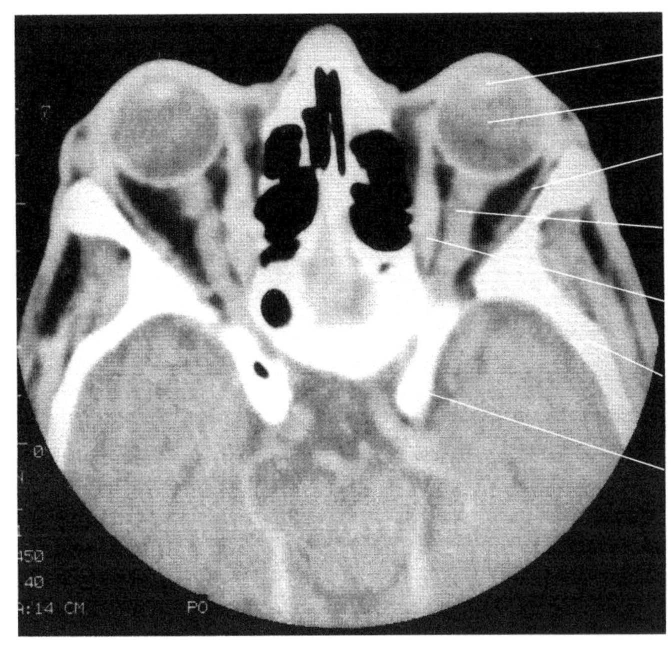

Fig. 2.26 CT scan through orbits

inferior orbital fissure. The superior orbital fissure separates the lateral wall and the roof, near the apex. Through it pass the oculomotor, trochlear and abducent nerves and the terminal branches of the ophthalmic nerve and the ophthalmic veins. The inferior orbital groove, in the floor of the orbit, becomes the infraorbital canal and opens on the anterior surface of the maxilla at the infraorbital foramen. Through it are transmitted the infraorbital nerves to supply the skin of the face adjoining the side of the nose.

The medial wall, which is very thin, is formed by the lacrimal bone and the orbital plate of the ethmoid bone. Anteriorly is the lacrimal groove which communicates with the nasal cavity through the nasolacrimal canal. The floor of the groove separates the orbital cavity from the nasal cavity.

The lateral wall is formed by the greater wing of sphenoid and the frontal process of the zygomatic bone.

NASAL CAVITY (Fig. 2.27)

The nasal cavity is an irregularly shaped cavity situated above the hard palate. It is divided into two halves, right and left, by a septum. The nasal septum has a bony part and a mobile, non-bony part. The bony part is formed by the perpendicular plate of the ethmoid bone superiorly and by the vomer inferiorly.

Each half of the nasal cavity consists of:

- the roof
- the floor
- the lateral wall
- the medial wall.

The roof of the nasal cavity comprises (a) an anterior sloping part, (b) an intermediate horizontal part, formed by the cribriform plate of the ethmoid bone, (c) a perpendicular part, anterior to the sphenoid sinus and (d) a curved part, below the sinus, which continues as the roof of the nasopharynx.

The **cribriform plate** separates the roof from the anterior cranial fossa and through its numerous openings pass the branches of the olfactory nerve from the mucous membrane in the upper part of the nasal cavity.

The floor is formed by the superior surface of the hard palate which separates the nasal cavity from the mouth.

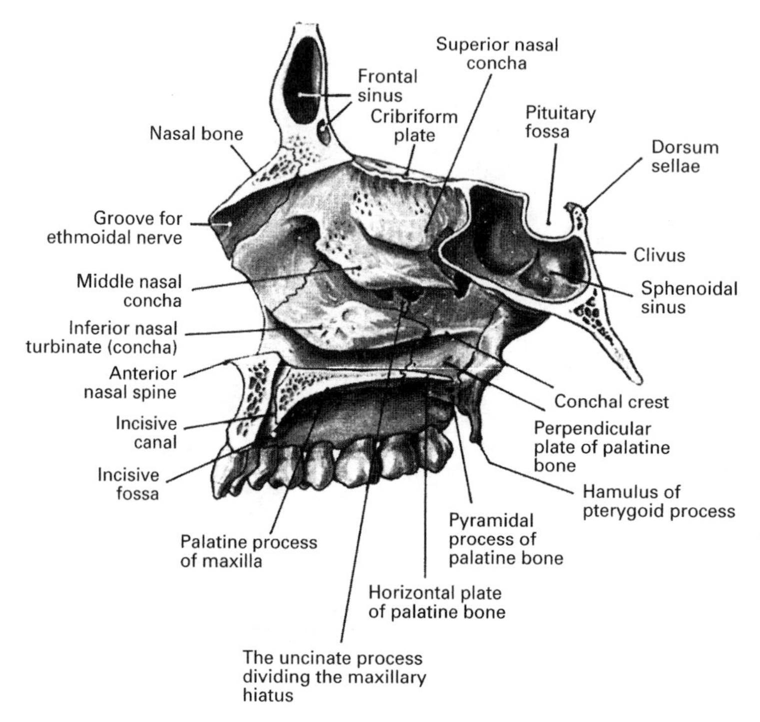

Fig. 2.27 Saggital section showing lateral wall of nasal cavity

The lateral wall (Fig. 2.27) is composed of a number of bones. The main features are three curved projections—the **superior, middle** and **inferior nasal turbinates** (conchae). Each consists of a thin, bony plate that curves downwards forming the roof of a meatus. The superior and middle turbinates are parts of the ethmoid bone and form the roof of the superior and middle meatuses into which the ethmoid sinuses open. The inferior turbinate is the largest. It is a separate bone and extends almost the full length of the lateral wall, overlying the inferior meatus into which the nasolacrimal canal opens. The meatus communis lies between the turbinates and the septum.

The medial wall is formed by the nasal septum.

The anterior nares—nostrils—form the anterior opening to the nasal cavity. The choanae form the opening between the nasal cavity and the nasopharynx.

Radiographic appearances of the cranial bones (Figs 2.28 to 2.35)

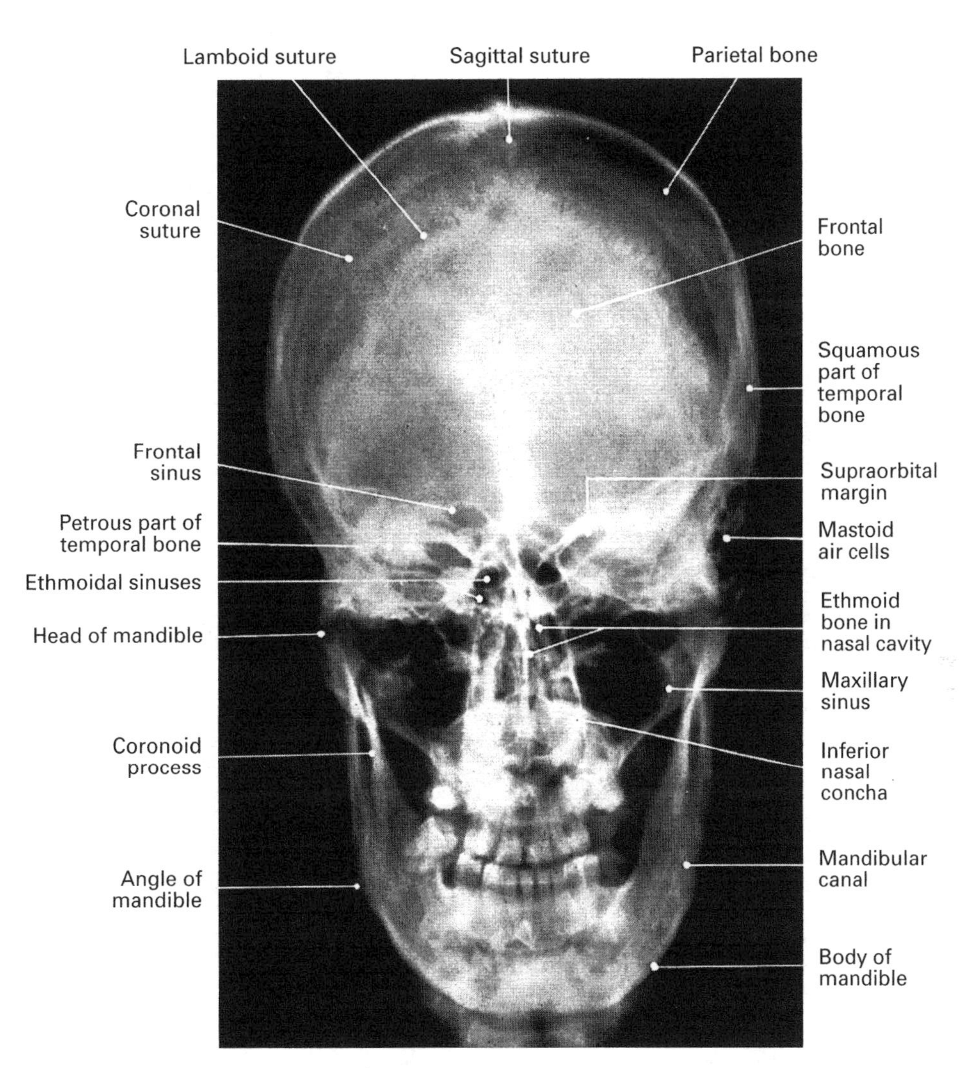

Fig. 2.28 Cranial bones: occipitofrontal view

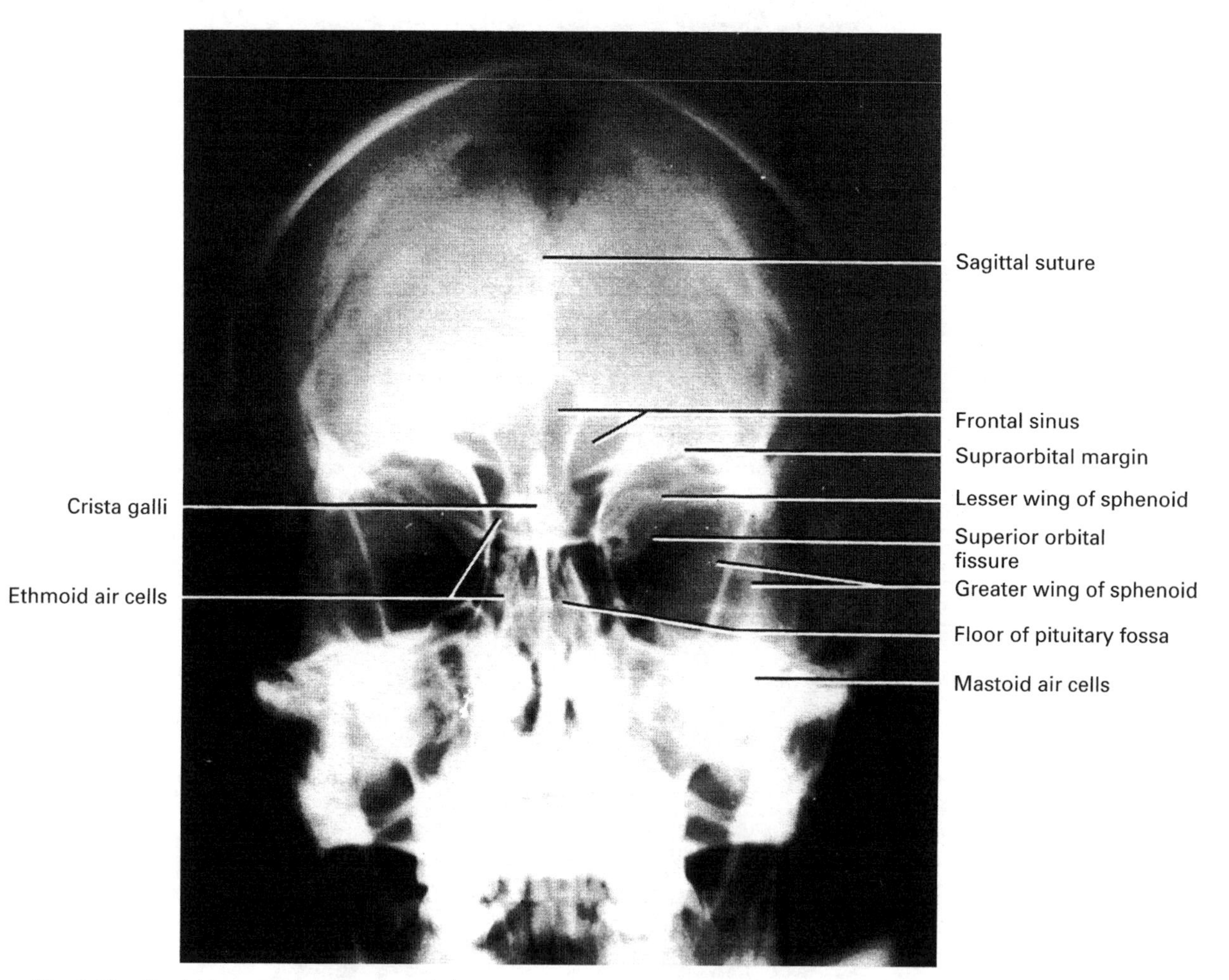

Fig. 2.29 Cranial bones: 20° occipitofrontal view

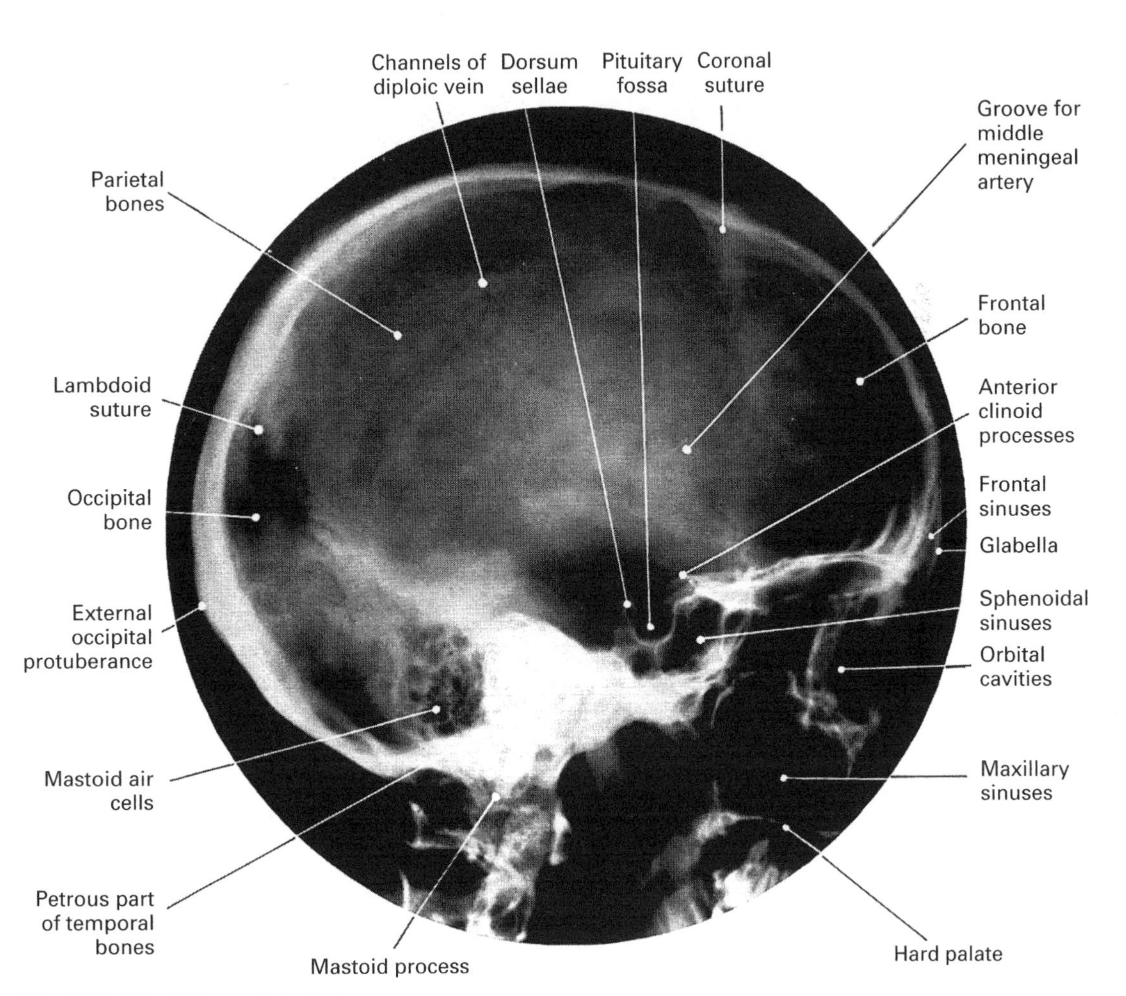

Fig. 2.30 Cranial bones: lateral view

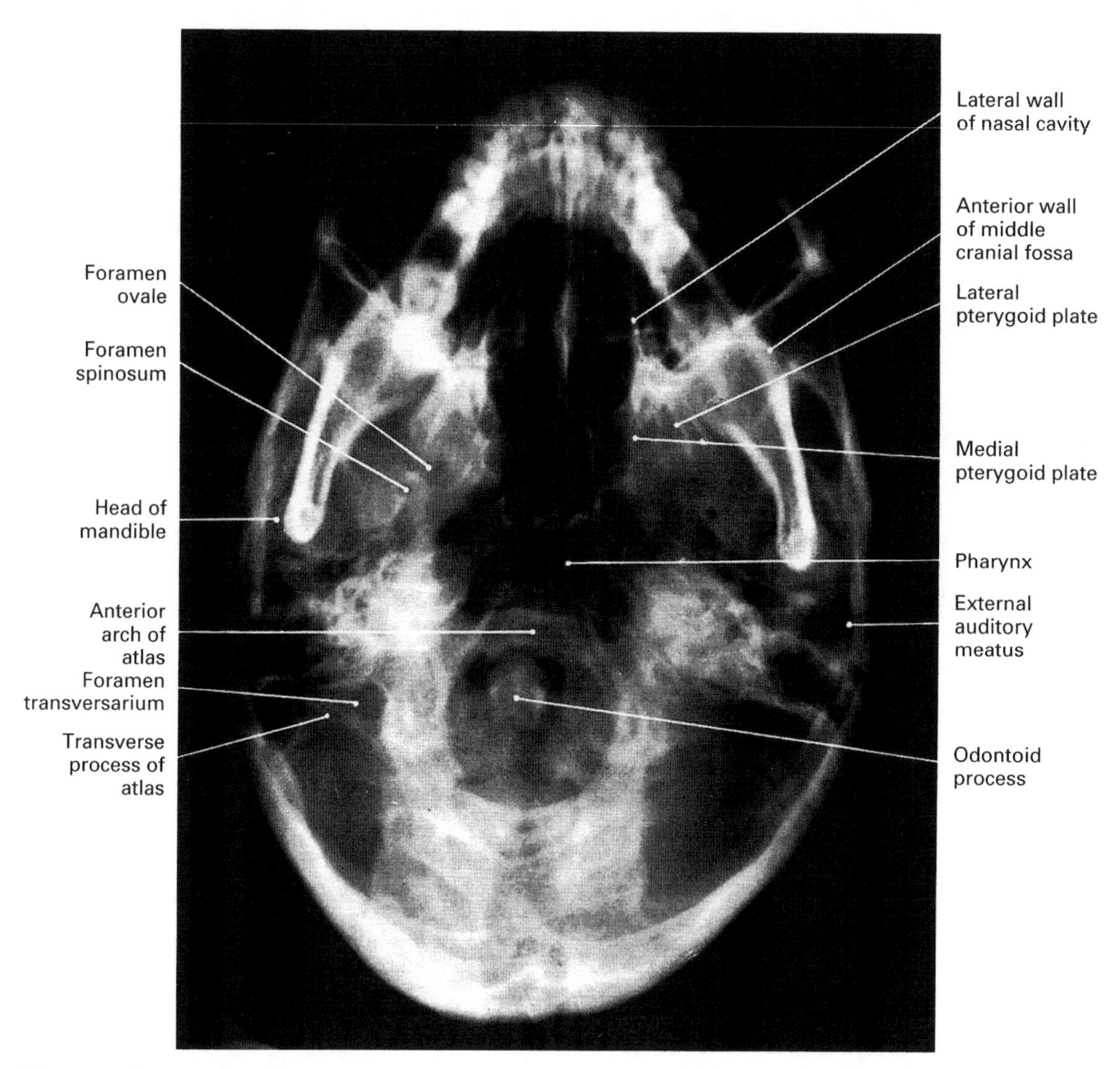

Fig. 2.31 Cranial bones: submentovertical view

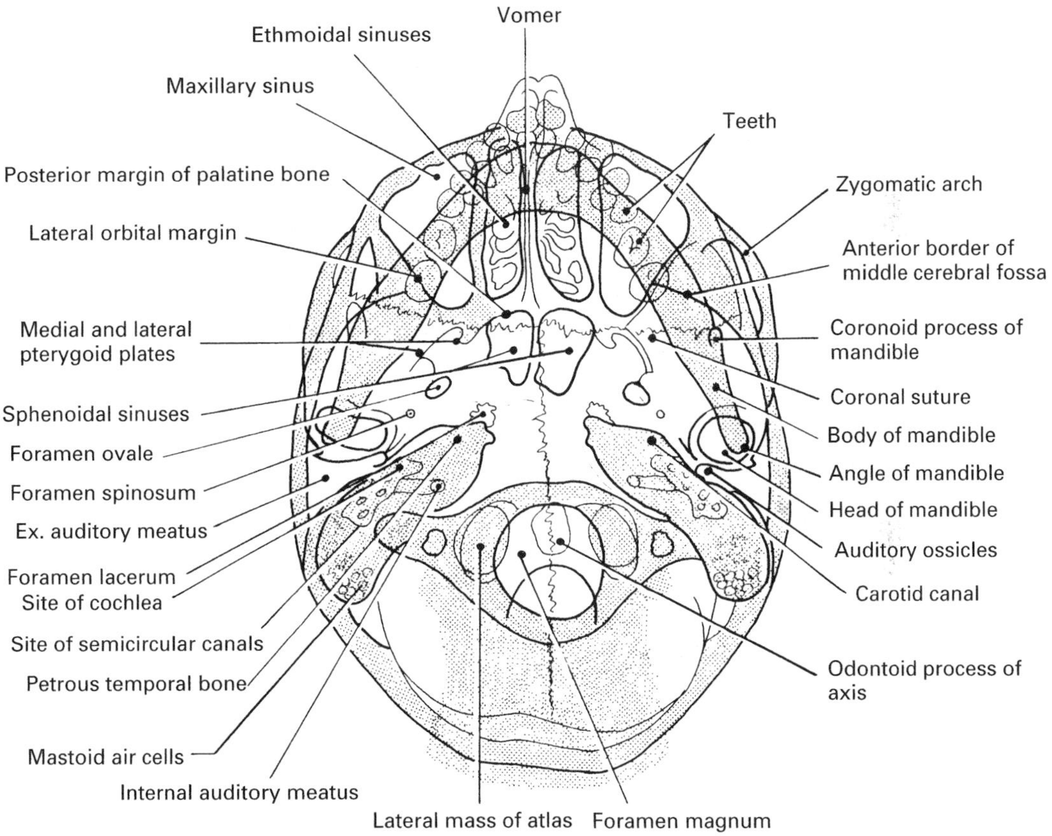

Fig. 2.32 Diagram of submentovertical view

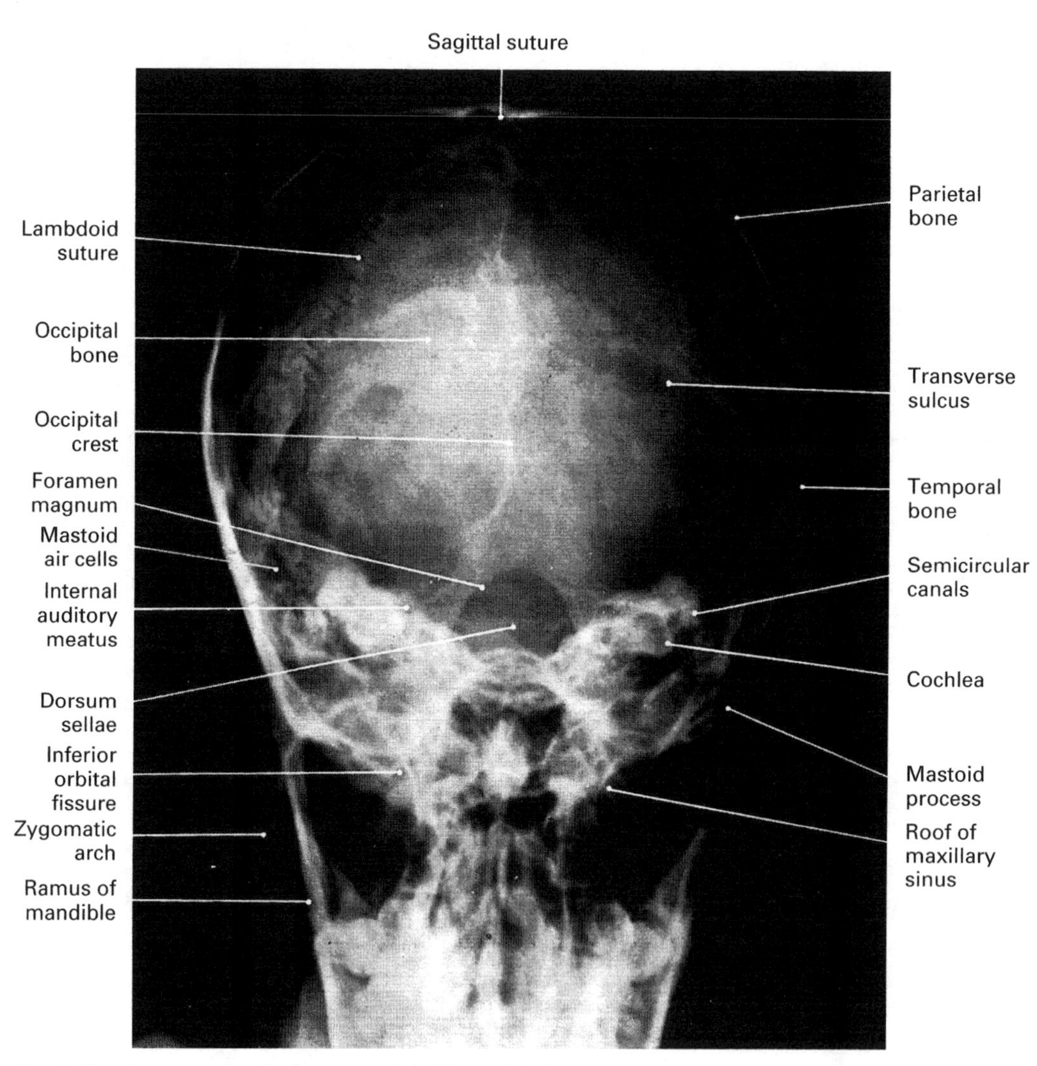

Fig. 2.33 Cranial bones: 30° frontooccipital ('Towne's') view

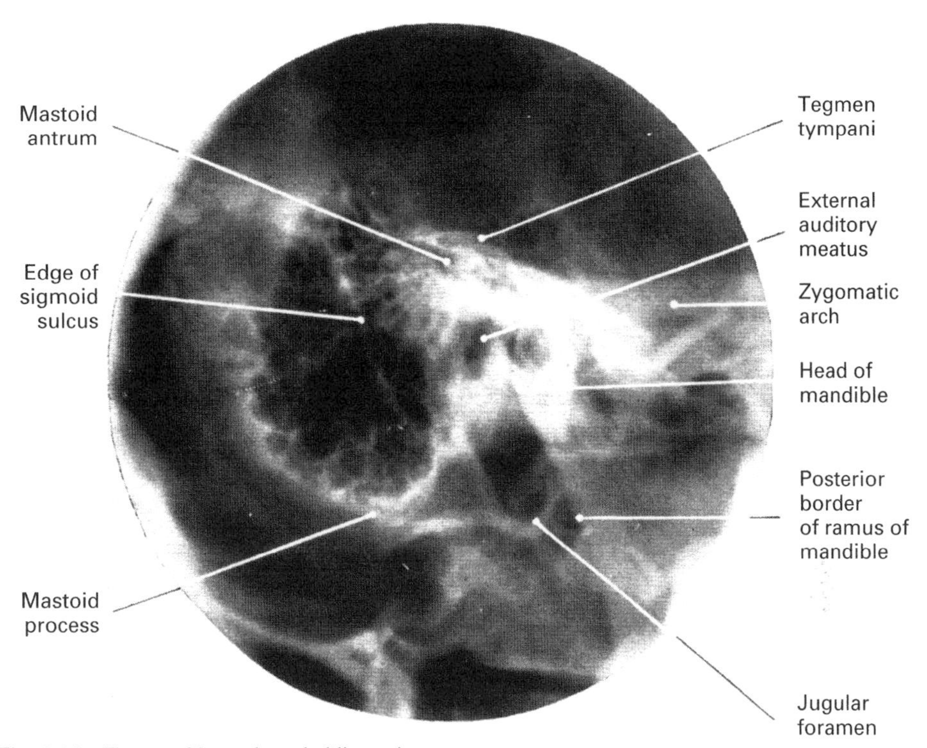

Fig. 2.34 Temporal bone: lateral oblique view

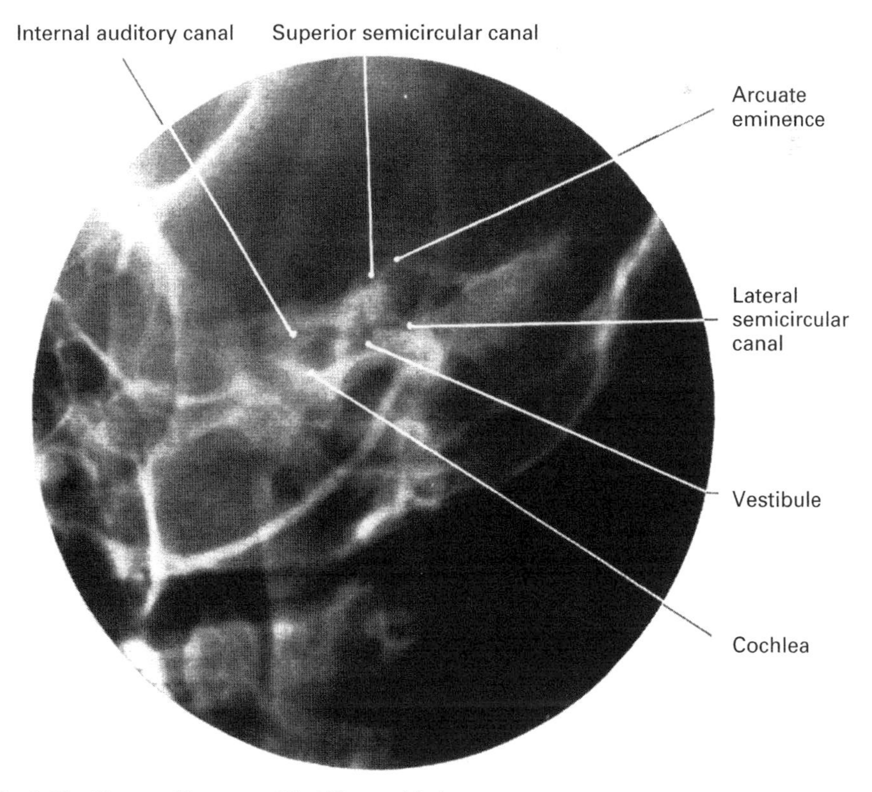

Fig. 2.35 Temporal bone: modified 'Stenver's' view

THE INDIVIDUAL BONES OF THE SKULL 2. FACIAL BONES

MAXILLA (PAIRED) (Fig. 2.36)

The maxillae form most of the bony structure of the face (Fig. 2.40), including the whole of the upper jaw, parts of the orbital and nasal cavities and the anterior two-thirds of the hard palate. Each maxilla consists of:

- the body
- the zygomatic process
- the frontal process
- the palatine process
- the alveolar process.

The body is pyramidal in shape and contains the large maxillary sinus (maxillary antrum).

The anterior surface faces anteriorly and laterally. It is separated from the superior, or orbital, surface by the infraorbital margin below which lies the infraorbital foramen. The inferior part of the anterior surface is marked by a number of ridges corresponding to the roots of the upper teeth. Beyond the last molar tooth is a rounded eminence—the maxillary tuberosity.

The superior surface is triangular in shape and forms the greater part of the floor of the orbit. It articulates with the lacrimal bone and with the orbital plate of the ethmoid. The posterior margin of this surface forms the anterior lip of the inferior orbital fissure.

The medial, or nasal, surface of the maxilla forms a large part of the lateral wall of the nasal cavity (Fig. 2.27 and p. 30) and exhibits an opening, the maxillary hiatus, which leads into the maxillary sinus. Anterior to the maxillary hiatus lies the lacrimal canal which transmits the naso-lacrimal duct. The posterior surface of the maxilla forms the anterior wall of the infratemporal fossa.

The zygomatic process is a roughened triangular area on the lateral aspect of the body. It articulates with the zygomatic bone.

The frontal process is triangular and projects upwards and slightly backwards. Its anterior

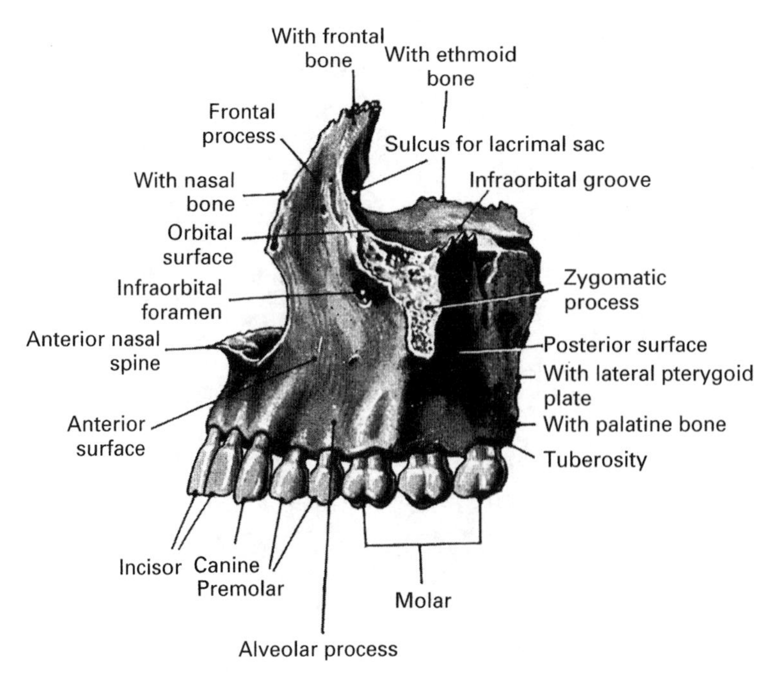

Fig. 2.36 Maxilla: lateral aspect

border articulates with the nasal bone, its posterior border with the lacrimal bone and its superior border with the nasal part of the frontal bone. On the medial surface is the ethmoid crest to which is attached the middle urbinate (concha).

The palatine process projects horizontally from the lower part of the nasal surface and forms, with the palatine process of the other side, the anterior two-thirds of the hard palate (Fig. 2.6, p. 13). In the midline the two palatine processes fuse to form the nasal crest which articulates with the lower border of the vomer and forms part of the nasal septum. The anterior end of the crest forms the nasal spine. On the inferior surface, behind the incisor teeth, is the incisive foramen which transmits the nasopalatine nerves and the terminal branches of the greater palatine vessels from the nasal cavity to the roof of the mouth.

The alveolar process projects downwards from the body of the maxilla to form, with the alveolar process of the other side, the upper dental arch in which are the sockets for the upper teeth. The alveolar process is broader posteriorly to accommodate the larger roots of the molar teeth (see also p. 55).

The maxillary sinus (antrum) is a large pyramidal cavity in the body of the maxilla. Its apex lies near the zygomatic bone, its base is formed by the lateral wall of the nasal cavity (see also p. 37).

ZYGOMATIC BONE (PAIRED)

The zygomatic bone is a small four-sided bone which forms the prominence of the cheek. It consists of:

- the body
- the maxillary process
- the frontal process
- the temporal process
- the orbital process.

The body has a slightly convex outer surface and in it are two foramina for the passage of nerves and vessels from the orbit to the soft tissues of the cheek.

The maxillary process articulates medially with the maxilla.

The frontal process articulates with the zygomatic process of the frontal bone.

The temporal process joins the zygomatic process of the temporal bone and so completes the zygomatic arch.

The orbital process forms part of the lateral and inferior walls of the orbit and articulates with the greater wing of sphenoid and the orbital surface of the maxilla.

PALATINE BONE (PAIRED)

The palatine bone is small and L-shaped. It consists of:

- the horizontal plate
- the vertical plate
- the pyramidal process
- the orbital process
- the sphenoidal process.

The horizontal plate forms, with that of the other side, most of the posterior quarter of the hard palate.

The vertical plate is a thin plate of bone in the posterior part of the lateral wall of the nasal cavity (p. 29). It articulates posteriorly with the medial pterygoid plate of the sphenoid bone. On the medial surface of the vertical plate is a horizontal ridge—the conchal crest—to which is attached the inferior nasal turbinate (concha).

The pyramidal process (or tubercle) fits in between the lower ends of the pterygoid plates of the sphenoid bone at the junction of the horizontal and vertical plates.

The orbital process forms part of the posterior wall of the orbit and articulates with the maxilla and the ethmoid bone.

The sphenoidal process is small and thin. It articulates with the medial pterygoid plate of the sphenoid bone.

The orbital and sphenoidal processes are separated by the sphenopalatine notch which is converted into the sphenopalatine foramen by the sphenoid bone. Through this foramen pass vessels and nerves to the mucous membrane of the nasal cavity.

LACRIMAL BONE (PAIRED)

The lacrimal bone is a small, thin bone situated in the medial wall of the orbit. It articulates with the frontal process of the maxilla, the orbital plate of the ethmoid bone, the orbital plate of the frontal bone and the orbital plate of the maxilla. The lacrimal sac lies in a deep groove—the lacrimal groove—which the lacrimal bone forms with the frontal process of the maxilla.

VOMER (SINGLE)

The vomer is a thin plate of bone which forms the posteroinferior part of the nasal septum (Fig. 2.7). The upper border of the vomer divides into two diverging plates of bone—the alae.

Superiorly the vomer articulates with the perpendicular plate of the sphenoid bone, the vertical plate of the ethmoid bone and with the septal cartilage.

NASAL BONE (PAIRED)

The nasal bones are small, oblong plates of bone which articulate with each other in midline to form the bridge of the nose (Fig. 2.27). The lateral border of each bone articulates with the frontal process of the maxilla. The short superior border articulates with the nasal part of the frontal bone and the lower border gives attachment to the nasal cartilages. (For nasal cavity, see p. 29.)

MANDIBLE (PAIRED)

The mandible (lower jaw) forms the skeleton of the lower part of the face. It consists of:

- the body
- two rami.

(Figs 2.28, 2.37, 2.38, 2.42 and 2.43).

The body is horseshoe shaped. Its superior surface—the alveolar process—contains the sockets for the roots of the lower teeth. Below the premolar teeth lies the mental foramen which transmits the mental nerve to supply the overlying skin. The median plane is marked by a ridge—the **symphysis menti**—which is the site of fusion of the two halves of the body. This fusion is complete by the second year. The inner surface of the body is divided obliquely by the mylohyoid line which gives attachment to the mylohyoid muscle. Above the line is the fossa for the sublingual salivary gland and below it, is the fossa for the submandibular gland. The mylohyoid muscle forms a large part of the floor of the mouth.

The rami. Each ramus projects upwards almost at right angles to the body. At the upper end are two projections—the **condylar process** and the **coronoid process**. They are separated by the mandibular notch through which nerves and vessels pass to the overlying masseter muscle—an important muscle of mastication. The site at which the posterior and inferior borders of the ramus join is called the angle of the jaw. On the inner surface of the ramus is the mandibular foramen—the opening of the mandibular canal—through which pass nerves and vessels supplying the lower teeth. Also on the inner surface is the mylohyoid groove which houses the mylohyoid nerve.

The condylar process consists of an expanded head and neck. The head articulates with the mandibular fossa of the temporal bone to form the temporomandibular joint (p. 48).

The coronoid process gives attachment to the temporal muscle.

HYOID BONE

The hyoid bone (Fig. 2.39) lies just above the larynx in the anterior part of the neck. It consists of:

- the body
- two greater cornua
- two lesser cornua.

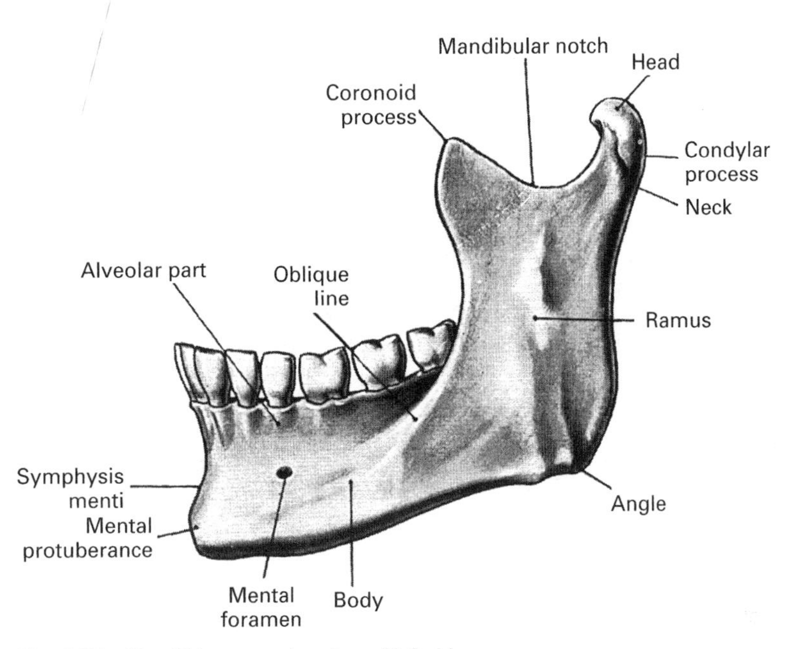

Fig. 2.37 Mandible: external surface of left side

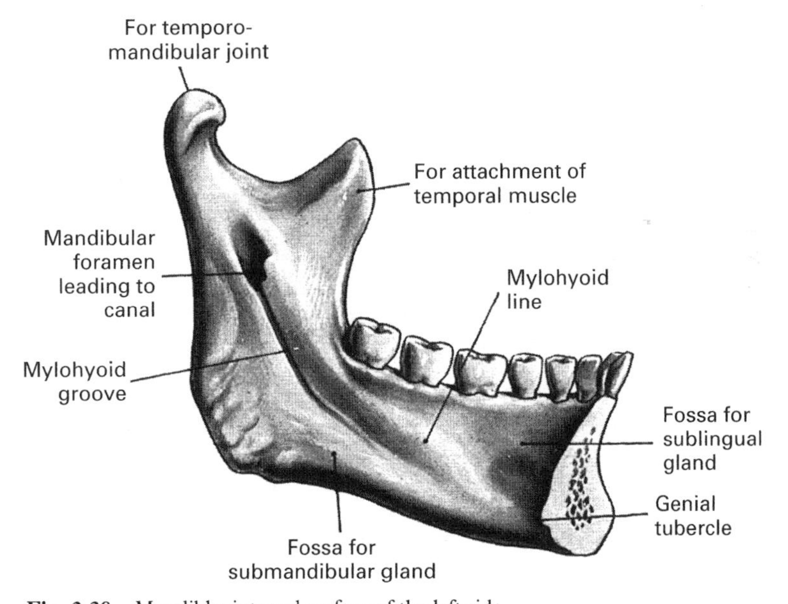

Fig. 2.38 Mandible: internal surface of the left side

The hyoid gives attachment above to some of the muscles of the floor of the mouth and to extrinsic muscles of the tongue and below to the infrahyoid muscles of the neck which depress the hyoid after it has been raised during swallowing. It also gives attachment to the middle constrictor muscles of the pharynx.

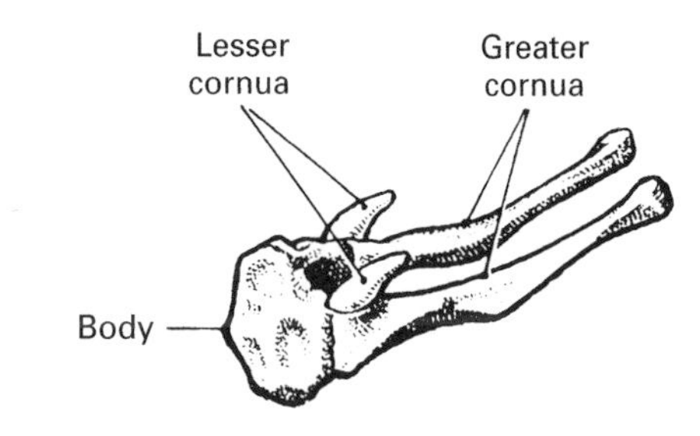

Fig. 2.39 Hyoid bone: left supero-lateral aspect

Radiographic appearances of the facial bones (Figs 2.40 to 2.43)

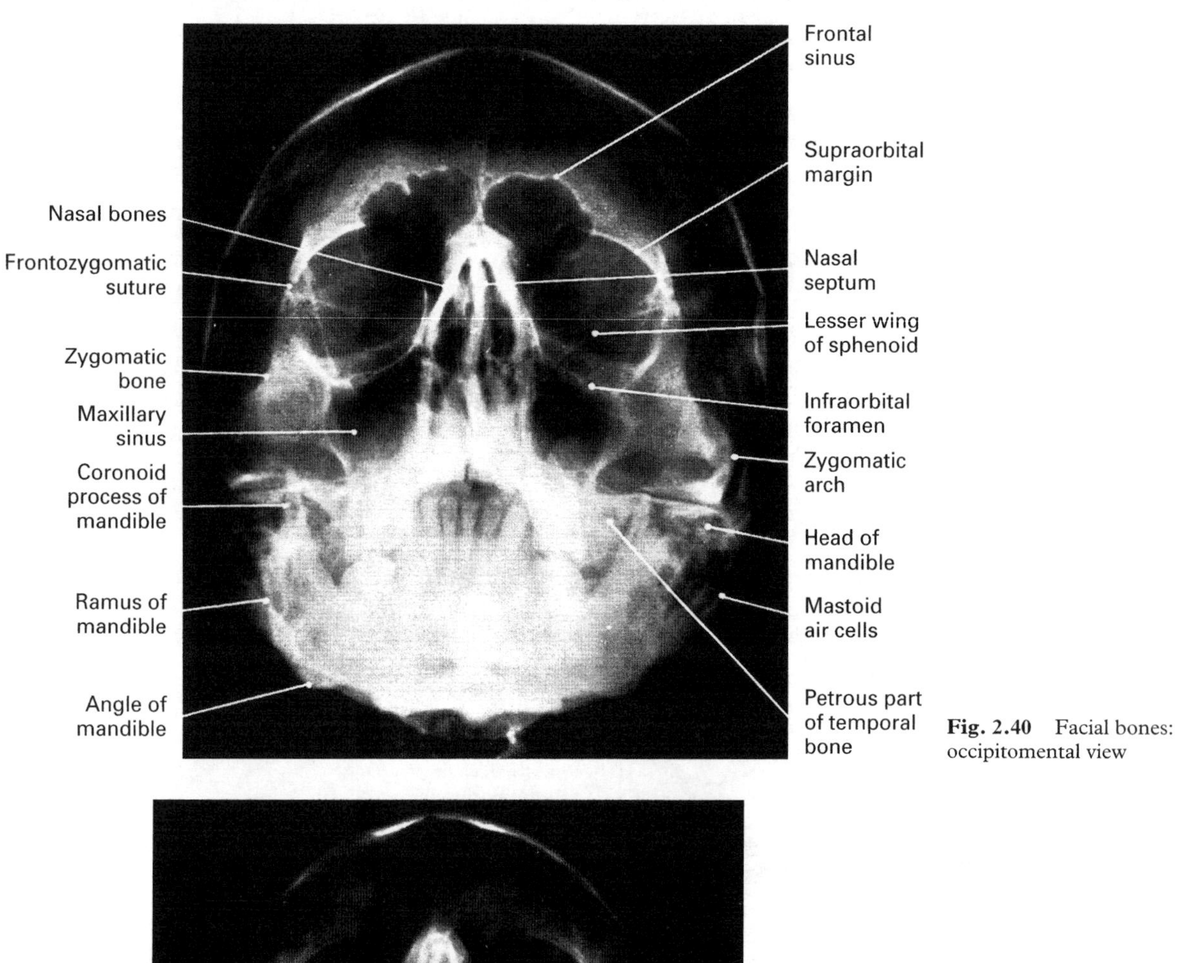

Fig. 2.40 Facial bones: occipitomental view

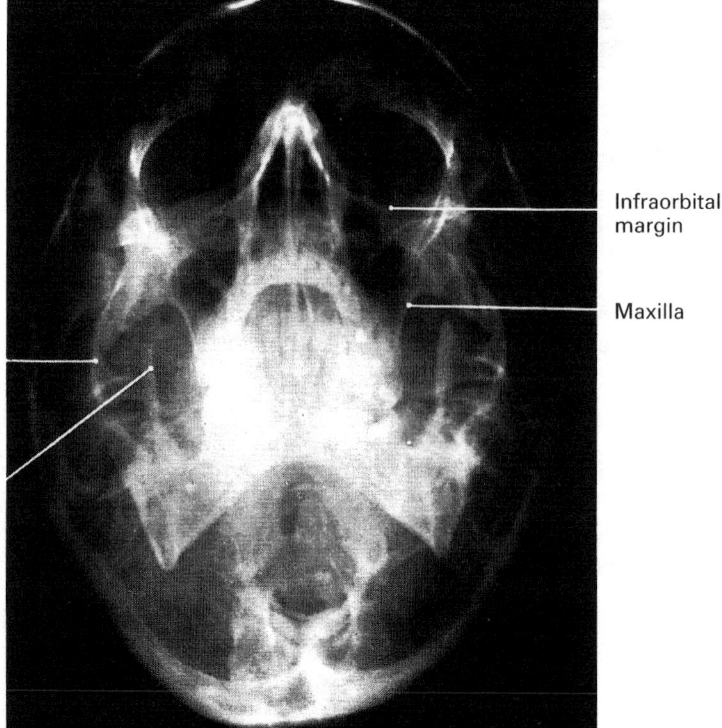

Fig. 2.41 Facial bones: 30° occipitomental view

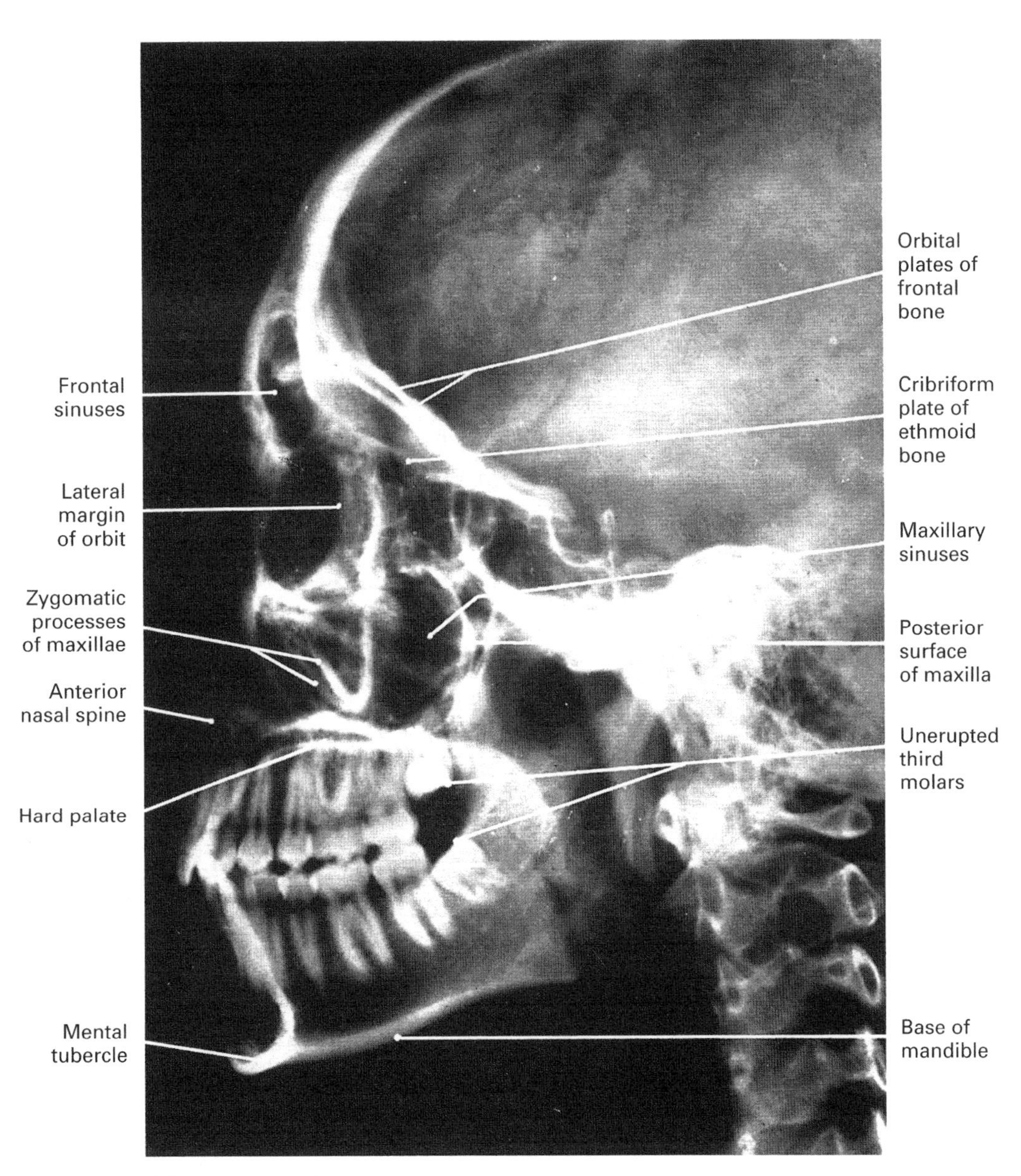

Fig. 2.42 Facial bones: lateral view

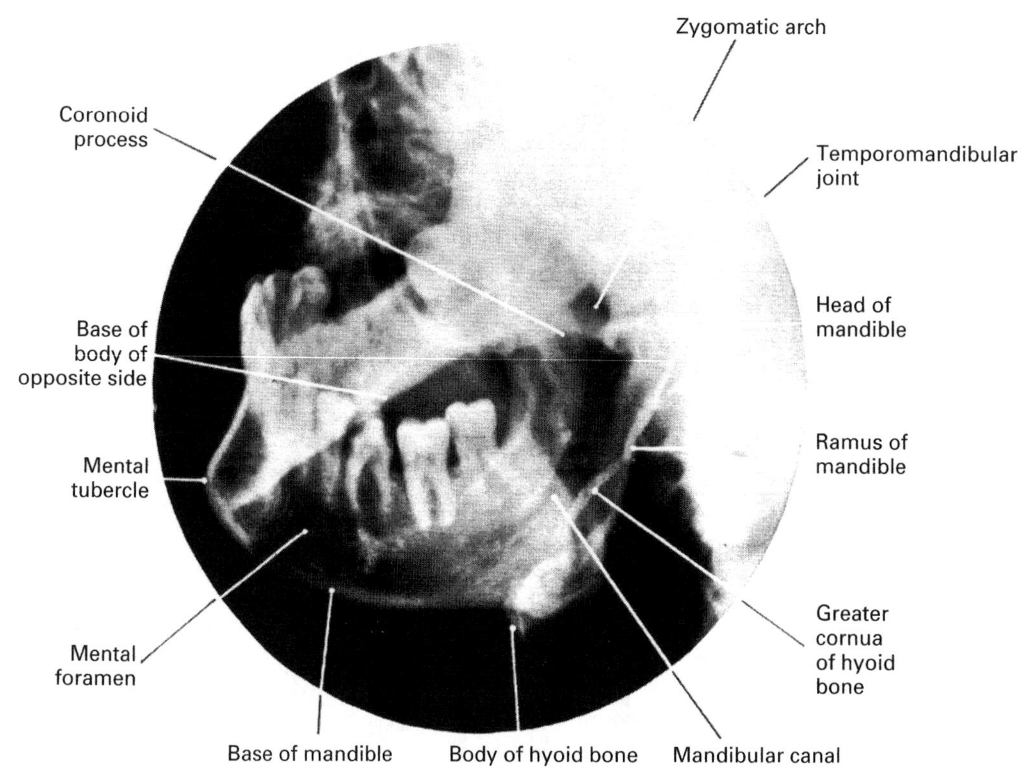

Fig. 2.43 Mandible: lateral oblique view

THE NASAL SINUSES (Figs 2.44 to 2.47)

There are four groups of nasal sinuses:

- the maxillary antra
- the sphenoidal sinuses
- the ethmoidal air cells
- the frontal sinuses.

They are paired but are often asymmetrical.

The maxillary antra are the largest of the paranasal air sinuses. Each antrum is a triangular-shaped cavity lying in the body of the maxilla and lying lateral to the lower half of the nasal cavity. The roof of the antrum is formed by the floor of the orbit, the base is formed by the lateral wall of the nasal cavity and the apex extends into the zygomatic process of the maxilla. The maxillary antrum opens into the middle meatus, the opening being near the roof of the antrum.

The sphenoidal sinuses occupy the body of the sphenoid bone. They are very variable in size and shape and are seldom symmetrical. They open into the sphenoethmoidal recesses, above and behind the superior turbinates (conchae).

The ethmoidal air cells (anterior, middle and posterior) consist of a variable number of thin-walled cavities lying between the upper part of the nasal cavity and the orbits, from which they are separated by the thin orbital plate of the ethmoid bone. The anterior and middle groups of cells open into the middle meatus and the posterior group into the superior meatus.

The frontal sinuses lie posterior to the supercilliary arches, between the outer and inner tables of the frontal bone (Fig. 2.7). They are divided by the median septum and are usually asymmetrical. Usually they open into the middle meatus directly but sometimes they share openings with the anterior ethmoid cells.

Radiographic appearances of the nasal sinuses (Figs 2.44 to 2.50).

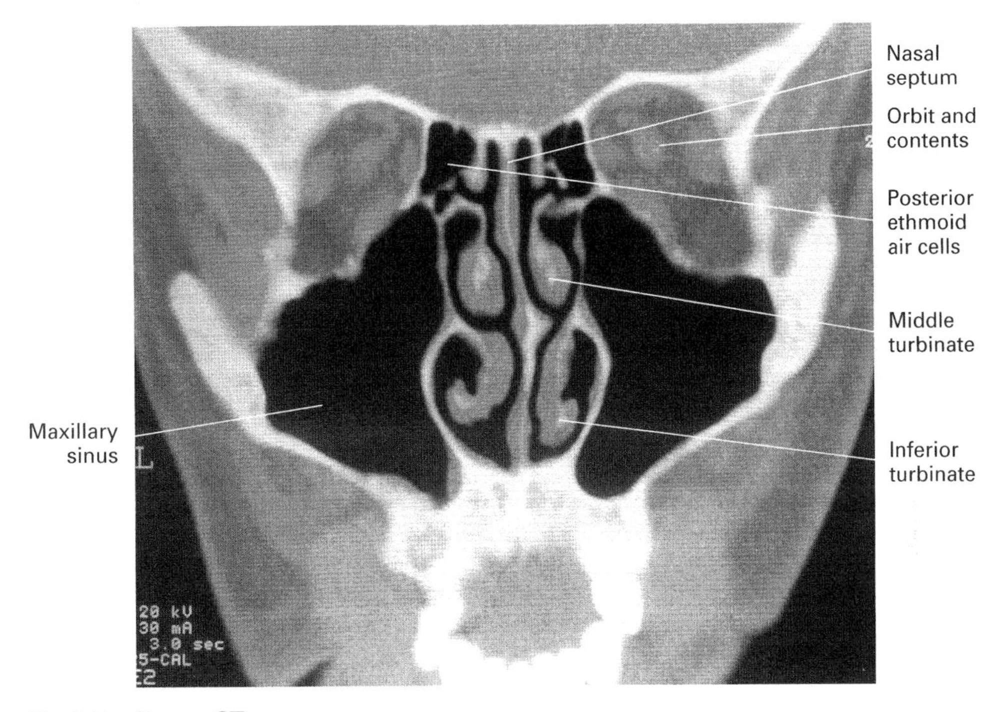

Fig. 2.44 Sinuses: CT scan

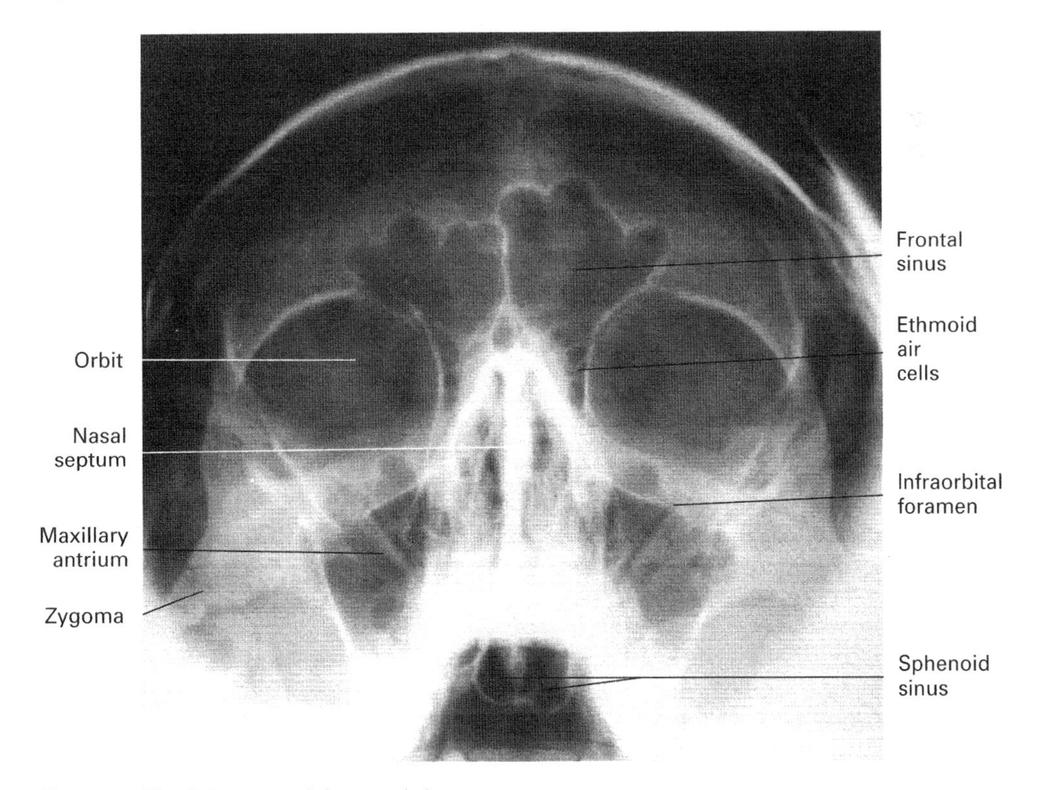

Fig. 2.45 Nasal sinuses: occipitomental view

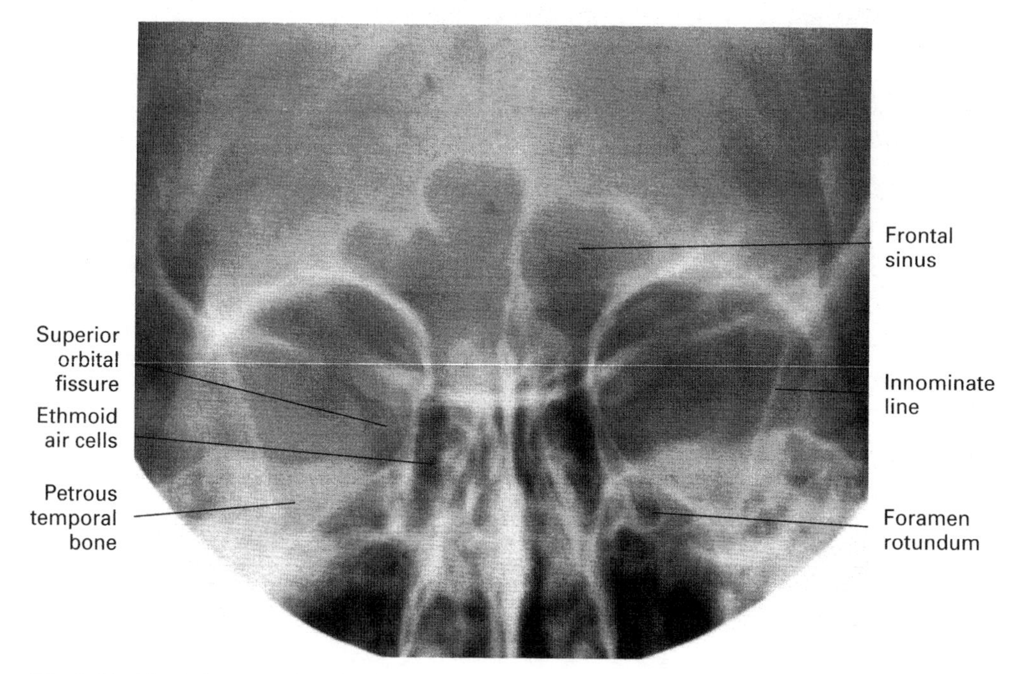

Fig. 2.46 Nasal sinuses: occipitofrontal view

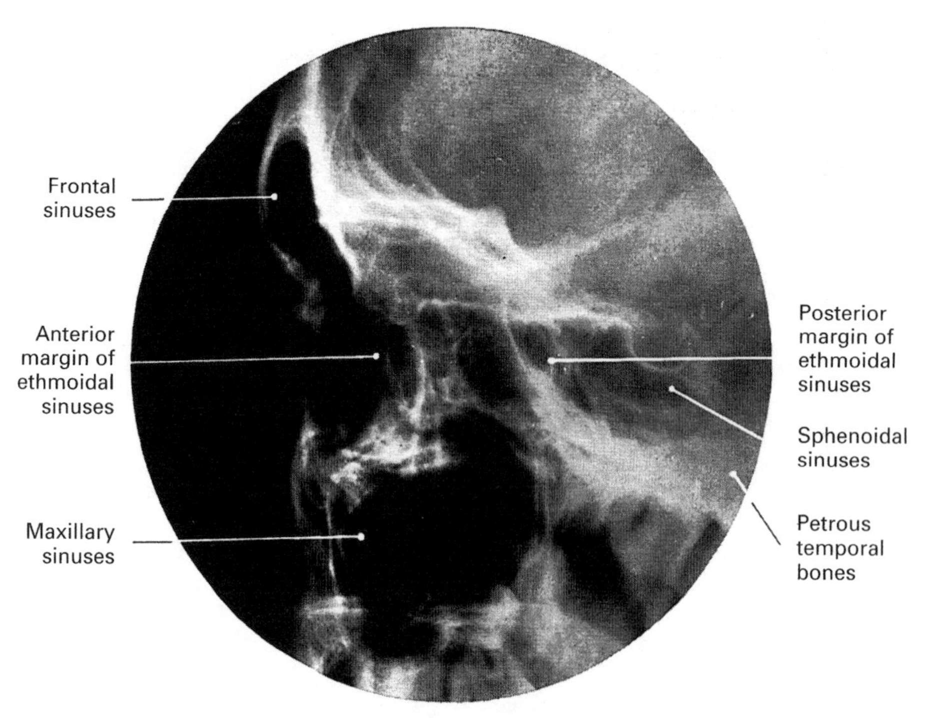

Fig. 2.47 Nasal sinuses: lateral view

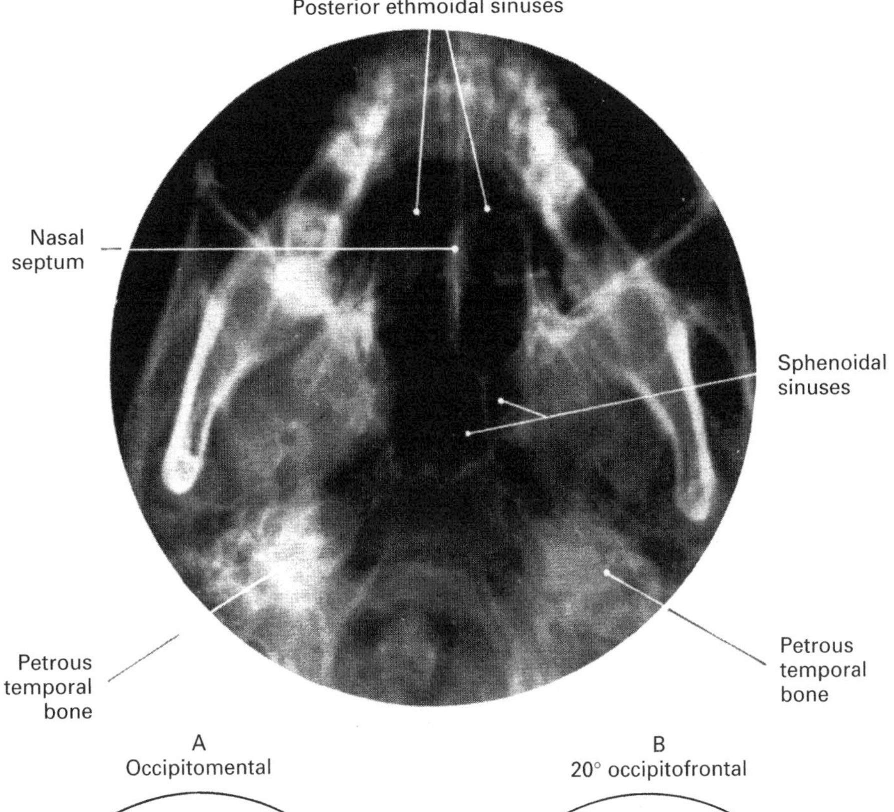

Fig. 2.48 Nasal sinuses: submentovertical view

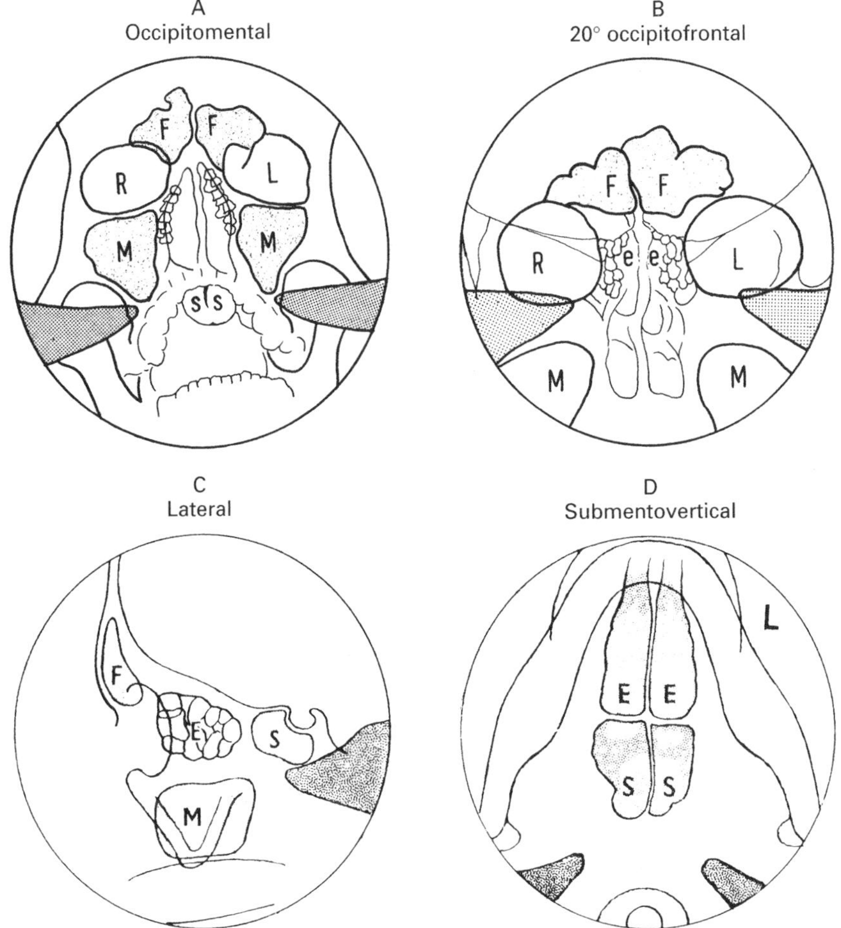

Fig. 2.49 Diagrams of the radiographic views of the nasal sinuses

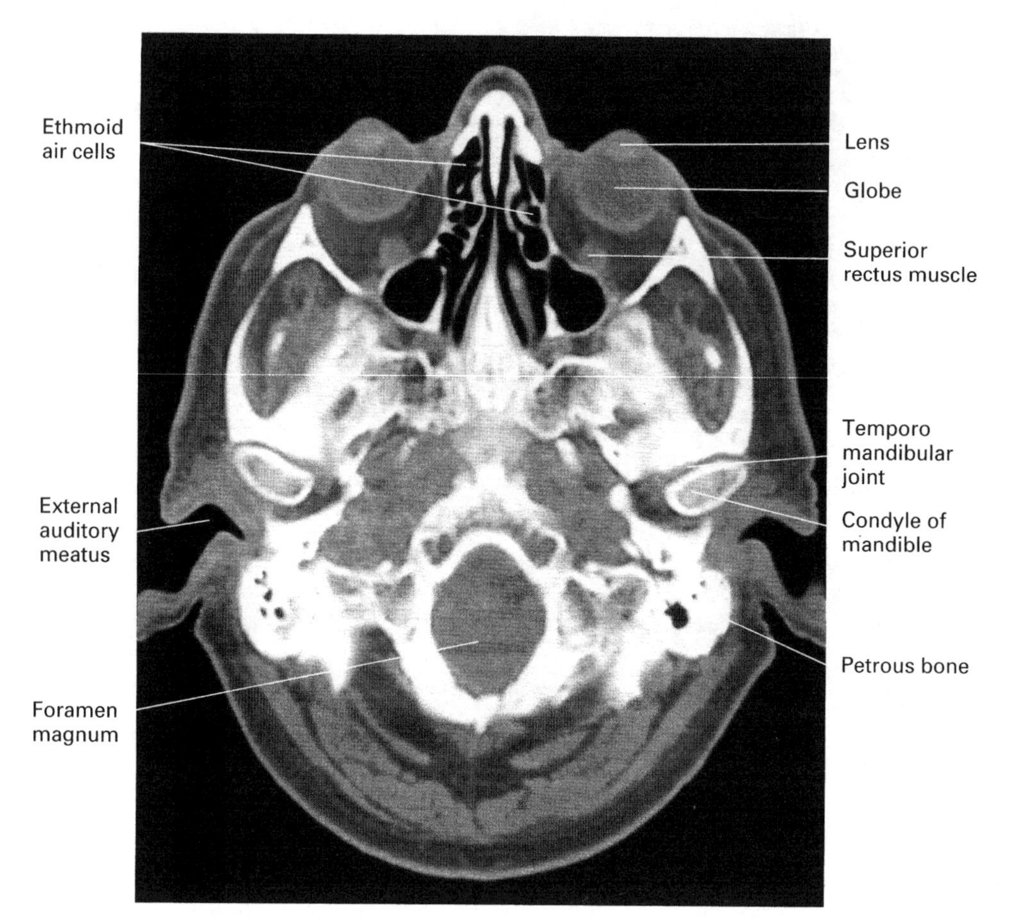

Fig. 2.50 CT scan at level of foramen magnum

TEMPOROMANDIBULAR JOINT

(Figs 2.51 to 2.53)

Type: Synovial, condylar. The joint is divided into two parts by an articular disc and each half has its synovial membrane.

Articular surfaces: Head of mandible and anterior part of mandibular fossa of the temporal bone.

Capsule: Attached to margins of mandibular fossa and below it to the neck of the mandible. Disc is attached to capsule which is loose above the disc (allowing forward movement) and tight below it.

Ligaments:

Lateral—temporomandibular. Thickened band on lateral side. Attached above to articular tubercle of temporal bone (at root of zygoma) and below to posterior border of mandibular neck. The fibres pass posteriorly and inferiorly and the ligament lies deep to the parotid gland.

Medial—sphenomandibular. Attached above to the spine of sphenoid bone and below to lingula of mandibular foramen. The ligament is thin above, becoming broader at the lingula.

Articular disc: Oval, fibrous disc, attached by its circumference to the capsule. It divides the joint into two separate synovial joints. The superior joint allows a forward gliding movement of the disc while the inferior joint allows both a hinge and a forward gliding movement. The superior surface of the disc is convex posteriorly and concave anteriorly. The

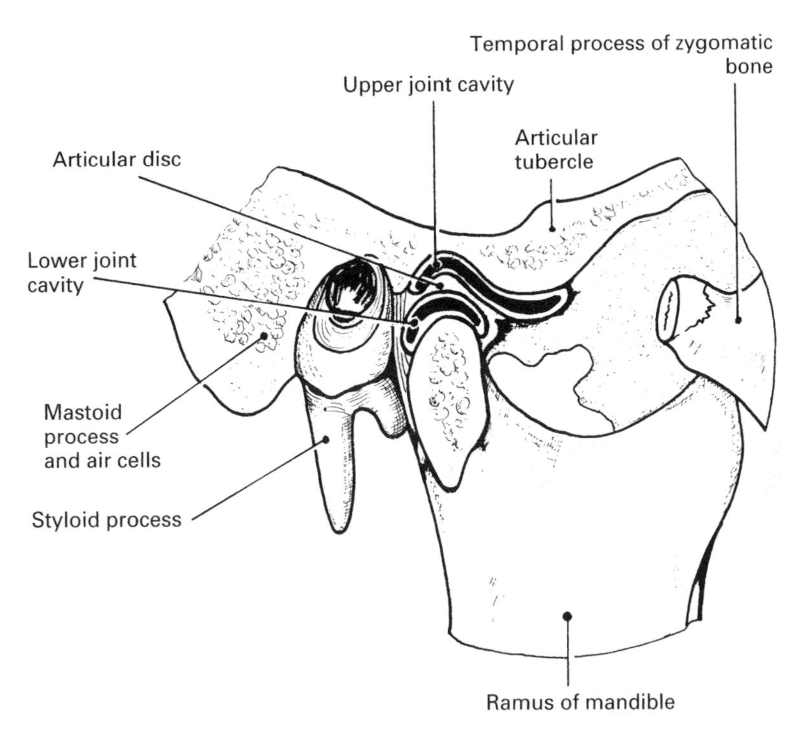

Fig. 2.51 Schematic diagram of temporomandibular joint

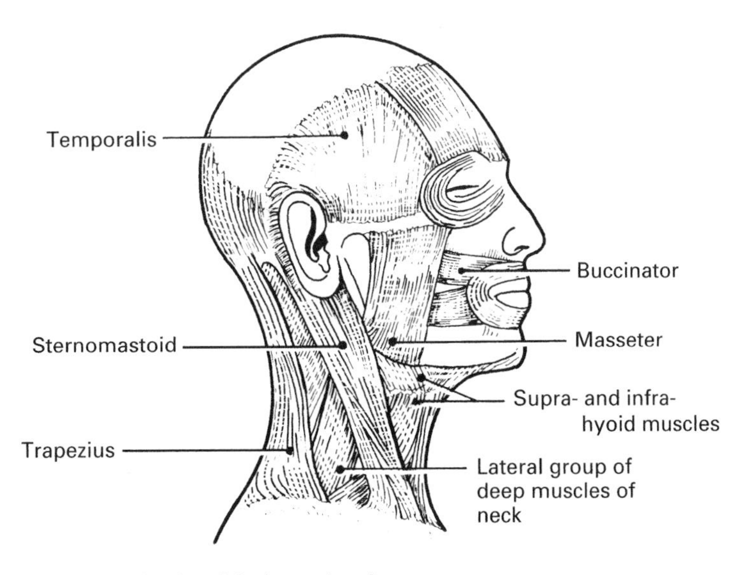

Fig. 2.52 Muscles of the jaw and neck

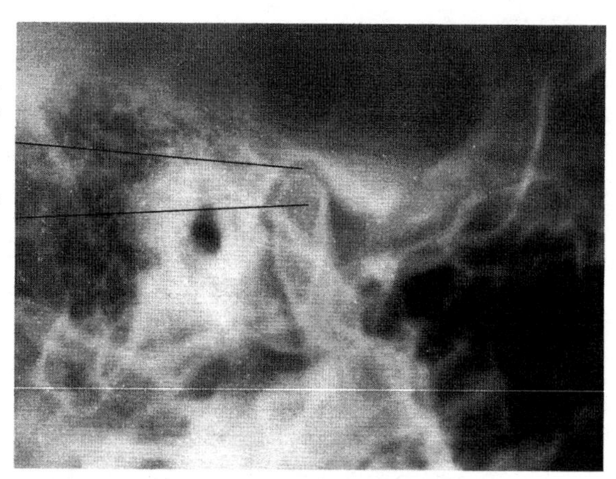

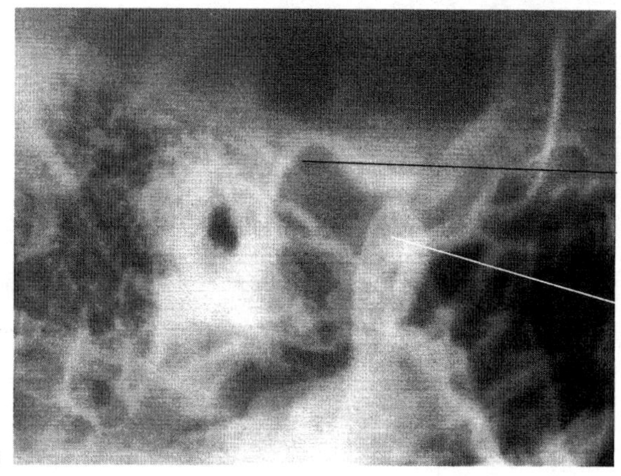

Fig. 2.53 A & B Temporomandibular joint: lateral oblique view. A: with mouth closed. B: with mouth open. When the mouth is closed, the head of the mandible lies in the mandibular fossa of the temporal bone. When the mouth is opened, the head of the mandible rotates and moves forward on the articular tubercle

head of the mandible articulates with the concave inferior surface of the disc and remains in intimate contact. The articular disc and mandibular head move together, the mandibular head first rotating about an horizontal axis (as in a hinge) and then gliding downwards and forwards over the convex anterior portion of the disc's lower surface. This movement brings about 'depression' of the mandible, i.e. opening the mouth. The converse movement brings about 'elevation', i.e. closing the mouth. Some lateral movement occurs also in chewing. The lateral pterygoid muscle is attached to the medial and anterior aspects of the mandibular head and to the articular disc; it pulls the disc forward in depression of the mandible.

Movements and muscles:

Depression—opening the mouth. Geniohyoid, mylohyoid, digastric, lateral pterygoid and platysma.

Elevation—closing the mouth. Temporalis (anterior fibres), masseter and medial pterygoid.

Protrusion—moving the jaw forwards. Medial and lateral pterygoids.

Retraction—moving the jaw backwards. Temporalis (posterior fibres), digastric, geniohyoid, and by middle parts of masseter in forced retraction.

Lateral movement—e.g. in chewing. Medial and lateral pterygoids, of each side, acting alternately producing an ovoid motion.

Geniohyoid:
- origin—inferior mental spine of mandible.
- insertion—body of hyoid bone.
- function—elevation of hyoid, depression of mandible.
- nerve supply—1st cervical nerve, via ansa cervicales.

Mylohyoid:
- origin—mylohyoid line of mandible.
- insertion—body of hyoid and with its fellow at the median raphé.
- function—elevation of hyoid, depression of mandible.
- nerve supply—mandibular branch of trigeminal (Vth cranial) nerve.

Digastric (two bellies):
- origin
  - anterior belly—inferior border of anterior mandible
  - posterior belly—mastoid notch of temporal bone (digastric fossa).
- insertion—the two bellies are joined by a tendon which passes through a sling of fascia attached to the hyoid bone.
- function—separately the anterior belly pulls the hyoid forwards and the posterior belly pulls backwards. The combined action is depression of the mandible.
- nerve supply
  - anterior belly—mandibular nerve
  - posterior belly—facial (VIIth cranial) nerve.

Platysma:
- origin—fascia over deltoid and pectoralis muscles.
- insertion—skin over lower face and lips, and to body of mandible.
- function—depression of mandible, tightening of skin over neck.
- nerve supply—facial (VIIth cranial) nerve.

Lateral pterygoid:
- origin—lateral aspect of lateral pterygoid plate of sphenoid.
- insertion—neck of mandibular condyle and articular disc.
- function—depression of mandible.
- nerve supply—mandibular branch of trigeminal (Vth cranial) nerve.

Medial pterygoid:
- origin—medial aspect of lateral pterygoid plate of sphenoid.
- insertion—medial surface of angle of mandible.
- function—elevation and protrusion of mandible.
- nerve supply—mandibular branch of trigeminal (Vth cranial) nerve.

Temporalis:
- origin—(fan-shaped), lateral aspect of skull (temporal fossa) and overlying fascia.
- insertion—coronoid process of mandible and its anterior surface.
- function—vertical fibres, elevation of mandible; horizontal fibres, retraction of mandible.
- nerve supply—mandibular branch of trigeminal (Vth cranial) nerve.

Masseter:
- origin—zygomatic arch (superficial and deep fibres).
- insertion—lateral aspect of ramus of mandible.
- function—elevation of mandible.
- nerve supply—mandibular branch of trigeminal (Vth cranial) nerve.

SUMMARY OF FORAMINA OF THE SKULL

Foramen	*Position*	*Transmits*
Anterior condylar canal	Occipital bone, above condyles	Hypoglossal nerve (12)
Carotid canal	Petrous part of temporal bone, posterolateral to foramen lacerum	Internal carotid artery
Foramen in cribriform plate	Ethmoid bone	Branches of olfactory nerve (1)
Foramen lacerum	Junction of petrous body and greater wing of sphenoid	Crossed by internal carotid artery, nerves and vein
Foramen magnum	Occipital bone	Spinal cord, vertebral arteries, spinal root of accessory nerve
Foramen ovale	Greater wing of sphenoid	Mandibular dividion of trigeminal nerve (5)
Foramen rotundum	Greater wing of sphenoid	Maxillary division of trigeminal nerve (5)
Foramen spinosum	Greater wing of sphenoid	Middle meningeal artery
Hypoglossal canal	Occipital condyle	Crossed by hypoglossal nerve (12)
Incisive canal	Palatine part of maxilla	Palatine nerves and vessels
Inferior orbital fissure	Between greater wing of sphenoid, maxilla, zygoma and palatine	Maxillary branch of trigeminal nerve (5) and infraorbital vessels
Infraorbital foramen	Maxilla, below infraorbital margin	Infraorbital nerve and vessels
Internal auditory meatus	Petrous part of temporal bone	Facial nerve (7), auditory nerve (8) and internal auditory artery
Jugular foramen	Between occiput and petrous part of temporal bone	Glossopharyngeal (9) vagus (10) accessory (11) nerves, internal jugular vein, inferior petrosal sinus
Mental foramen	Mandible	Mental nerve and artery
Optic foramen	Sphenoid bone, between greater and lesser wings	Optic nerve (2), ophthalmic artery
Stylomastoid foramen	Temporal bone, base of styloid process	Facial nerve (7)
Superior orbital fissure	Sphenoid bone, between greater and lesser wings	Oculomotor nerve (3), trochlear (4), ophthalmic branch of trigeminal (5), abducent (6), ophthalmic nerves and veins
Supraorbital foramen	Frontal bone	Supra orbital vessels and nerves

OSSIFICATION AND BONY DEVELOPMENT OF THE SKULL

At birth the cranial vault is large in relation to the body (Fig. 1.3). It is large in relation to the facial bones also—mainly because the upper and lower jaws are small, the teeth are unerupted and the air sinuses are rudimentary (Fig. 2.54).

Gaps, called fontanelles, are present in the skull of an infant. The anterior and posterior fontanelles, situated at the ends of the sagittal suture, are the largest of these fontanelles. The posterior fontanelle closes soon after birth by growth of the surrounding bone but the anterior fontanelle remains open until the second year.

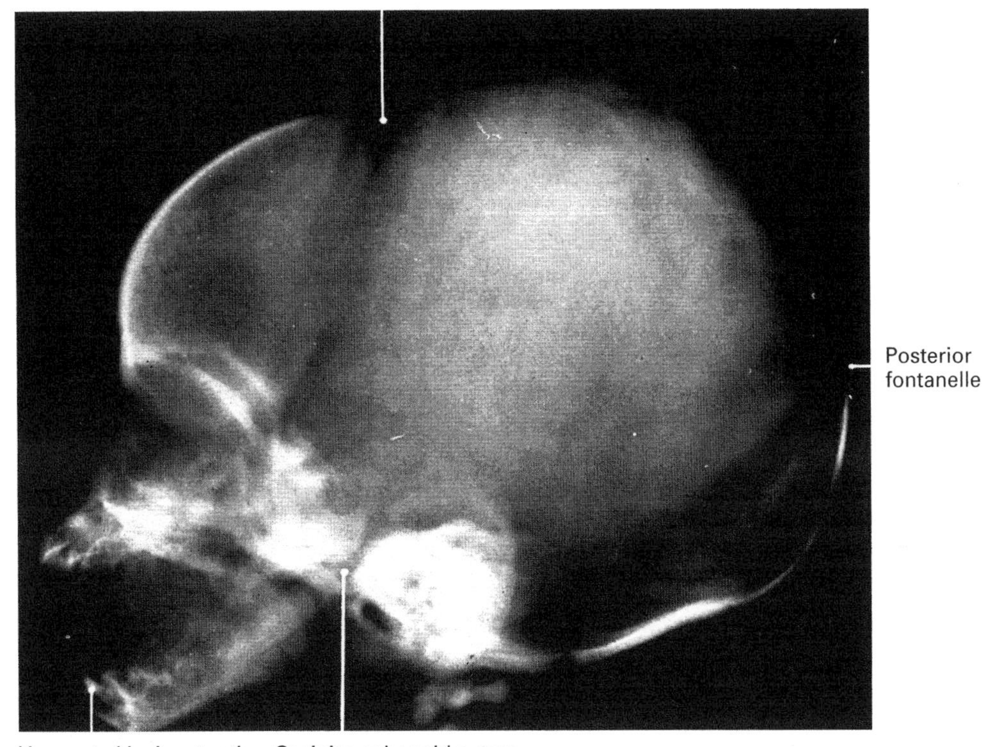

Fig. 2.54 The skull at birth: lateral view

Ossification of the bones of the skull is incomplete at birth and some bones, e.g. the mandible, consist of separate parts which do not fuse until after birth. At birth the two halves of the frontal bone are separated in the midline by a suture—the metopic suture. The two halves usually fuse at about the eighth year.

Ossification of the bones of the vault of the cranium and of the face takes place in membrane whereas the bones of the base of the skull ossify in cartilage. Thus some bones that form part of the base and the vault, e.g. occipital bone, are ossified partly in cartilage and partly in membrane from separate primary centres.

During postnatal life, growth of the facial bones exceeds that of the cranial vault. At birth the capacity of the cranial vault is two-thirds that of the adult. In the adult the volume occupied by the face is about the same as that occupied by the cranium whereas at birth the volume occupied by the face is only about one-eighth of that occupied by the cranium. Growth of the cranium and facial bones is rapid up to the 7th year, when the petrous part of the temporal bones, the sinuses, the orbits and the foramen magnum have reached almost adult size. Between the 7th year and puberty, growth is slower, but after puberty there is a further period of rapid growth, coincident with the eruption of the permanent dentition.

The sutures begin to close between 30 and 40 years of age but sometimes they remain unfused throughout life.

TEETH

The teeth consist of two dentitions, the deciduous and the permanent. The deciduous teeth erupt during the first 2 years of childhood and the permanent teeth replace them as the jaws grow to full maturity; this covers the period from the 6th year to some time between the 17th to the 25th year.

Each tooth consists of three main parts, the crown, the neck and the root (Fig. 2.55). Enclosed down the centre of each is a canal—the pulp

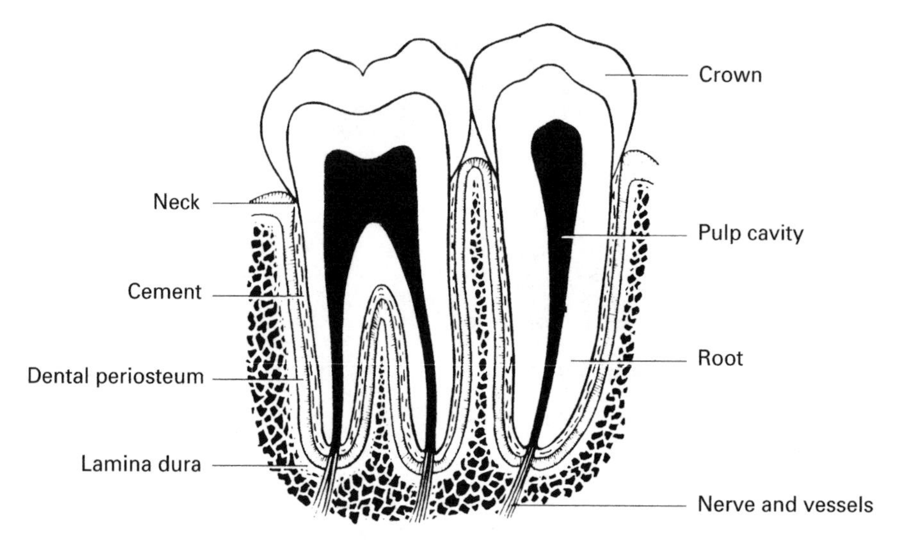

Fig. 2.55 Sectional diagrram of teeth in their sockets

cavity—containing the blood vessels and nerve of the tooth which enter the cavity through the apex of the root.

The main structure of the tooth is of a hard substance called dentine. This is covered at the crown by an even denser layer of enamel which forms a protective cap over the exposed part of the tooth. The enamel encases the tooth as far as the neck, where it is superseded by the cement covering of root.

The socket for the tooth, contained in the alveolar process of the jaw, is lined with a compact layer of bone called the lamina dura. Fibrous periodontal membrane (dental periosteum) forms the attachment between the lamina dura and the tooth root, which is a peg-and-socket type of fibrous joint (gomphosis).

Deciduous (or milk) teeth

The deciduous dentition consists of 20 teeth, 10 in each jaw.

In each jaw the two central incisors occupy the medial positions in the dental arches at the front of the mouth, with a lateral incisor on either side of them. These teeth have single roots, and their crowns have bevelled cutting edges.

Next to the incisors are the canine teeth, two in each arch, with long single roots. Their crowns carry single conical tubercles or cusps.

Beyond each canine, at the back of the dental arches, are two molar teeth, with roots having three prongs in the upper jaw and two prongs in the lower jaw. These roots are splayed in order to give space between them for development of the permanent premolar teeth which will eventually replace them.

The crowns of the molar teeth are somewhat square in shape, with bulging sides that overhang the necks. On their grinding surfaces there are three to five cusps which interlock with their opposite numbers when the jaws are occluded.

Permanent teeth

The permanent dentition consists of 32 teeth, 16 in each jaw. The incisors—medial and lateral—and the canines are similar to those of the deciduous dentition but they are larger.

In addition, there are eight premolars (bicuspids), two distal to each canine tooth, occupying the position previously filled by the deciduous molars. These teeth usually have single roots which are grooved; occasionally a root may be bifurcated down the line of the groove. The premolars have crowns with two cusps, one labial (cheek side) and the other lingual (toward the tongue).

Beyond each premolar, where previously there were no teeth in the deciduous dentition, are three permanent molars. The molar teeth of the upper jaw have three roots. The roots of the smaller third molars are often partly fused and pyramidal in shape. The lower molars have two roots, curved with a slight backward inclination. The crowns have from three to five cusps and are larger than the molars of the upper jaw.

The teeth are numbered according to established dental formulae. Conventionally, those of the upper and lower jaws of each side are numbered from 1 to 8 for permanent teeth and from *a* to *e* for deciduous teeth. Thus:

R	8 7 6 5 4 3 2 1	1 2 3 4 5 6 7 8	L
	8 7 6 5 4 3 2 1	1 2 3 4 5 6 7 8	

Adult teeth

R	e d c b a	a b c d e	L
	e d c b a	a b c d e	

Deciduous teeth

By this formula, individual teeth are indicated thus: $\overline{8|}$ for adult right lower third molar, $\underline{|3}$ for adult left upper canine, $\overline{|c}$ for deciduous left lower canine.

In the two-digit dental recording system, instituted by the Fédération Dentaire Internationale, the formula is easier to write and type and, unlike the conventional formula, can be relayed by Telex because it dispenses with the quadrant sign. Thus:

R	18 17 16 15 14 13 12 11	21 22 23 24 25 26 27 28	L
	48 47 46 45 44 43 42 41	31 32 33 34 35 36 37 38	

Adult teeth

R	55 54 53 52 51	61 62 63 64 65	L
	85 84 83 82 81	71 72 73 74 75	

Deciduous teeth

In this formula, the first digit indicates the quadrant and also distinguishes between adult and deciduous teeth. For example, individual teeth are indicated thus: 48 for adult right lower third molar, 23 for adult left upper canine, 73 for deciduous left lower canine.

Development of the teeth

The teeth begin to develop as early as the sixth week of intrauterine life and at birth calcification of the deciduous teeth is well advanced. Beside the lingual aspect of each deciduous tooth lies the bud of the permanent tooth, which will eventually replace it, and at this stage, tooth buds of the first permanent molars are also present beyond the area of the deciduous teeth.

Each tooth, deciduous or permanent, develops in its own erupting sac, or crypt, until it is hard enough to bear the pressure which will be exerted on it during mastication. When this stage has been reached it erupts through the fibrous tissues of the gum and becomes visible. During this growing period of the permanent teeth the roots of the deciduous teeth are absorbed, leaving the crowns to be pushed off by the pressure of the developing tooth below.

The following tables give the normal dates of eruption of both dentitions but there is considerable variation, and in some subjects the permanent third molars may never develop at all.

Eruption of deciduous teeth

Lower central incisors	6 to 9 months
Upper four incisors	8 to 10 months
Lower lateral incisors	15 to 21 months
First molars	15 to 21 months
Canines	16 to 20 months
Second molars	20 to 24 months

Eruption of permanent teeth

First molars	6th year
Central incisors	7th year
Lateral incisors	8th year
First premolars	9th year
Second premolars	10th year
Canines	11th–12th year
Second molars	12th–13th year
Third molars	17th–25th year

It should be noted that the first permanent molars are in use before the loss of any deciduous teeth, giving the stability of a grinding surface during the changeover from one set of teeth to the other.

Radiographic appearances of the teeth
(Figs 2.56 and 2.57)

Due to the curved shape of the jaws, it is necessary to take either a series of small films or a panoral tomogram to obtain radiographic demonstration of a full set of teeth. On any one intraoral view several teeth may be demonstrated but usually only two are seen clearly with the least possible distortion or overlap (Fig. 2.56). On a panoral tomogram (Fig. 2.57) a view of both dental arches is obtained on one radiograph.

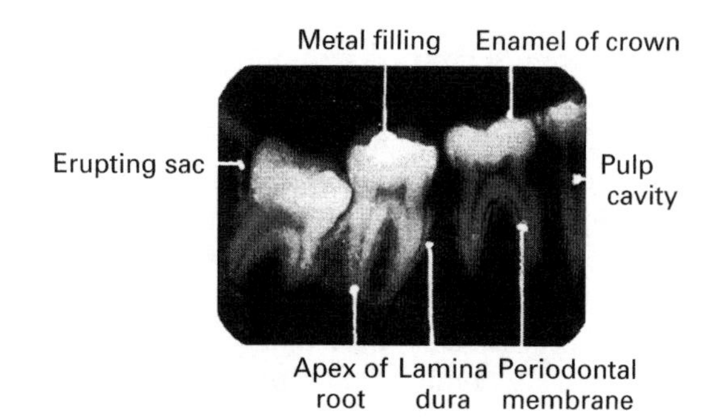

Fig. 2.56 Teeth: intraoral view

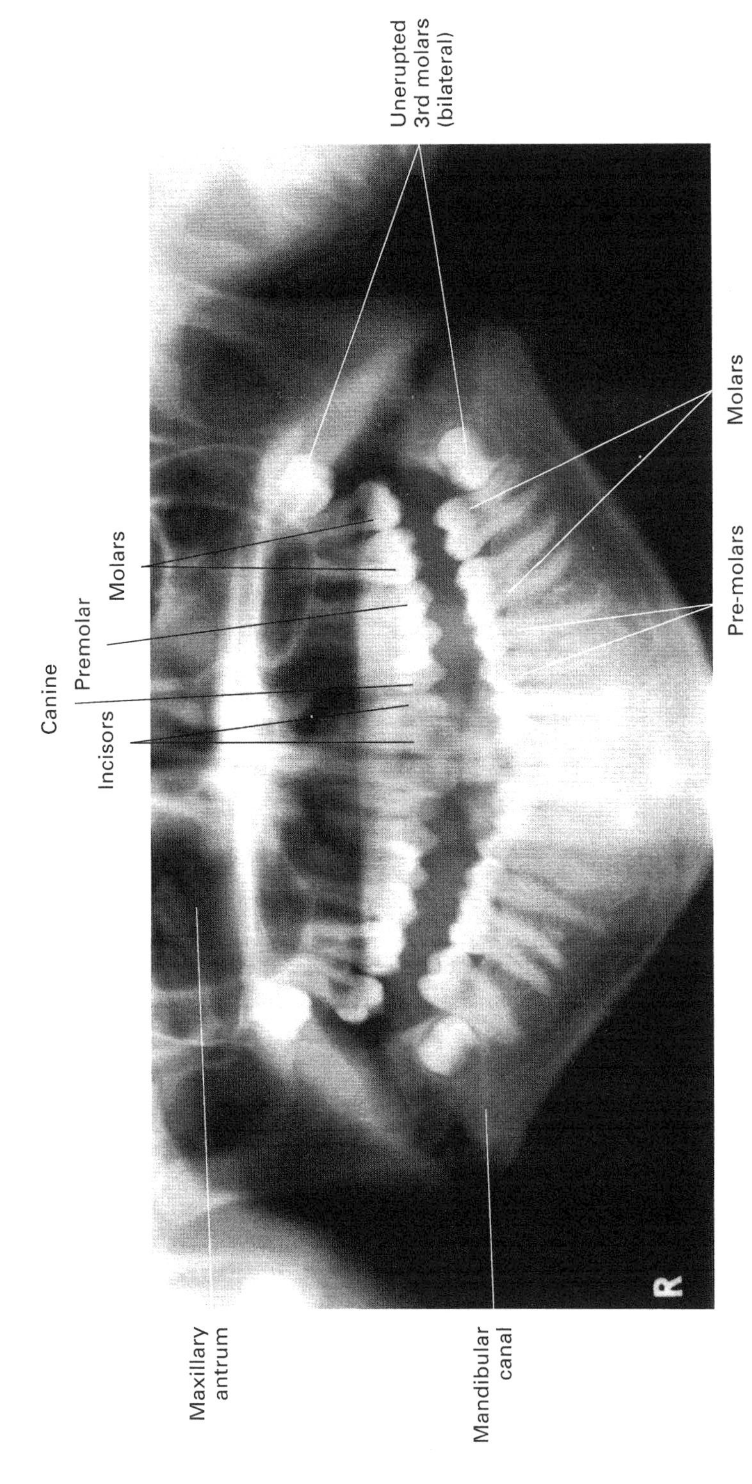

Fig. 2.57 Teeth: panoral tomograph

3. The vertebral column

The vertebral column forms the central bony axis of the body. It consists of 33 individual bones (vertebrae) placed one upon another. The vertebrae are divided into five groups, each group showing certain modifications of a general structural pattern. The groups are:

- cervical
- thoracic
- lumbar
- sacral
- coccygeal.

There are 7 vertebrae in the cervical region, 12 in the thoracic, 5 in the lumbar, 5 in the sacral and 4 in the coccygeal region. Both the sacral and coccygeal vertebrae fuse to form the sacrum and coccyx about puberty and are referred to as the fixed vertebrae, whereas the other remain as moveable vertebrae throughout life.

A TYPICAL VERTEBRA

A typical vertebra (Figs 3.1 and 3.2) consists of:

- the body (anterior part)
- the vertebral arch (posterior part).

The two parts together enclose the **vertebral foramen**, which contains the spinal cord, the meninges and associated vessels. The body of one vertebra is bound to the bodies of the vertebrae above and below by fibrocartilaginous discs.

The body is shaped like a short cylinder and is formed of spongy bone thinly covered by compact bone. The anterior surface is convex and is slightly roughened. The posterior surface is slightly concave and is perforated by one or two openings for veins. The upper and lower surfaces are flat and roughened.

The size of the body varies with the position of the vertebra. It is smallest in the cervical region and becomes progressively larger towards the lumbar end of the vertebral column.

The vertebral arch has a vertical anterior part on each side, the pedicle, and a broader posterior

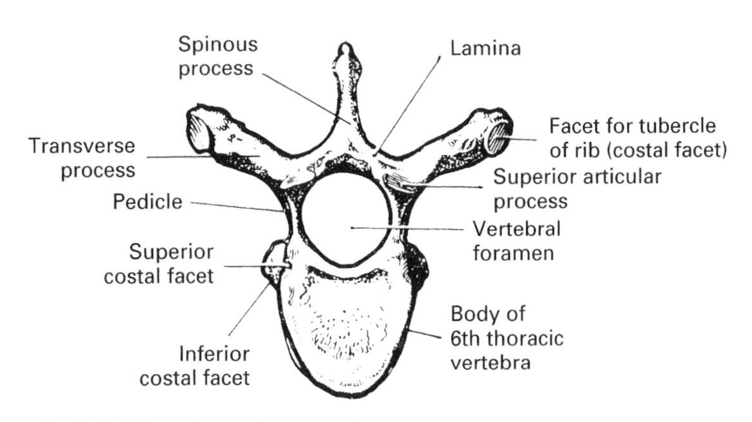

Fig. 3.1 Thoracic vertebra: superior aspect

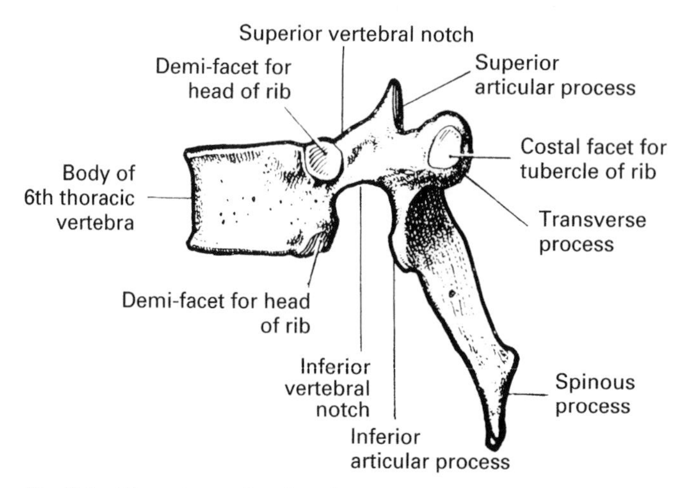

Fig. 3.2 Thoracic vertebra: lateral aspect

part, the lamina. Projecting from the vertebral arch, are seven processes—paired transverse and superior and inferior articular processes, and a single spinous process.

The pedicles are short, rounded processes which project backwards from either side of the body at the junction of its lateral and dorsal surfaces.

The laminae are flat plates of bone, continuous with the pedicles. They project backwards and medially from the ends of the pedicles to meet and fuse in the midline posteriorly—thus completing the vertebral foramen. The anterior part of the vertebral foramen is formed by the posterior surface of the vertebral body.

The spinous process projects backwards in the midline from the junction of the laminae and acts as a lever for the muscles of the back.

The transverse processes project laterally from the junction of the pedicles and laminae and give attachment to ligaments and muscles.

The articular processes are arranged in pairs, a superior and an inferior process on each side of the arch. The superior articular processes bear facets which face backwards and articulate with the forward facing inferior facets of the vertebra above.

The pars interarticularis is a slender segment of bone between the superior and inferior articular processes.

The intervertebral foramina lie between the pedicles of adjacent vertebrae and transmit the spinal nerves and vessels. The concavities above and below the pedicles are called the superior and inferior vertebral notches.

All the vertebrae, with certain exceptions, conform to the general plan, but the vertebrae of each particular region show characteristic features. A typical vertebrae of each group is situated in the middle of the group; the upper and lower vertebrae of that group are modified to provide a gradual transition from one group to another.

CERVICAL VERTEBRAE (Figs 3.3 to 3.17)

The cervical vertebrae are the smallest of the movable vertebrae. The 1st, 2nd and 7th have special features which distinguish them from the others.

The body of a typical cervical vertebra (Figs 3.3, 3.4 and 3.5) is small and is wider from side to side than from back to front. The lateral margins of the upper surface are upturned and articulate with the lateral margins of the vertebra above. The small synovial joints (uncovertebral joints) so formed are found only in the cervical region.

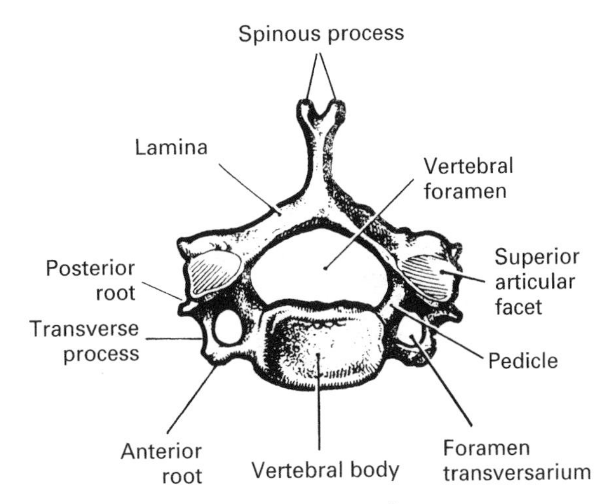

Fig. 3.3 Fifth cervical vertebra: superior aspect

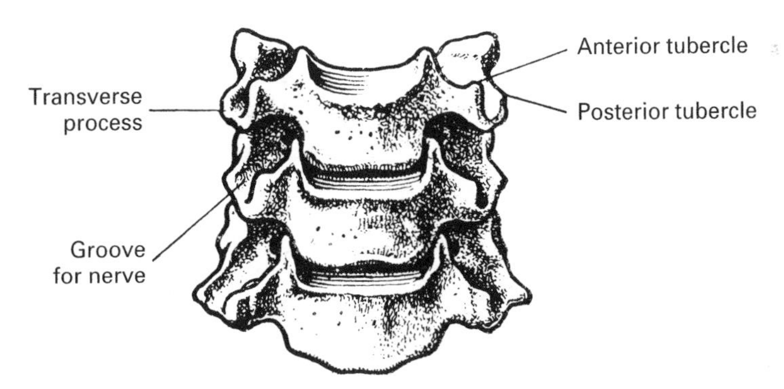

Fig. 3.4 Fourth to sixth cervical vertebrae: anterior aspect

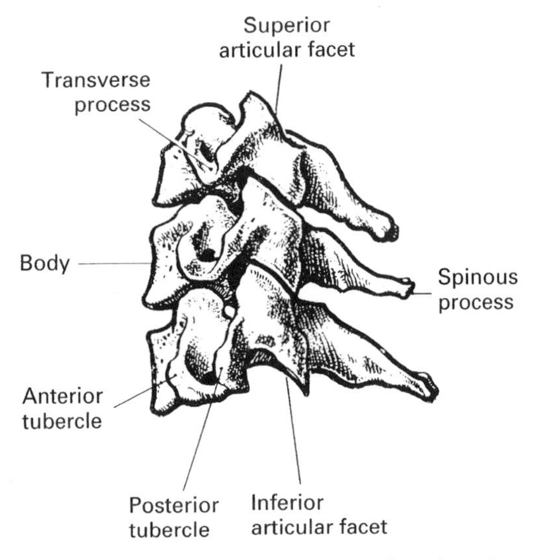

Fig. 3.5 Fourth to sixth cervical vertebrae: lateral aspect

The vertebral arch as a whole is large in comparison with the body and the vertebral foramen is also large and triangular. The transverse processes are each pierced by a foramen (the foramen transversarium) for the passage of the vertebral vessels. The posterior root of the transverse process represents the true transverse process, whilst the anterior root represents the costal element, which becomes separate and forms a rib in the thoracic region; each root ends in a tubercle.

The articular processes form a bony pillar on each side, at the junction of the pedicles and laminae, behind the transverse processes. The articular facets are oval and flat. The superior facets face upward and backward, and the inferior facets downward and forward. The spinous processes are short and bifid.

The intervertebral foramina in the cervical spine face forwards and outwards at an angle of 45° to the median plane. The size of the intervertebral foramina varies—the largest being at C5/6 and C6/7 because of the large motor roots for the brachial plexus.

ATLAS (Figs 3.6 and 3.7)

The name atlas is given to the 1st cervical vertebra because it supports the head. It has no body and no spinous process and is shaped like a ring.

It consists of:

- an anterior arch
- a posterior arch
- two lateral masses.

The anterior arch is the smaller part of the ring. A small projection, the anterior tubercle, is present on its anterior surface, whilst on the posterior surface is a small facet for articulation with the odontoid process of the second cervical vertebra (axis).

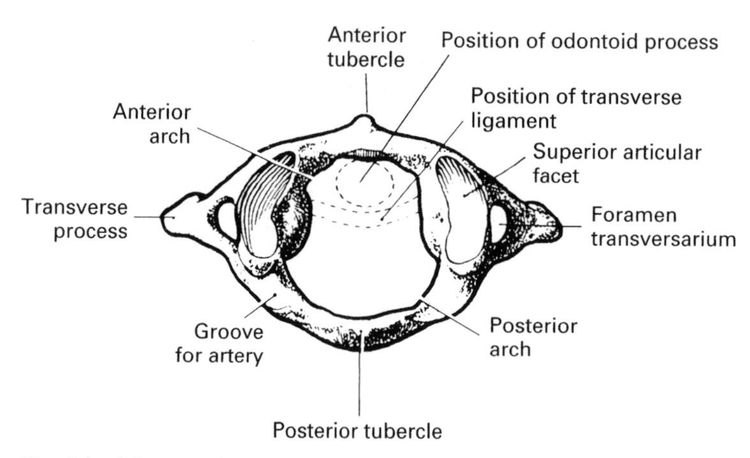

Fig. 3.6 Atlas: superior aspect

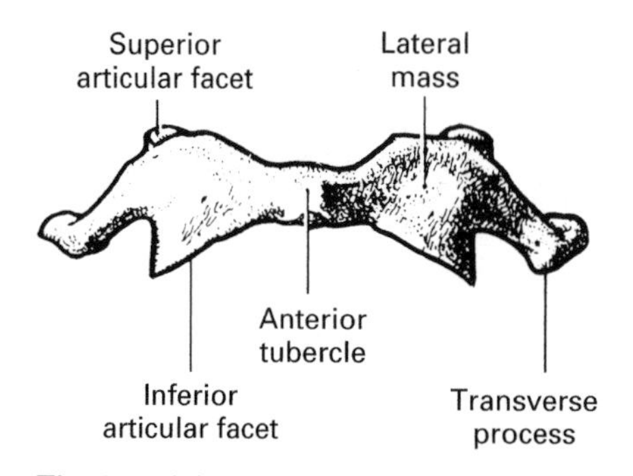

Fig. 3.7 Atlas: anterior aspect

The posterior arch forms the larger part of the ring: on its posterior surface is the posterior tubercle which represents a much reduced spinous process. The obliquely set lateral masses have articular facets on their upper and lower surfaces. The upper facets are oval and concave and articulate with the occipital condyles of the skull to form the atlanto-occipital joints. The inferior facets are round and flat and articulate with the superior facets of the 2nd cervical vertebra (axis).

On the inner side of each lateral mass is a prominent tubercle to which the transverse ligament is attached. This ligament divides the ring into two parts; the anterior smaller part is occupied by the odontoid process of the axis, whilst the spinal cord occupies the larger posterior part.

The perforated transverse processes are longer than those of the other cervical vertebrae and act as levers for the rotatory muscles of the head. Behind the articular facets on the upper surface of the atlas are shallow grooves for the vertebral arteries which enter the cranial cavity through the foramen magnum.

AXIS (Figs 3.8 and 3.9)

The 2nd cervical vertebra—the axis—consists of a body and a vertebral arch. Its chief feature is the odontoid process (odontoid peg) which projects upward from the upper surface of the body. This process articulates with the posterior surface of the anterior arch of the atlas and is held in place by the transverse ligament of the atlas. It forms the pivot round which the atlas and head rotate, and represents the body of the atlas fused with that of the axis.

The spinous process is large and bifid. To it are attached the muscles which control rotation and extension of the head on the vertebral column.

The superior articular facets are borne on the lateral part of the body and adjacent part of the pedicles and articulate with the lateral masses of the atlas. The transverse processes are small.

SEVENTH CERVICAL VERTEBRA

This vertebra has a prominent spinous process which ends in a single tubercle and which is easily palpable ('vertebra prominens'). The transverse process is large—the anterior part (the costal element) may develop as a cervical rib (see also p. 102). The foramen transversarium is small and may even be absent.

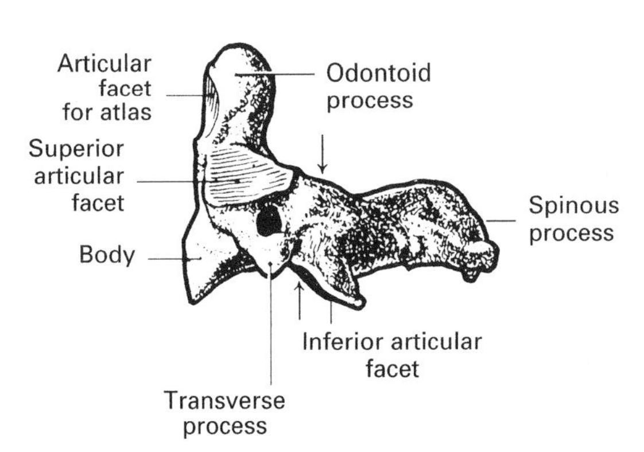

Fig. 3.8 Axis: lateral aspect

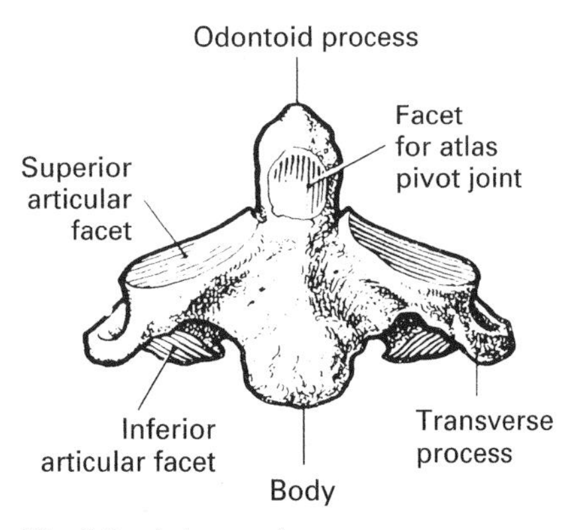

Fig. 3.9 Axis: anterior aspect

Radiographic appearances of the cervical vertebrae (Figs 3.10 to 3.17)

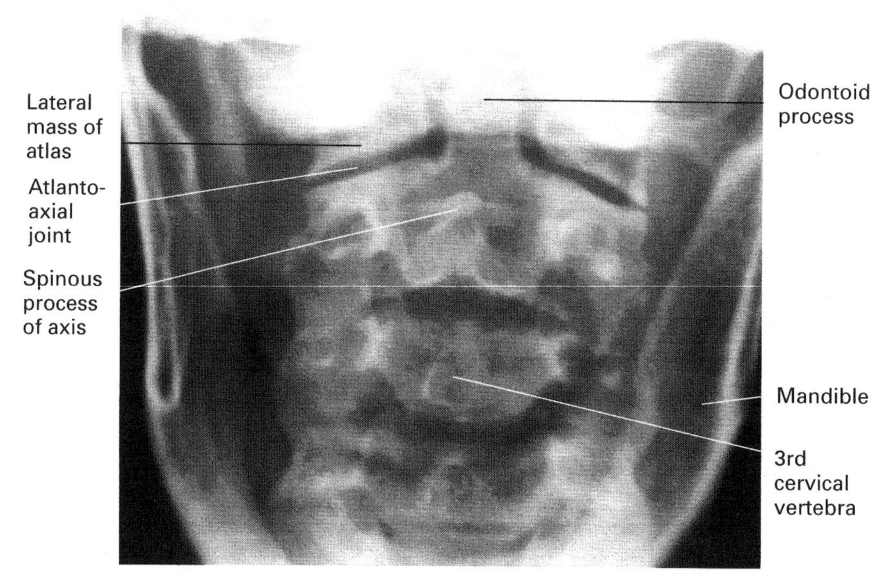

Fig. 3.10 Upper cervical vertebrae: anteroposterior view taken with mouth open

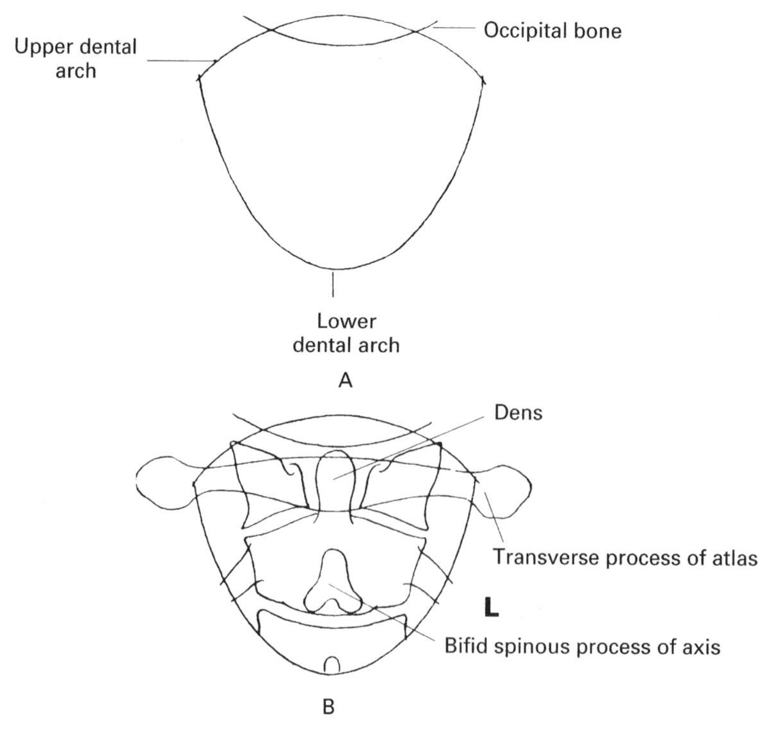

Fig. 3.11 Atlas and axis: diagram of anteroposterior view

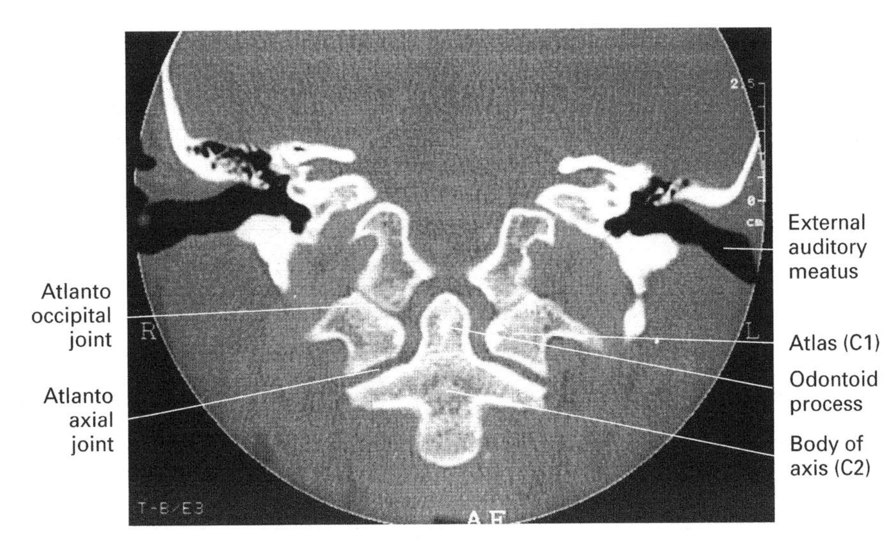

Fig. 3.12 Upper cervical spine: CT scan

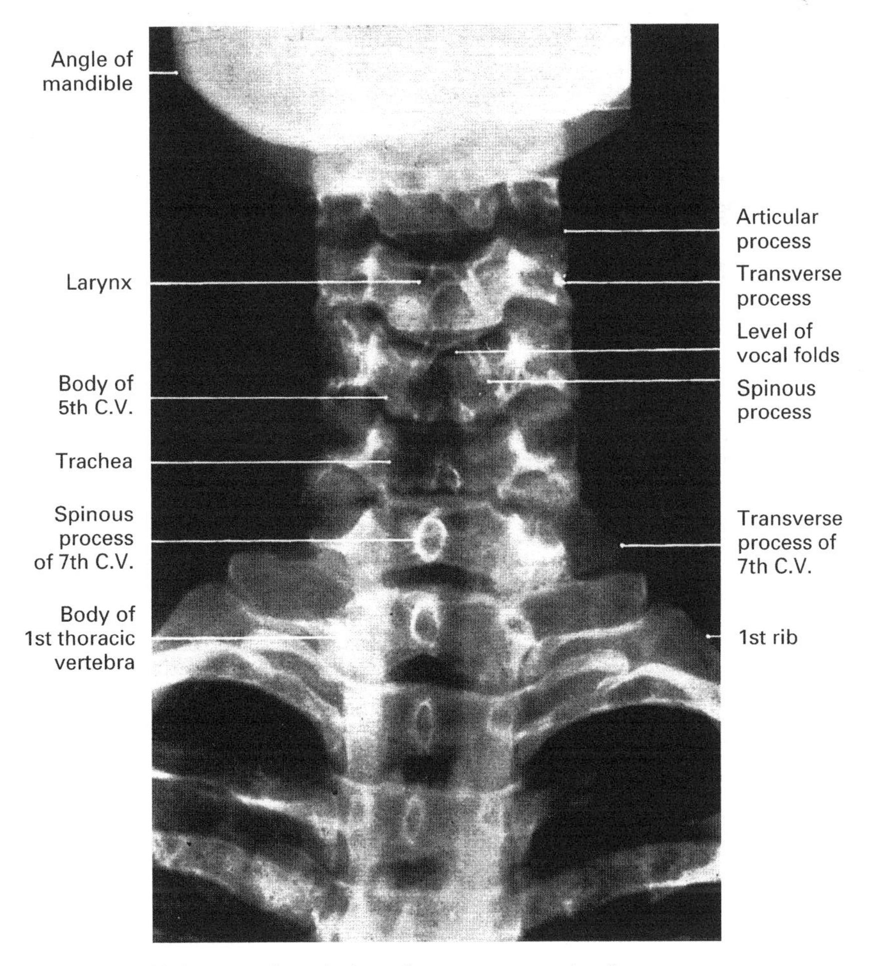

Fig. 3.13 Third to seventh cervical vertebrae: anteroposterior view

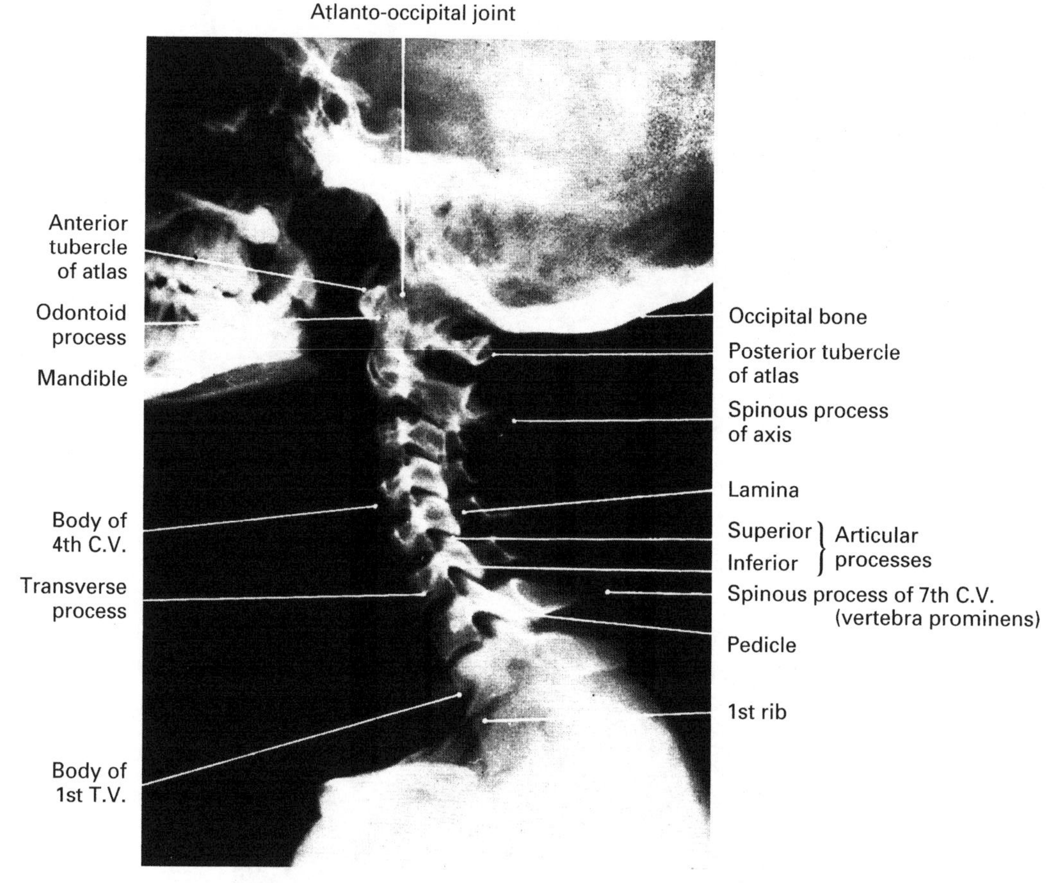

Fig. 3.14 Cervical vertebrae: lateral view

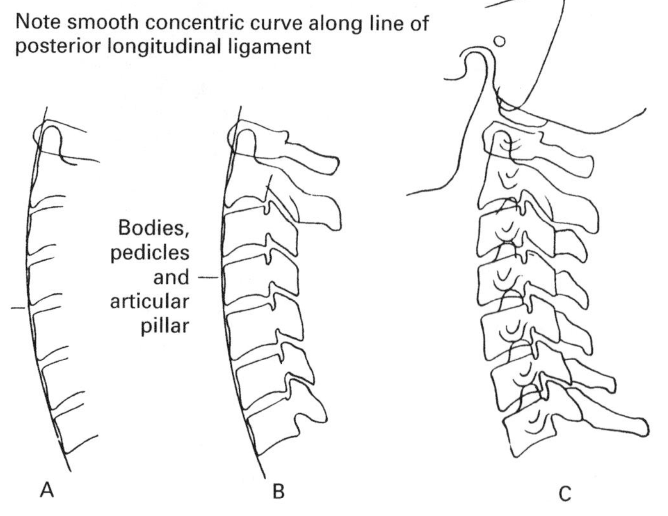

Fig. 3.15 Cervical vertebrae: construction of diagram of the lateral view

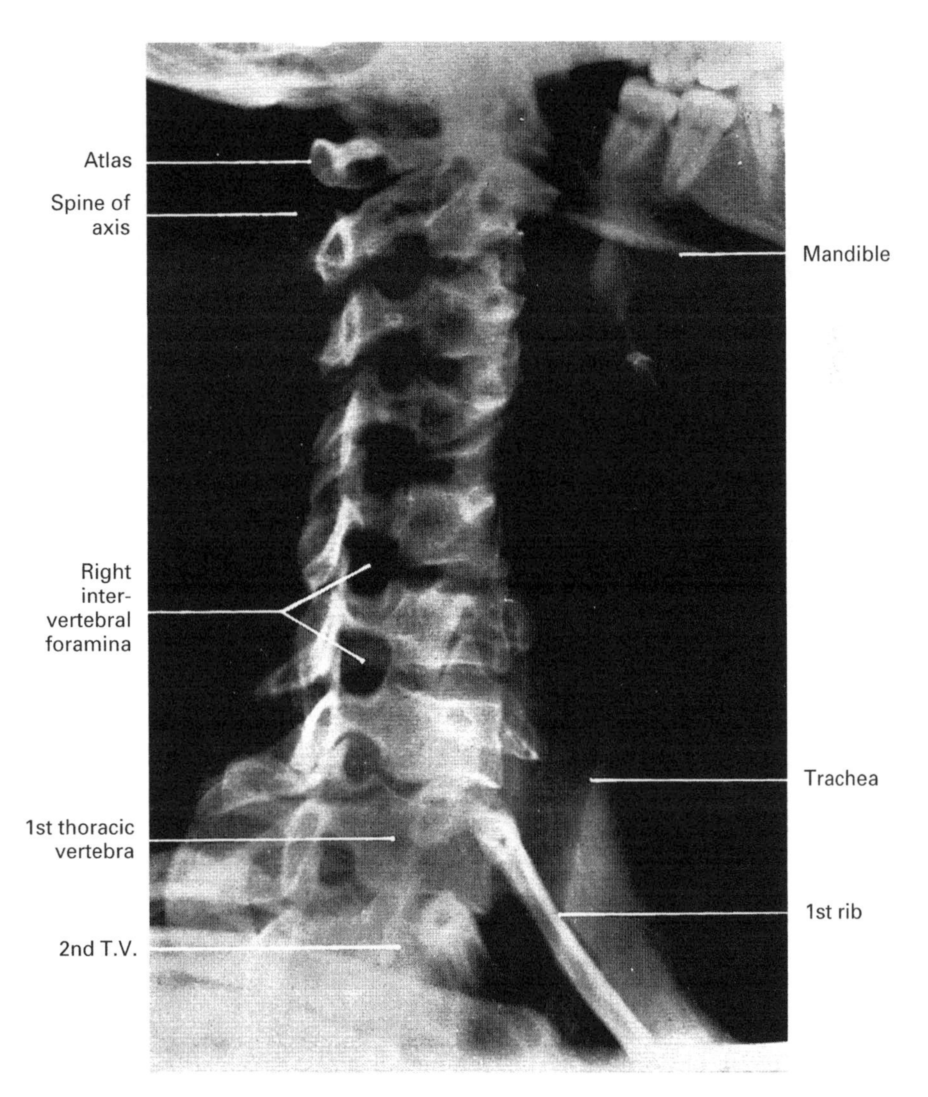

Fig. 3.16 Cervical vertebrae: oblique anteroposterior view: right foramina shown

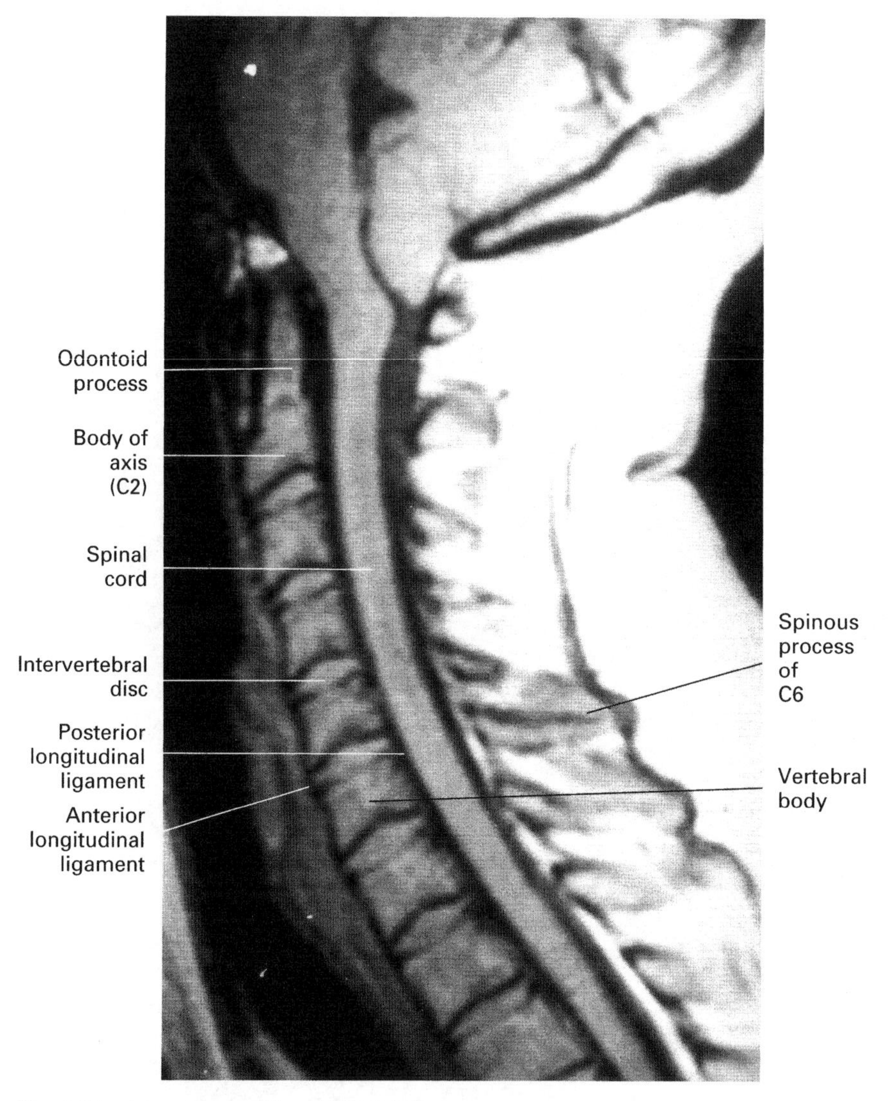

Fig. 3.17 Atlas and axis (C1 and C2): M. R. scan

THORACIC VERTEBRAE (Figs 3.18 to 3.24)

The 12 thoracic vertebrae can be distinguished by the presence of small articular facets on the lateral sides of the bodies and, with the exception of the 11th and 12th vertebrae, on the transverse processes also. These facets articulate with the heads and tubercles of the ribs. The 1st, 9th, 10th, 11th and 12th vertebrae differ slightly from the others and will be considered separately.

The body of a typical thoracic vertebra (Figs 3.1 and 3.2, pp. 59 and 60) is heart-shaped. The height and width are about equal, but the bodies are a little deeper behind than in front, thus causing the normal slight anterior concavity (kyphosis) of the thoracic spine as a whole. The bodies decrease slightly in size from the 1st to the 3rd vertebra, and then increase progressively in size to the 12th. On either side of the body are two demi-facets, superior and inferior, for articulation with the heads of the ribs. Each rib therefore articulates with two vertebral bodies (Fig. 4.13, p. 104).

The pedicles are short and project backward from the body to meet the short deep laminae. The vertebral foramen is circular and it is smaller than in the cervical region.

The transverse processes are thick and strong. On the anterior surface of each is a facet for articulation with the tubercle of a rib.

The articular processes are almost vertical. The superior facets face posteriorly and the inferior facets anteriorly.

The spinous processes are long and slender and are directed downwards and posteriorly, those of T5–T8 being the most nearly vertical.

The intervertebral foramina are formed by a relatively deep inferior vertebral notch and open laterally, in front of the transverse processes.

FIRST THORACIC VERTEBRA (Fig. 3.18)

On each side of the body is a whole superior articular facet for the head of the 1st rib, and demi-facet below for the head of the 2nd rib.

NINTH THORACIC VERTEBRA (Fig. 3.20)

The body has demifacets above; the lower facets are very small or absent.

TENTH THORACIC VERTEBRA (Fig. 3.20)

Complete superior facets are present; the lower facets are absent.

ELEVENTH AND TWELFTH THORACIC VERTEBRA (Fig. 3.20)

Complete articular facets are present on the sides of the bodies; no facets are present on the transverse processes which are small. The bodies

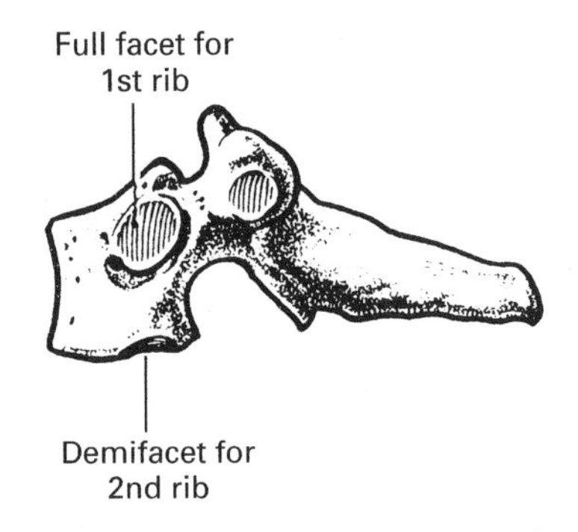

Fig. 3.18 First thoracic vertebrae: lateral aspect

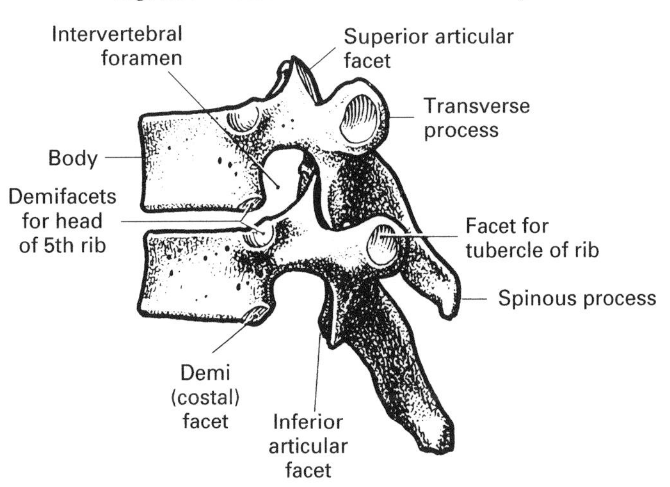

Fig. 3.19 Fourth and fifth thoracic vertebrae: lateral aspect

are large and the superior articular facets are beginning to face laterally as in the lumbar region. The inferior articular facets of the 12th thoracic vertebra are lumbar in morphology.

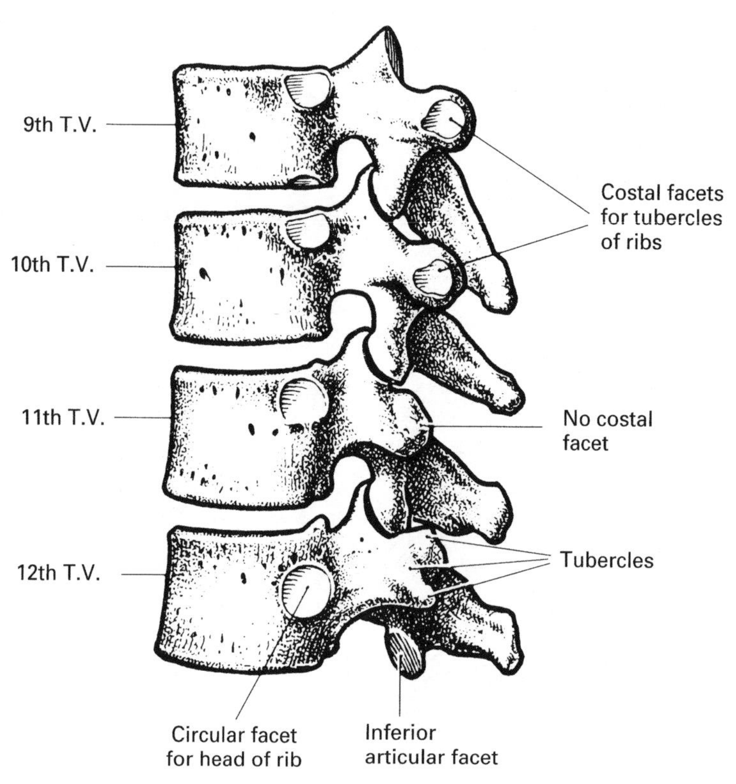

Fig. 3.20 Ninth to twelfth thoracic vertebrae: lateral aspect

Radiographic appearances of the thoracic vertebrae (Figs 3.21 to 3.25)

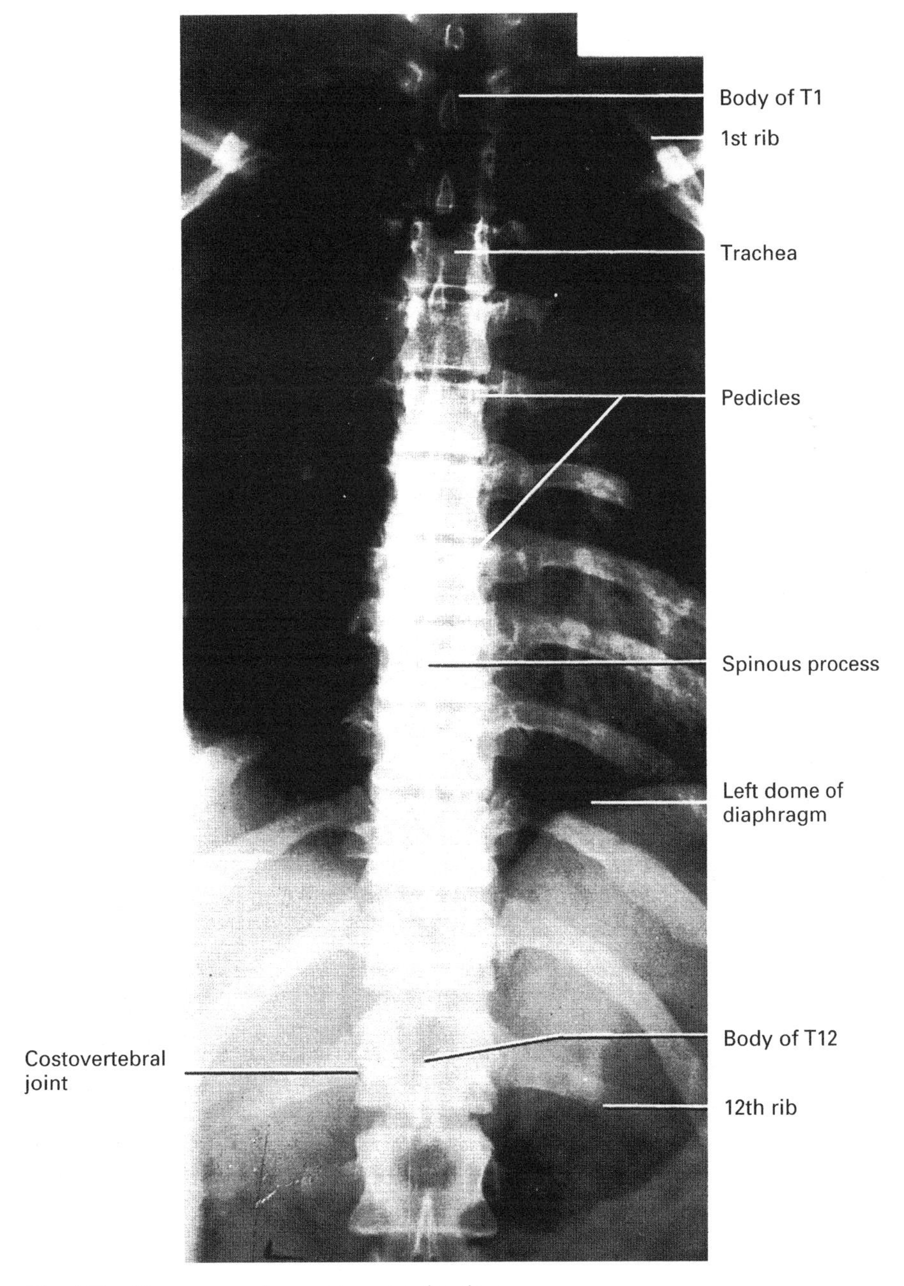

Fig. 3.21 Thoracic vertebrae: anteroposterior view

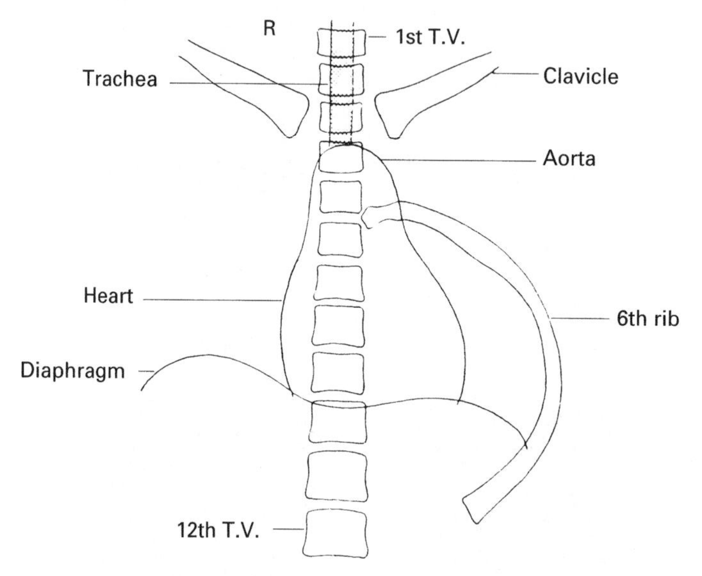

Fig. 3.22 Thoracic vertebrae: a simplified diagram of the anteroposterior view

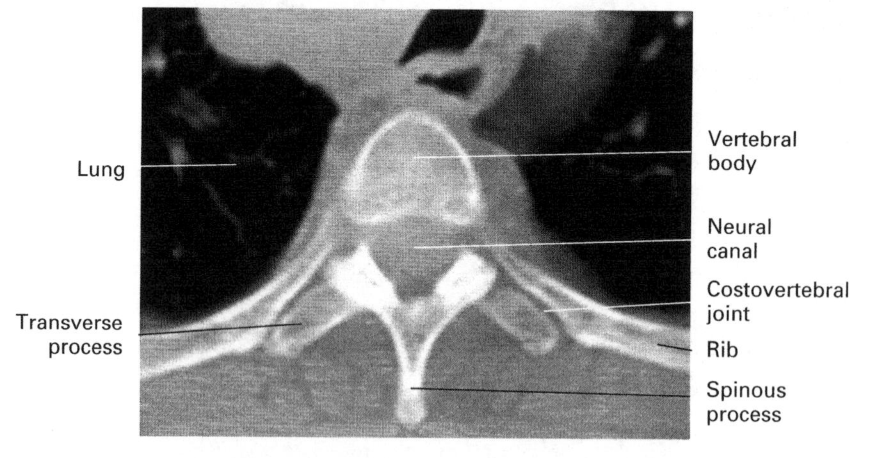

Fig. 3.23 Thoracic vertebra: CT scan

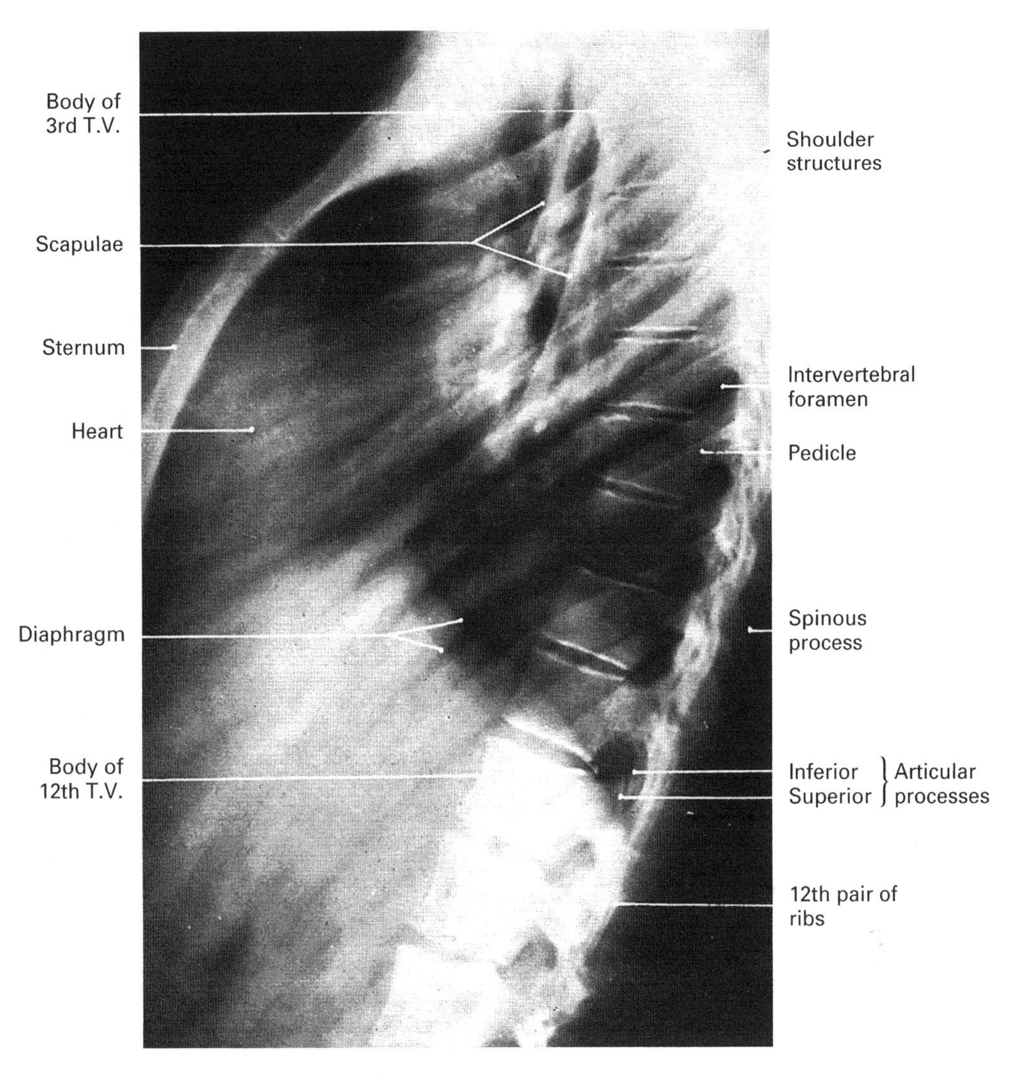

Fig. 3.24 Thoracic vertebrae: lateral view

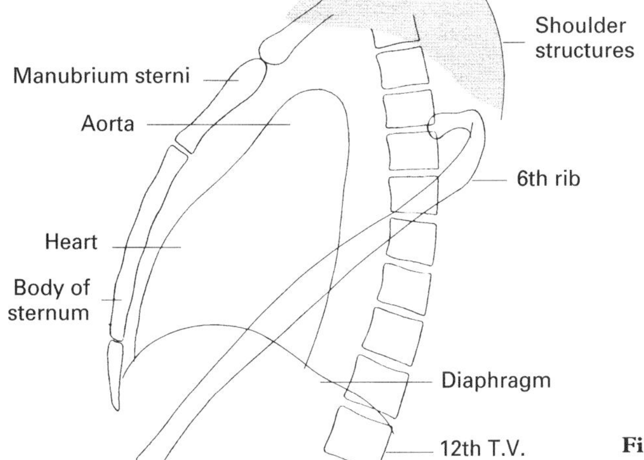

Fig. 3.25 Thoracic vertebrae: a simplified diagram of the lateral view

LUMBAR VERTEBRAE (Figs 3.26 to 3.31)

The five lumbar vertebrae are large and strongly built because they have to support considerable weight. The 5th vertebra differs slightly from the others.

A TYPICAL LUMBAR VERTEBRA

The body of a typical lumbar vertebra (Figs 3.26 and 3.27) has wide, kidney-shaped upper and lower surfaces and is a little deeper anteriorly than posteriorly.

The pedicles are short and thick. The vertebral foramen is relatively small and is triangular in shape.

The transverse processes are long and thin. They project laterally, slightly upwards and backwards.

The spinous processes are thick and broad and are horizontal in direction. The articular processes are large; the superior facets face backwards and inwards and the inferior facets face forwards and outwards.

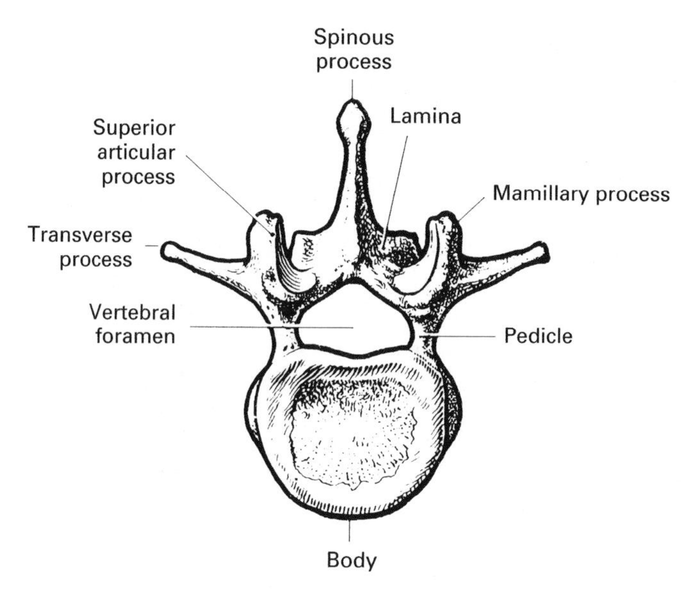

Fig. 3.26 Third lumbar vertebra: superior aspect

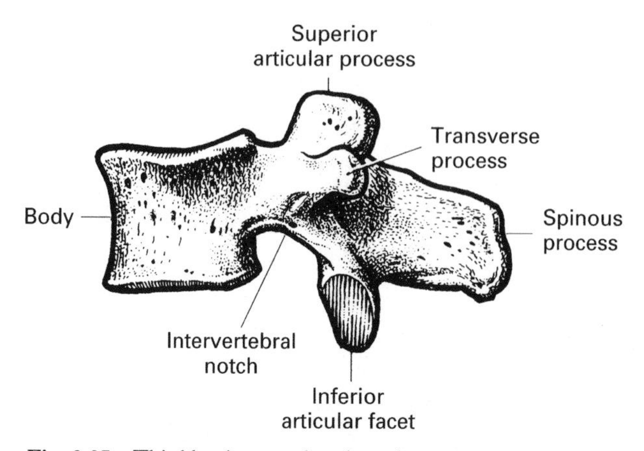

Fig. 3.27 Third lumbar vertebra: lateral aspect

FIFTH LUMBAR VERTEBRA (Fig. 3.28)

This vertebra forms the lumbosacral angle with the sacrum. The body is appreciably deeper anteriorly than posteriorly. The transverse processes are short and they arise from the sides of the vertebral body as well as from the pedicles. They incline upwards slightly, and the iliolumbar ligament attaches here.

Radiographic appearances of the lumbar vertebrae (Figs 3.28 to 3.31)

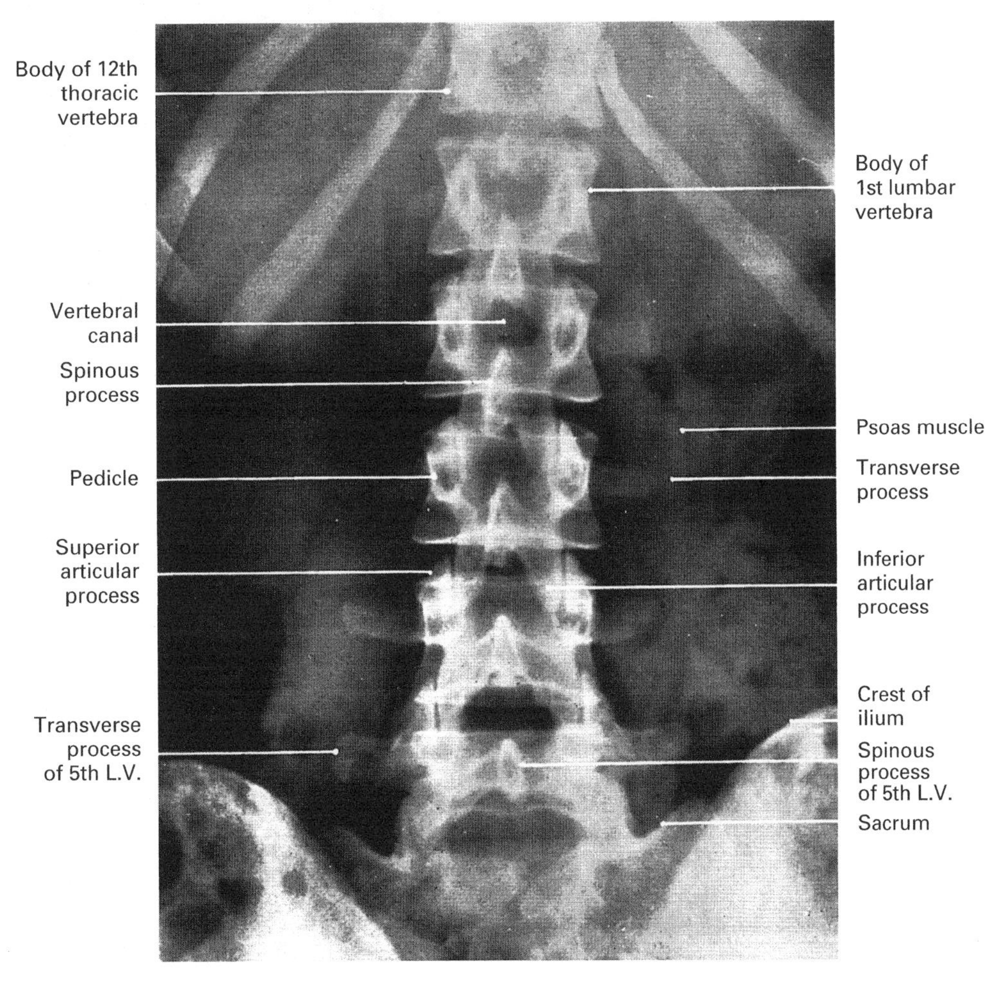

Fig. 3.28 Lumbar vertebrae: anteroposterior view

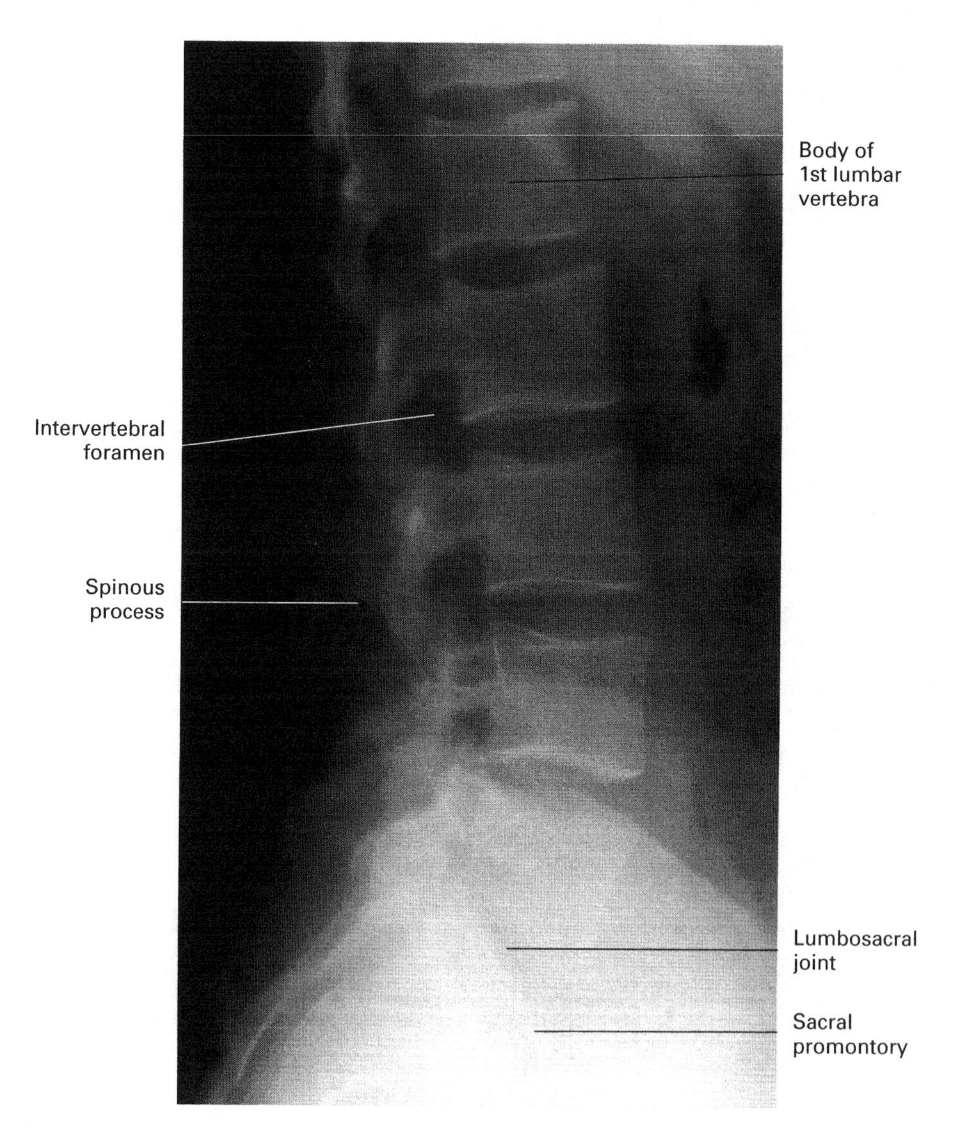

Fig. 3.29 Lumbar vertebrae: lateral view

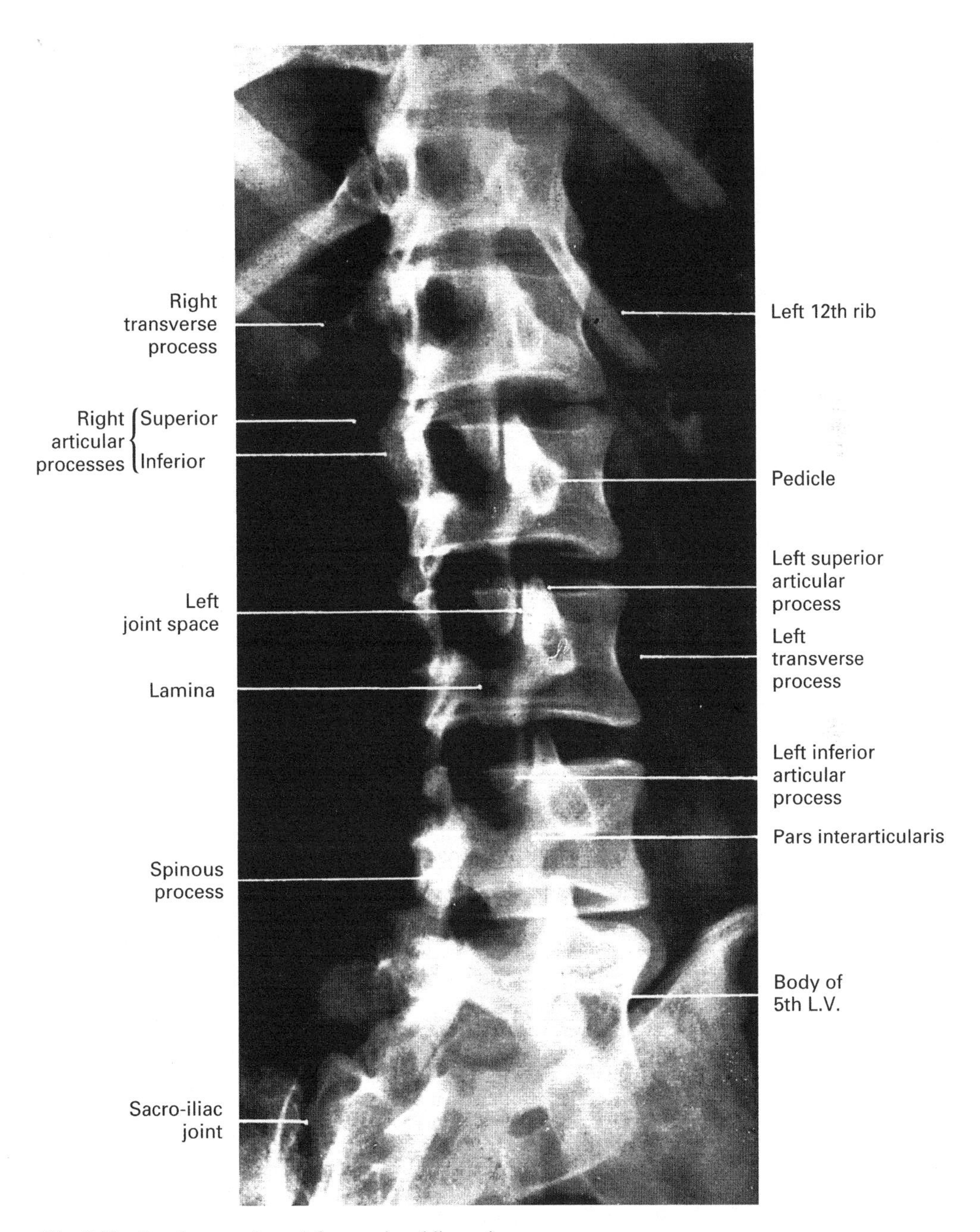

Fig. 3.30 Lumbar vertebrae: left posterior oblique view

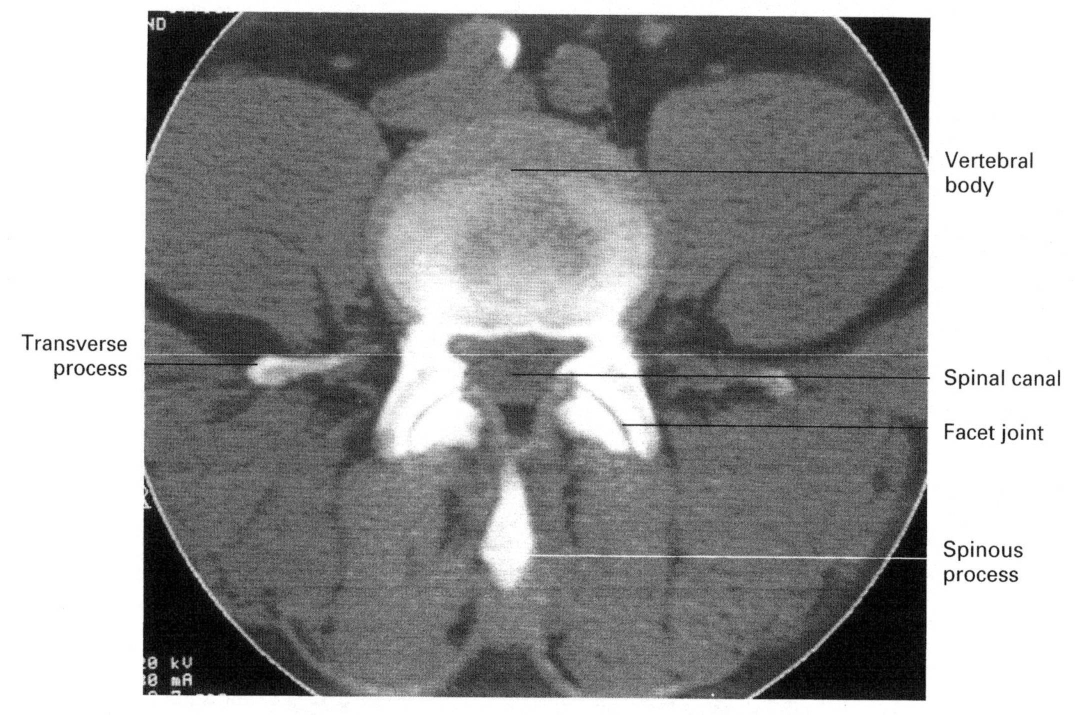

Fig. 3.31 Lumbar spine: CT scan

SACRUM (Figs 3.32–3.35 and 3.37)

There are five sacral vertebrae (sacral segments) which diminish progressively in size. They fuse in adolescence to become a single wedge-shaped bone with a marked anterior concavity. The base articulates with the 5th lumbar vertebra to form the lumbosacral joint (p. 91). The apex articulates

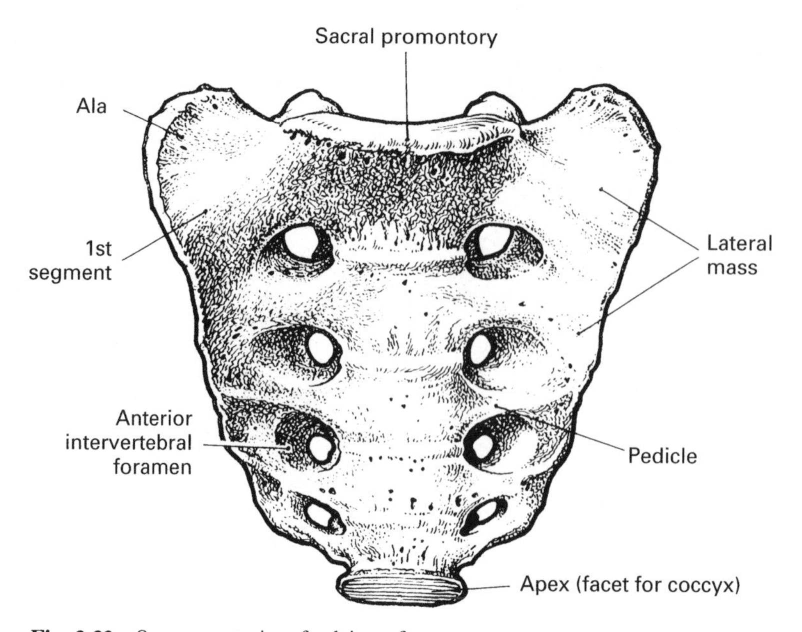

Fig. 3.32 Sacrum: anterior of pelvic surface

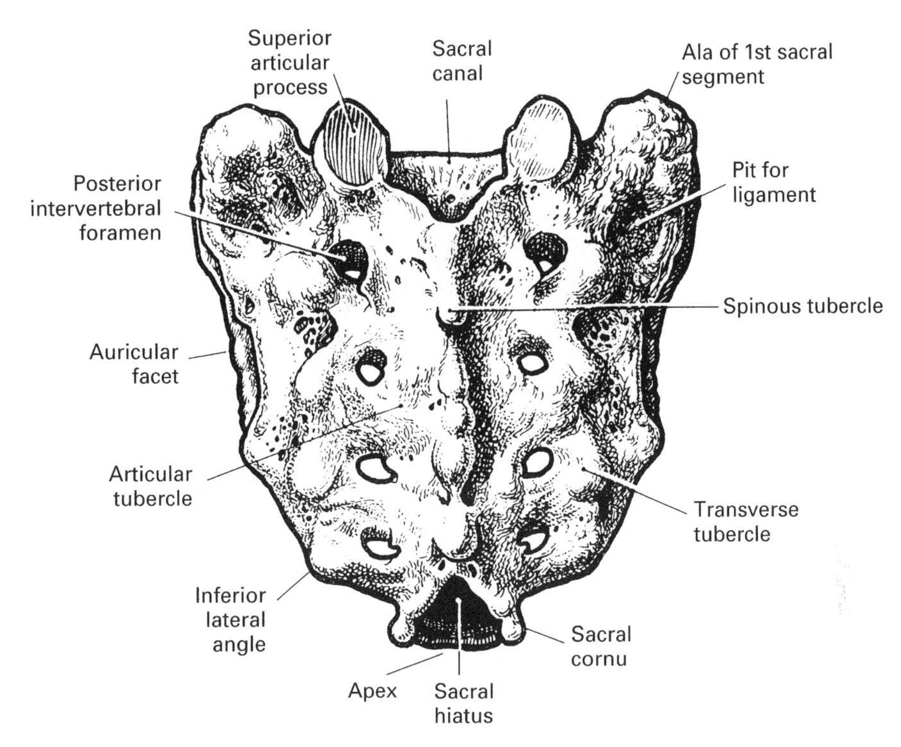

Fig. 3.33 Sacrum: dorsal surface

with the coccyx and the auricular (ear-shaped) surfaces articulate with the iliac bones at the sacroiliac joints (p. 178).

The basic components of a typical vertebra are still present in a sacral segment but in a modified form. The part lateral to the anterior sacral foramina is called the lateral mass (ala) and consists of fused transverse processes and costal elements. The sacral canal is a continuation of the vertebral canal.

The anterior (pelvic) surface of the sacrum is concave (Fig. 3.32). Four transverse ridges are present in the centre of the bone and mark the site of the fusion of the individual sacral bodies. The ridges terminate laterally at the anterior intervertebral foramina, of which there are four on each side. The foramina transmit the anterior primary rami of the sacral nerves from the sacral canal to the pelvis. The lateral mass of the sacrum is lateral to these foramina. The upper part of the lateral mass is large and is formed by the upper three sacral vertebrae. On the lateral aspect is the auricular surface for articulation with the iliac bone. The degree of the curvature varies between the sexes and between persons of the same sex.

The dorsal surface is convex. Extending down the centre of the bone is a prominent crest, bearing 4 spinous tubercles, which represent spinous processes. Below the last spinous tubercle there is a small gap—the sacral hiatus. On either side of the central crest are 4 intervertebral foramina which transmit the posterior primary rami of the sacral nerves from the sacral canal. The superior articular processes of the 1st sacral segment only bear articular facets. The articular processes below this are fused and are represented by small bony prominences—articular tubercles—which are medial to the dorsal foramina. The inferior articular processes of the last sacral segment form small bony projections—the sacral cornua—on either side of the sacral hiatus. The inferior articular surface of the 5th sacral vertebra is called the apex.

The lateral surface of the sacrum is wide above and narrow below. The auricular surface for

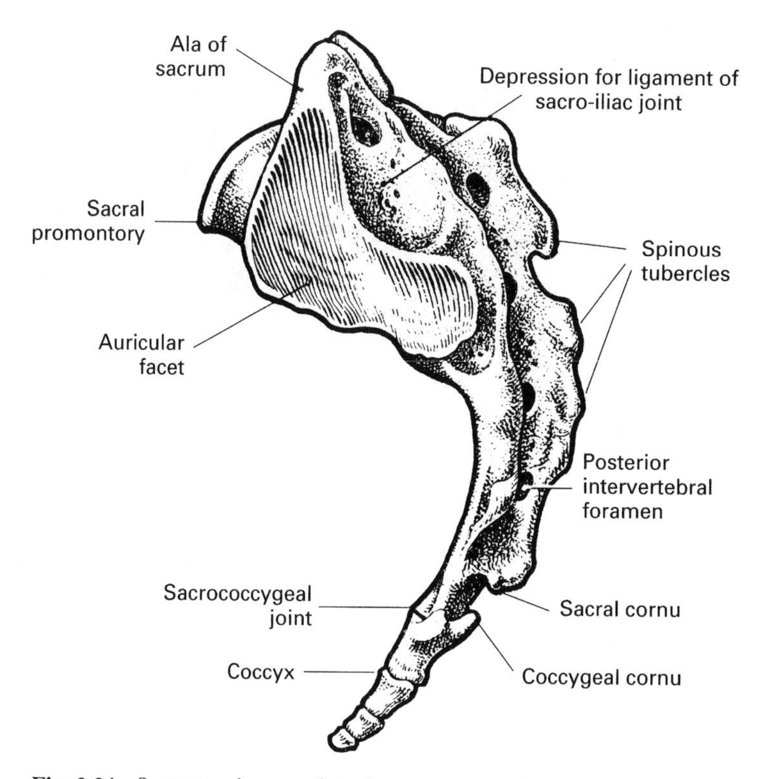

Fig. 3.34 Sacrum and coccyx: lateral aspect

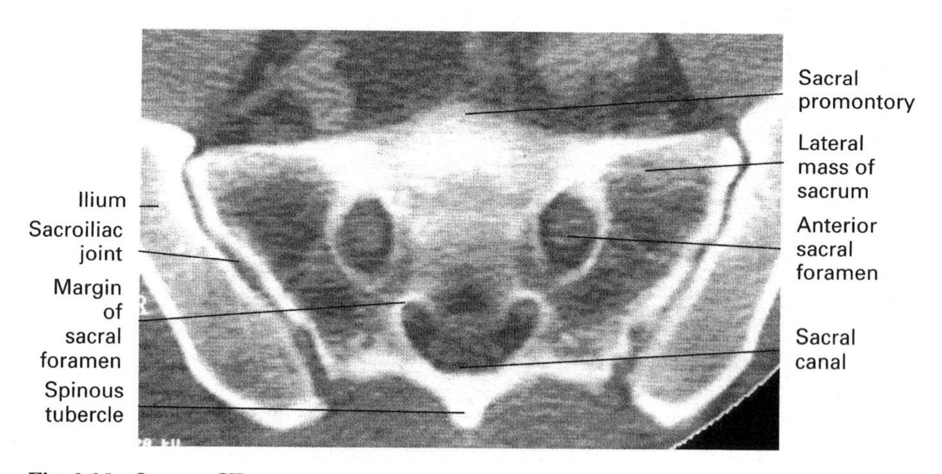

Fig. 3.35 Sacrum: CT scan

articulation with the ilium at the sacroiliac joint is situated toward the anterior upper part and behind the articular area the bone surface is roughened for attachment of ligaments. The lower part of the lateral surface is thin and gives attachment to the sacrotuberous ligament.

The base of the sacrum is formed by the upper surface of the first sacral vertebra. The anterior margin of the upper surface of the first vertebral body has a well-marked edge—the sacral promontory—which is an important landmark in obstetrics. Behind the body, the short pedicles and laminae enclose the triangular sacral canal. The superior articular facets face inward and backward and articulate with the inferior facets of the 5th lumbar vertebra.

The apex of the sacrum bears a small oval articular facet for articulation with the coccyx at the sacrococcygeal joint.

COCCYX (Figs 3.34, 3.36–3.39)

This small triangular bone (Fig. 3.34) consists of 4 segments, usually fused together.

The 1st segment possesses a body with rudimentary transverse processes and superior articular processes (the coccygeal cornua). The remaining segments consist of rudimentary bodies only, which decrease in size progressively.

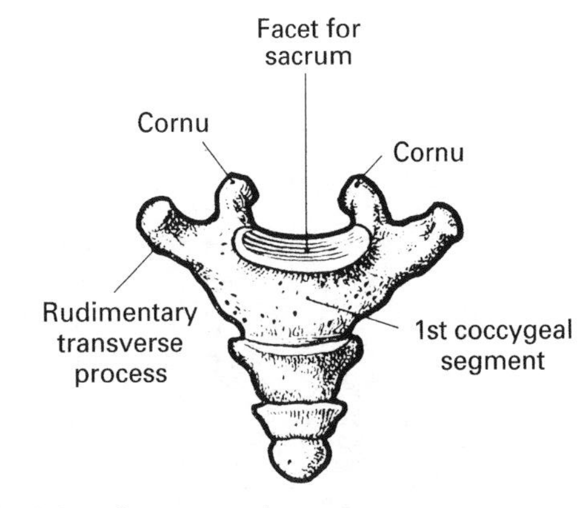

Fig. 3.36 Coccyx: anterior aspect

Radiographic appearances of the sacrum and coccyx (Figs 3.37 to 3.39)

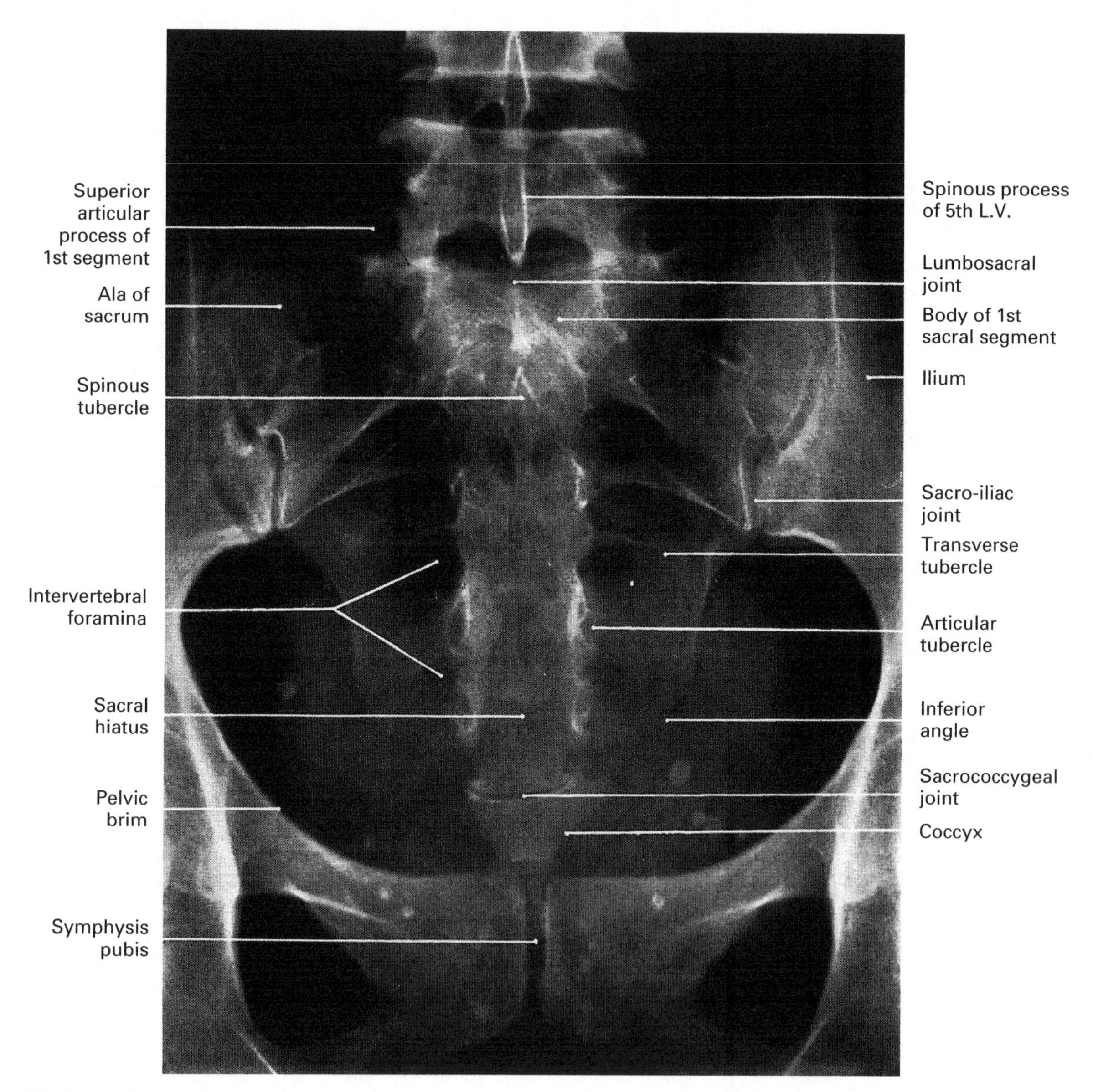

Fig. 3.37 Sacrum: anteroposterior view

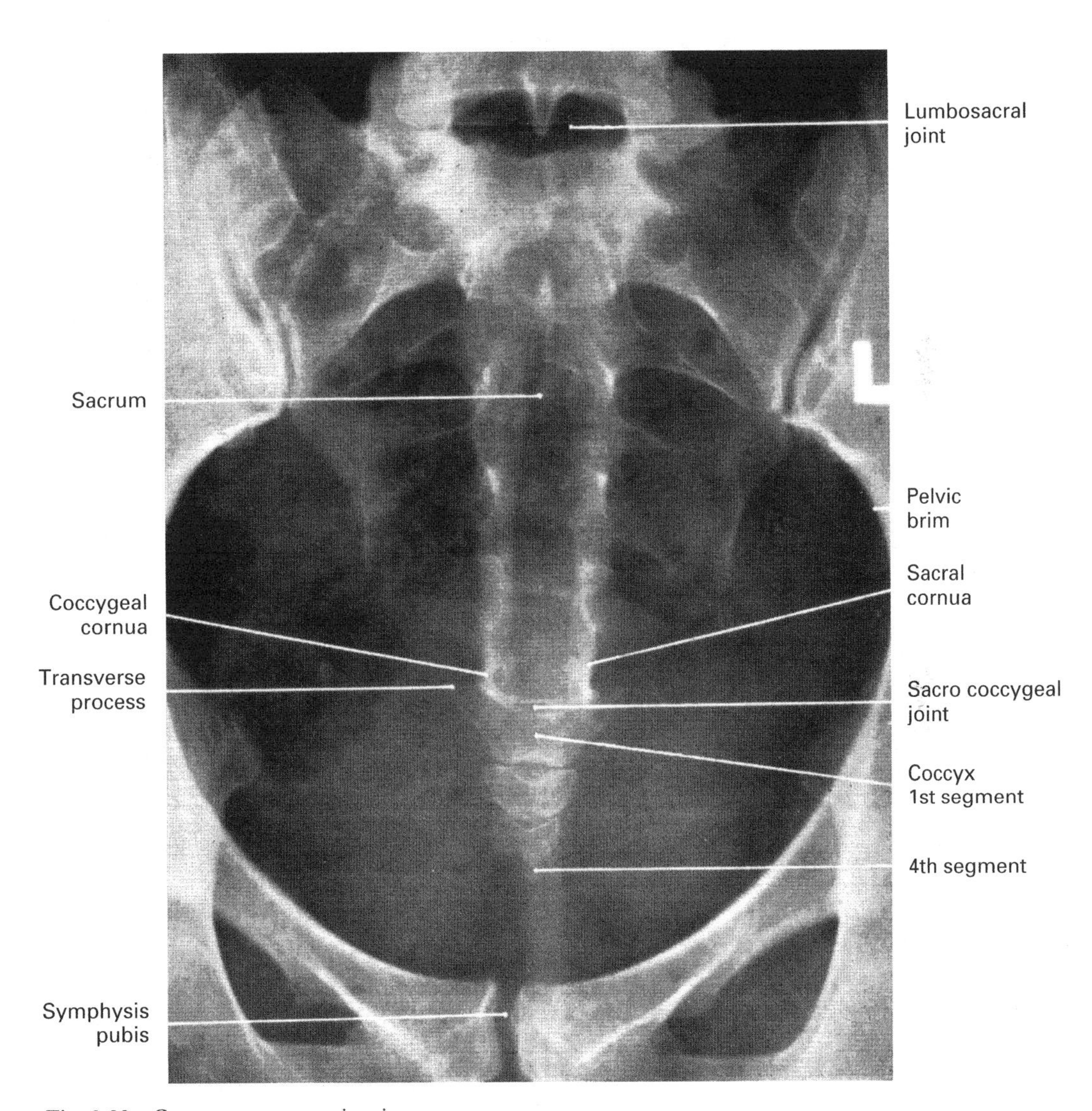

Fig. 3.38 Coccyx: anteroposterior view

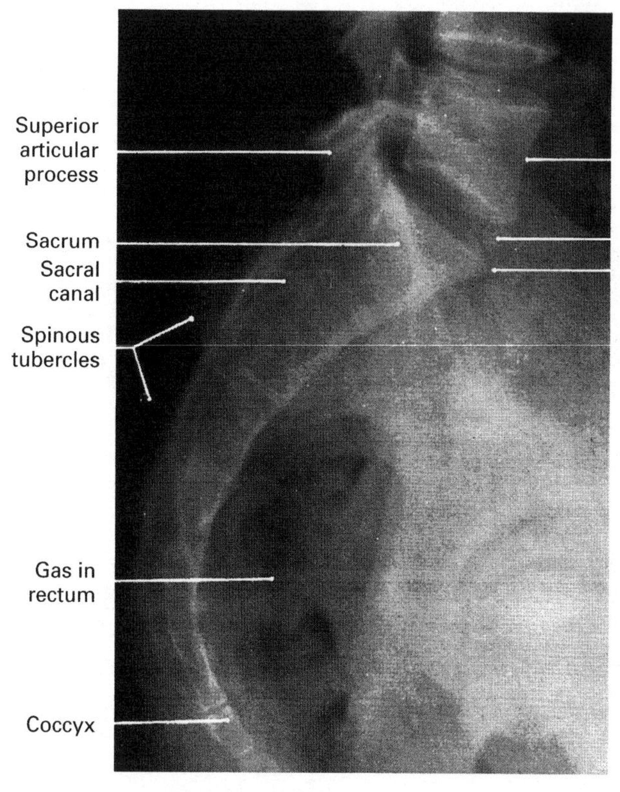

Fig. 3.39 Sacrum and coccyx: lateral view

OSSIFICATION OF THE VERTEBRAE
(Figs 3.40 and 3.41)

Primary centres (typical vertebra)

- for the body
- for each half of the vertebral arch.

Both appear early in intrauterine life. The two halves of the arch unite during the first year, commencing in the lumbar region and extending upward. Fusion of the bodies and the complete arches takes place between the third and sixth years.

Secondary centres

- for the tip of the spinous process
- for the tip of each transverse process
- for the upper surface of the vertebral body
- for the lower surface of the vertebral body.

All four appear at about 14 years. The epiphyses unite with the rest of the bone about the 25th year.

Exceptions

Atlas—ossifies from three primary centres—one for the anterior arch and one for each lateral mass. The centres for the lateral masses appear in intrauterine life but that for the anterior arch appears in the first year.

Axis—five primary centres, additional ones for each side of the odontoid process.

Seventh cervical vertebra—two additional centres for the anterior part of the transverse processes. These usually fuse with the rest of the bone about the fifth year but they may remain separate and grow to form cervical ribs.

Sacrum—has additional secondary centres, including one for each auricular surface.

Coccyx—each segment ossifies from one primary centre.

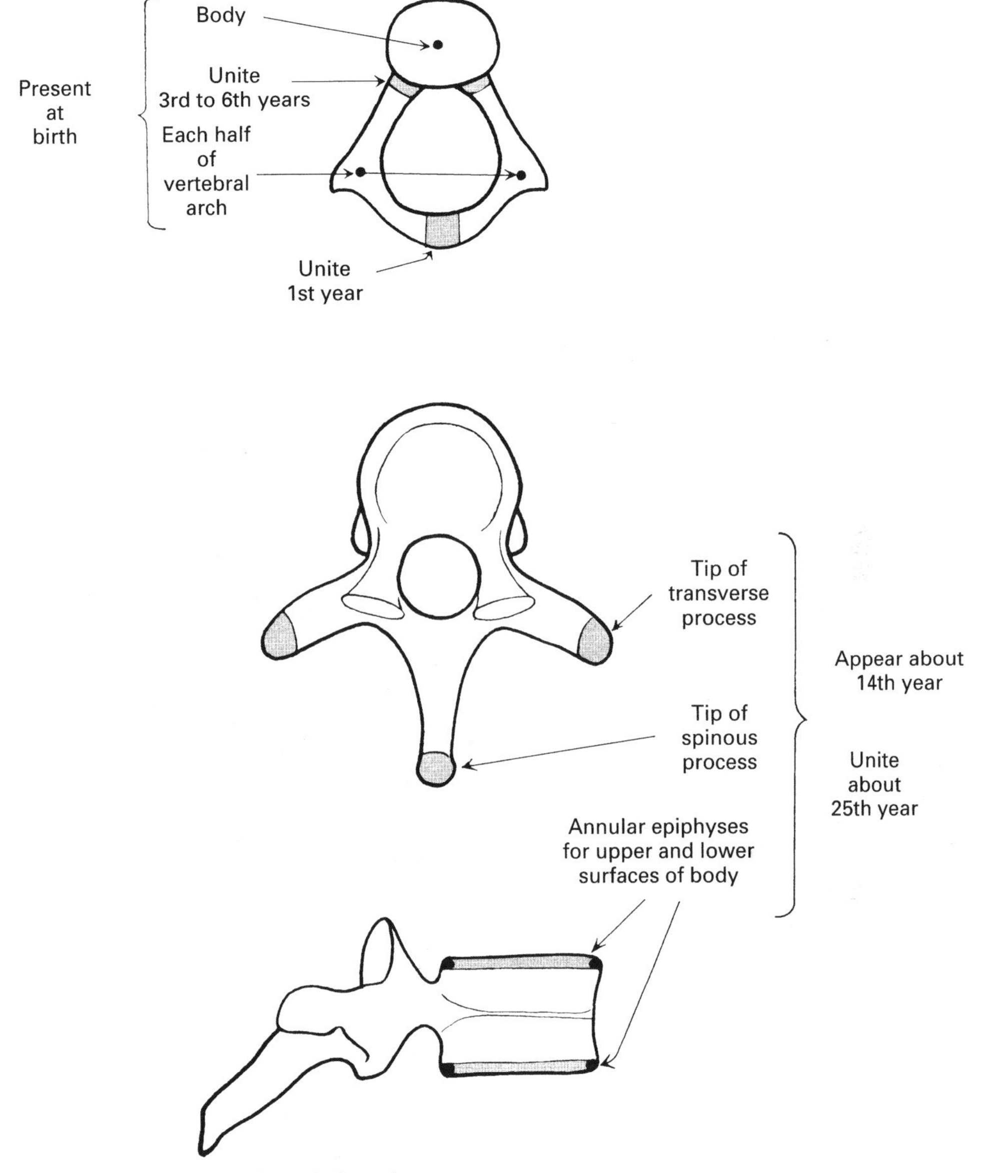

Fig. 3.40 Ossification of a typical vertebra

CURVES OF THE VERTEBRAL COLUMN (Fig. 3.42)

The vertebral column of the fetus has two primary curves—thoracic and pelvic. Both these curves are concave forward. The cervical and lumbar curves are secondary and are convex forward. The cervical curve develops during the first year of life when the child learns to hold up its head. The lumbar curve develops in the second year when the child begins to walk.

SPINAL CORD

In the early stages of fetal life the vertebral column and the spinal cord are of equal length but, owing to the more rapid growth of the vertebral

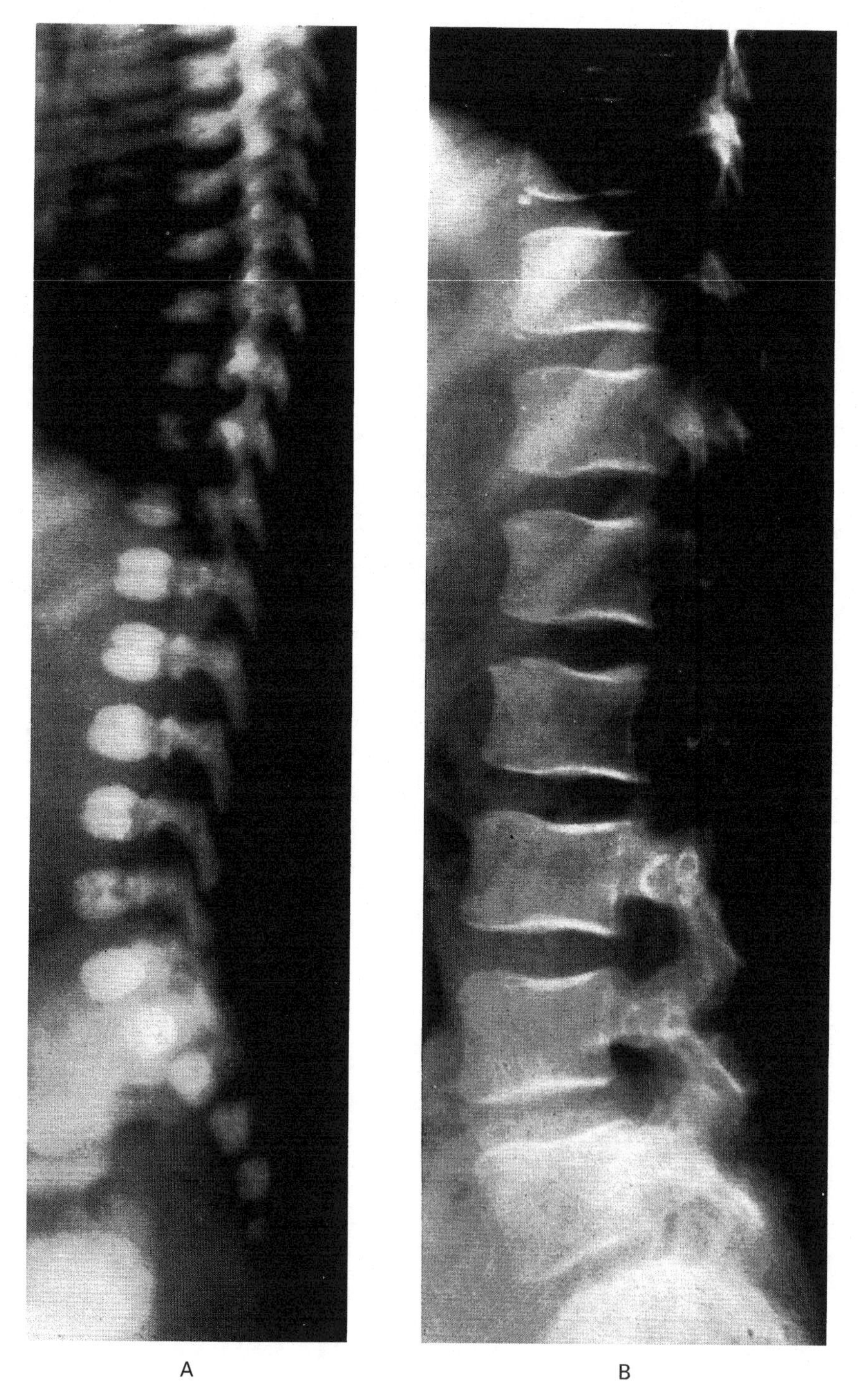

Fig. 3.41 Ossification of the vertebral column. A. Infancy B. About 14 years

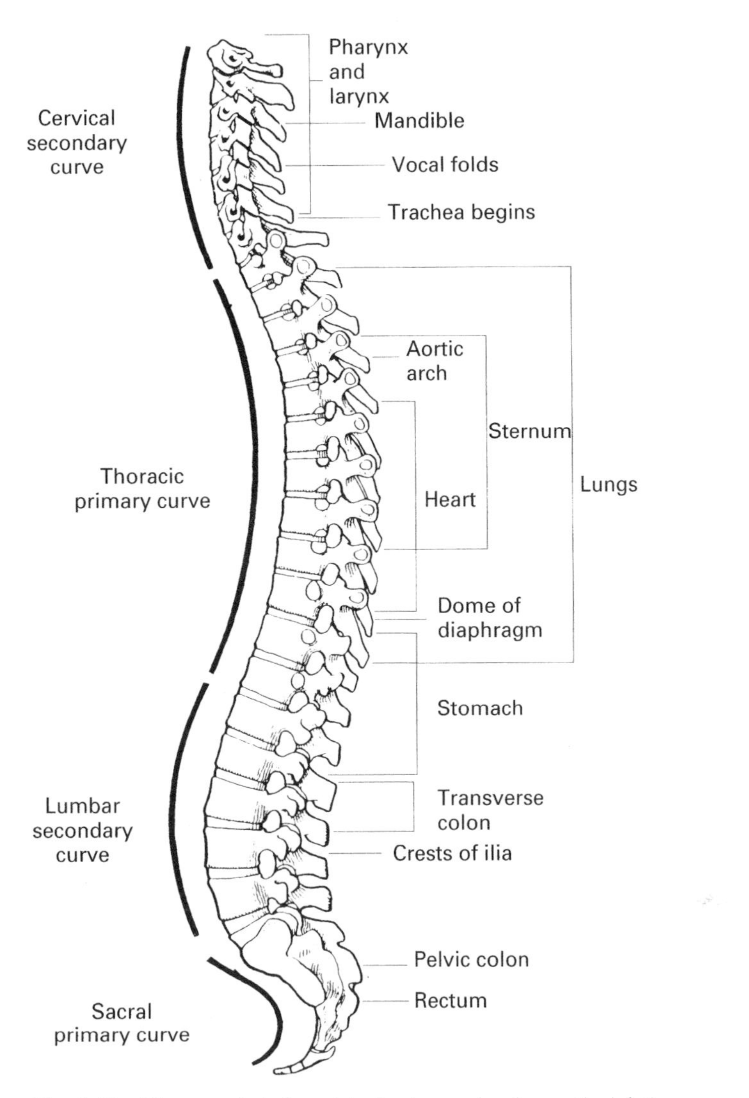

Fig. 3.42 Diagram of whole vertebral column, showing on the left the primary and secondary curves and on the right the approximate vertebral levels of other body structures

column, the cord terminates at the level of the third lumbar vertebra at birth, and at the level of the intervertebral disc between the first and second lumbar vertebrae in the adult.

The spinal cord occupies only part of the vertebral canal and is surrounded by three membranes—the pia mater, the arachnoid and the dura mater. These membranes help to suspend the cord in the vertebral canal and to protect it, while allowing a small degree of movement to take place during movements of the vertebral column. The arachnoid membrane is separated from the pia mater, which closely invests the cord, by the subarachnoid space; this space contains cerebrospinal fluid and is continuous with the cavities of the ventricles of the brain where the fluid is secreted.

Two areas of enlargement of the cord are present—the cervical and the lumbar. They correspond to the areas of innervation of the limb buds of the upper and lower limbs respectively. The lower end of the cord tapers to a conical extremity, the conus medullaris, which lies at the level of the intervertebral disc between the first and second lumbar vertebrae. A thin thread, the filum terminale (which does not contain nervous tissue) extends from the apex of the conus medullaris to be attached to the first segment of the coccyx. Although the cord ends high up in the lumbar region, the subarachnoid space extends down to the second sacral segment. Thus, puncture of the membranes in the lower lumbar region for the purpose of examining the cerebrospinal fluid can be undertaken without risk of damage to the cord.

31 pairs of spinal nerves arise from the spinal cord, roughly corresponding to the number of vertebrae, and leave the vertebral canal through the intervertebral foramina. The 1st cervical nerve emerges above the atlas. The 8th cervical nerve emerges below the 7th cervical vertebra. The remaining spinal nerves emerge below their respective vertebrae.

As the spinal cord is considerably shorter than the vertebral column, successive spinal nerves pass more and more obliquely downward to reach their respective intervertebral foramina. Below the termination of the cord these nerves form a large sheaf, the cauda equina, so called on account of its supposed resemblance to a horse's tail.

FUNCTIONS OF THE VERTEBRAL COLUMN

- It forms the main axis of the trunk; it gives attachment to the ribs, shoulder girdle and the upper limbs and to the pelvis and lower limbs.
- It supports the skull.
- It forms a strong protection for the spinal cord which lies in the vertebral canal formed by the vertebral foramina.
- The intervertebral foramina allow the spinal nerves to emerge from each vertebral canal.

THE JOINTS OF THE VERTEBRAL COLUMN

Adjacent vertebrae from the second cervical vertebra to the first sacral segment articulate with each other by:

- fibrocartilaginous joints between the vertebral bodies
- four synovial joints between the articular processes of the vertebral arches
- paired uncovertebral joints in the cervical region only
- The joints are reinforced by a number of ligaments (p. 93 and Fig. 3.44, 3.45).

JOINTS OF THE VERTEBRAL BODIES

These are secondary cartilaginous joints or amphiarthroses (symphyses). The upper and lower surfaces of the vertebral bodies are covered with a thin layer of hyaline cartilage. The opposing surfaces of two vertebrae are joined by a fibrocartilaginous intervertebral disc and by strong ligaments.

The intervertebral disc is composed of two parts:

- the anulus fibrosus—an outer, fibrosus ring
- the nucleus pulposus—an inner, gelatinous layer.

The disc adheres firmly to the layer of hyaline cartilage covering the surface of the vertebral bodies. Except for their peripheral parts the discs are avascular. They form about one-fifth of the length of the vertebral column and they increase in thickness from the upper thoracic region downwards. The cervical and lumbar discs are thicker anteriorly, producing the anterior convexity (lordosis) in these regions. The anterior concavity of the thoracic spine is largely due to the smaller anterior vertebral body height.

JOINTS OF THE VERTEBRAL ARCHES

(Zygapophyses—also known as facet joints)

These are synovial, plane joints between the inferior and superior articular processes of adjacent vertebrae.

Each joint has a fibrous capsule attached to the joint margins. In the cervical region the capsular ligaments are loose to allow a greater degree of movement in that region. The laminae, spines and transverse processes of the vertebrae are connected by the ligamenta flava, interspinous and intertransverse ligaments respectively. In the neck, the ligamentum nuchae joins the spinous processes of the cervical vertebrae; it is thickened posteriorly and passes to the occipital bone.

ATLANTO-OCCIPITAL JOINTS

There are two joints between the atlas and the occiput.

Type: Synovial, ellipsoid

Articular surfaces: Oval, convex articular facets of the occipital condyles articulate with the superior articular facets of the atlas.

Capsule: Fibrous, surrounds the occipital condyles and the corresponding articular facets of the atlas.

Ligaments:

Anterior atlanto-occipital membrane—a broad, fibrous sheet extending from the anterior margin of the foramen magnum to the upper border of the anterior arch of the atlas. The continuation of the anterior longitudinal ligament fuses with this membrane in the midline, joining the base of the occiput and the tubercle of the anterior arch of the atlas.

Posterior atlanto-occipital membrane—a broad, thin fibrous sheet extending from the posterior margin of the foramen magnum and the upper border of the posterior arch of the atlas. Laterally it arches over the vertebral artery allowing the latter to pass to the foramen magnum.

Movements and muscles:

Flexion—rectus capitis anterior, longus capitis.
Extension—splenius capitis, semispinalis capitis, trapezius, obliquus capitis superior, rectus capitis posteriores major and minor.
Lateral flexion—sternomastoid, trapezius, semispinalis capitis, splenius capitis, rectus capitis lateralis.

Suboccipital muscles (Fig. 3.43)

Obliquus capitis superior:
origin—transverse process of atlas.
insertion—occipital bone between superior and inferior nuchal lines.
function—extension and lateral flexion of head.

Obliquus capitis inferior:
origin—spine and lamina of axis.
insertion—transverse process of atlas.
function—rotation of head to same side.

Rectus capitis posterior major:
origin—spine of axis.
insertion—inferior nuchal line of occiput laterally.

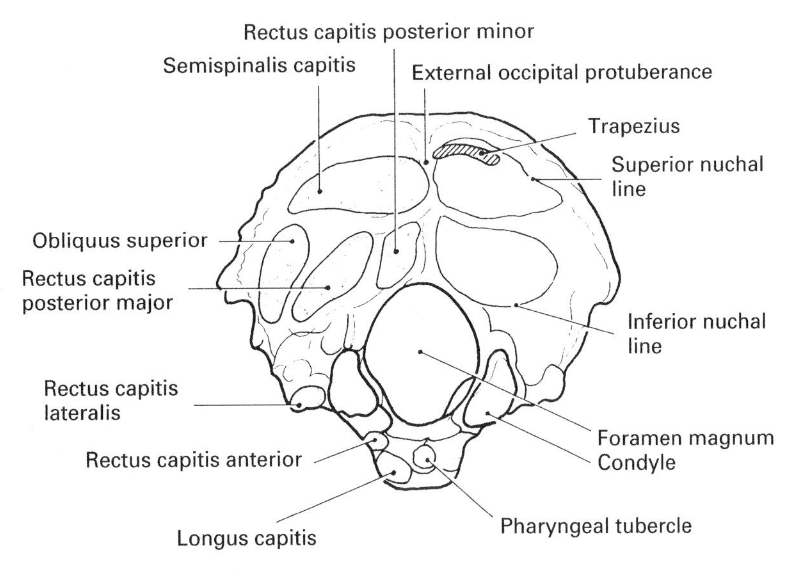

Fig. 3.43 Occipital bone: muscle attachments

function—extension and rotation of head to same side.

These three muscles form a triangle (suboccipital triangle) between the occipital bone, spine of axis and transverse process of atlas.

Rectus capitis minor:
origin—lies medial to rectus capitis major and takes origin from tubercle of posterior arch of atlas.
insertion—inferior nuchal line of occiput.
function—extension of head.
nerve supply—the suboccipital muscles are supplied by the posterior ramus of the first cervical nerve.

Rectus capitis anterior and lateralis belong to the prevertebral group of muscles.

Rectus capitis anterior:
origin—transverse process and lateral mass of atlas, anterior surface.
insertion—occipital bone, anterior to the condyle.
function—flexion of the head.
nerve supply—anterior ramus of first cervical nerve.

Rectus capitis lateralis:
origin—transverse process of atlas.
insertion—occipital bone, jugular process.
function—lateral flexion of the head to same side.
nerve supply—anterior ramus of first cervical nerve.

Trapezius: See page 125.

ATLANTOAXIAL JOINTS

There are three joints between the atlas and the axis:

- two lateral joints between the inferior articular facets of the atlas and the superior articular facets of the axis
- a midline pivot joint between the odontoid process (dens) of the axis and the articular facet on the posterior surface of the anterior arch of the atlas.

Type:

- lateral joints—plane.
- midline joint—pivot.

Capsule: a thin, loose capsule surrounds each joint.

Ligaments:

1. *Atlantoaxial*

Anterior longitudinal—joins surface of body of axis to anterior arch of atlas.

Ligamenta flava—join posterior arch of atlas to laminae of axis.

Ligamentum nuchae—interconnects to posterior tubercle of atlas with neural spine of axis and with occiput.

Transverse—attached to medial surface of each lateral mass of atlas. (Keeps odontoid process in position against anterior arch of atlas).

2. *Occipitoaxial*

Alar—attached to each side of odontoid process and to medial surface of each occipital condyle.

Apical—attached to apex of odontoid process and to anterior margin of foramen magnum.

Cruciate—so called because the longitudinal bands form vertical portion of a cross along with transverse ligament. The superior longitudinal band joins transverse ligament and basi-occiput at the anterior margin of foramen magnum. The inferior longitudinal band runs from transverse ligament to the posterior surface of the body of axis.

Tectorial membrane—a wide, thin ligament lying posterior to cruciate ligament. Joins posterior surface of axial body and basi-occiput just inside foramen magnum. It is continuous with posterior longitudinal ligament.

Movements and muscles:

Rotation—pivoting around odontoid process and at lateral joints simultaneously. Contraction of posterior deep and superficial muscles of neck on one side assisted by sternomastoid muscle of the opposite side. Limited by tightening of alar ligaments.

LUMBOSACRAL JOINT

This joint is similar to the other intervertebral joints. The intervertebral disc is wedge-shaped, being thicker anteriorly and it lies at the angle between the lumbar and sacral vertebrae. The disc transmits the weight of the vertebral column to the pelvis. The posterior joints between the vertebral arches (zygapophyses) are wider than those in the other vertebrae.

Iliolumbar ligament—the horizontal portion joins the transverse process of the fifth lumbar vertebra to the iliac crest. The vertical portion passes to the lateral part of the sacrum below. The horizontal part gives origin to the quadratus lumborum muscle (p. 92).

SACROCOCCYGEAL JOINT

Type: symphysis.

Articular surfaces: Fifth lumbar vertebra (apex) and first coccygeal vertebra, separated by an articular disc. Rarely the joint may be synovial in type thus allowing some movement.

Ligaments:

Anterior and deep posterior—join the respective anterior and posterior surfaces.

Superficial posterior—joins the margins of the sacral hiatus, converting it into a canal.

Lateral—lateral angle of sacrum inferiorly to the transverse process of first coccygeal vertebra.

Intercornual—a pair of small posterior ligaments joining the cornua of the sacrum and the coccyx.

INTERCOCCYGEAL JOINTS

The joints between the coccygeal vertebrae are symphyses with articular discs intervening. The ligaments associated are continuous with those of the vertebrae above.

MOVEMENTS AND MUSCLES OF THE VERTEBRAL COLUMN

Movement between individual vertebrae is very limited because of the spine's function in protection of the spinal cord. However movements of the column as a whole are considerable. Movement is maximum in the cervical region and minimum in the thoracic region.

Flexion—longus cervicis, sternomastoid, scalene, psoas major and rectus abdominis (paired muscles acting together). Movement limited by posterior spinal ligaments.

Extension—erector spinae (sacrospinalis), semispinalis capitis, splenius capitis.

Lateral flexion—erector spinae, quadratus lumborum, psoas major, scalene, splenius capitis, semispinalis capitis.

Rotation—sternomastoid, scalene, splenius cervicis, multifidus, rotatores. Maximum rotation occurs in thoracic region.

Superficial muscles

Erector spinae: A group of three paired muscles—iliocostalis (lateral), longissimus (or longus) and spinalis (medial). They form over the posterior thorax having taken origin from the sacral spinous processes, iliac crests and angles of the ribs. They course upwards and insert into the cervical and thoracic transverse and spinous processes and the ribs. The three muscles are subdivided further by the site of their insertion.
- Function—extension
- Nerve supply—posterior primary rami of spinal nerves.

Longus cervicis: (part of erector spinae):
- origin—transverse processes of T1–T6.
- insertion—transverse processes of C2–C6.
- function—extension and lateral flexion of neck.
- nerve supply—posterior primary rami of cervical and thoracic nerves.

Splenius capitis:
- origin—ligamentum nuchae, spinous processes of lower cervical and upper thoracic vertebrae.
- insertion—superior nuchal line of occipital bone, mastoid part of temporal bone.
- function—extension, lateral flexion and rotation of head.
- nerve supply—posterior rami of 4th–8th cervical spinal nerves.

Splenius cervicis:
origin—spines of upper thoracic vertebrae.
insertion—transverse processes of C1–C4.
function—extension (with splenius capitis), lateral flexion of head and neck.
nerve supply—posterior rami of 4th–8th spinal nerves.

Deep muscles

Semispinalis capitis: (part of a deep muscle which is divided into capitis, cervicis and thoracic portions):
origin—transverse processes of lower cervical and upper six thoracic vertebrae.
insertion—occipital bone.
function—extension and rotation of head.
nerve supply—posterior primary rami of cervical and thoracic spinal nerves.

Multifidus (has cervical, thoracic, lumbar and sacral portions):
origin—posterior sacrum, iliac crest, posterior lumbar vertebrae, thoracic transverse processes and lower cervical articular processes.
insertion—spinous processes of C2–C5.
function—extension of vertebral column.
nerve supply—posterior rami of spinal nerves (cervical-sacral).

Rotatores (lies deep to multifidus):
origin—transverse processes of thoracic vertebrae.
insertion—the vertebral lamina above.
function—extension and rotation of thoracic spine.
nerve supply—posterior rami of thoracic spinal nerves.

Other muscles acting upon vertebral column

Sternomastoid (sternocleidomastoid)—two heads:
origin—(a) sternal head—manubrium (anterolateral aspect).
(b) clavicular head—superior surface of medial third of clavicle.
insertion—lateral aspect of mastoid process of temporal bone and lateral third of superior nuchal line of occiput.
function—flexion and rotation of neck, rotation of head to opposite side.
nerve supply—accessory (XIth cranial nerve).

Scalene—(anterior, medius and posterior):
origin—(a) Anterior—anterior tubercles of transverse processes of C3–C6.
(b) Medius—posterior tubercles of all cervical vertebrae.
(c) Posterior—posterior tubercles of C4–C6.
insertion—(a) Anterior —scalene tubercle on superior of 1st rib (separates subclavian vein in front from subclavian artery behind).
(b) Medius—superior surface of 1st rib posterior to subclavian artery.
(c) Posterior—superior surface of 2nd rib.
function—lateral flexion of neck, stabilises 1st and 2nd ribs.
nerve supply—anterior primary rami of 3rd–7th cervical spinal nerves.

Psoas major
origin—lateral aspect of vertebral bodies and intervertebral discs of T12–L5 and spinous processes of L1–L5.
insertion—joins with tendon of iliacus muscle, inserting at lesser trochanter of femur.
function—flexion and lateral rotation of hip; flexion and lateral flexion of spine.
nerve supply—anterior rami of 2nd–4th lumbar spinal nerves.

Quadratus lumborum (quadrilateral shape):
origin—posterior of iliac crest and iliolumbar ligament, transverse processes of L1–L4.
insertion—inferior border of 12th rib.
function—lateral flexion of spine. Stabilizes 12th rib during respiratory movements.
nerve supply—anterior rami of 12th thoracic—1st lumbar spinal nerves.

Rectus abdominus (formed anteriorly by a pair of straight muscles either side of the midline of the abdomen which are enclosed by a fascial sheath. The recti are united in the median plane by a white tendinous band (linea alba) extending from the symphysis pubis to the xiphoid process of the sternum):

origin—pubic crest and anterior ligament of symphysis.
insertion—extends superiorly to the xiphoid process of sternum and costal cartilages of 5th–7th ribs.
function—flexes spine, strengthens anterior abdominal wall, increases intra-abdominal pressure against fixed diaphragm.
nerve supply—muscular branches of intercostal nerves T7–T12.

LIGAMENTS OF THE VERTEBRAL COLUMN (Figs 3.44 and 3.45)

The spinous processes, the laminae and the transverse processes of the arches are connected by a number of ligaments:

Supraspinous—extending from the apex of the spinous process of the 7th cervical vertebra to the sacrum. It consists of a tough fibrous cord which is thicker in the lumbar region. The superficial fibres skip between, or loop over, several vertebrae. The deep fibres are continuous with the interspinous ligament.

Interspinous—superficially continuous with the supraspinous ligament and deeply with the ligamentum flavum. It is thin and weaker than the supraspinous ligament.

Intertransverse—extending between the transverse processes of adjacent vertebrae and are most developed in the thoracic region. Muscles largely take over their function in the cervical and lumbar regions.

Anterior longitudinal—extending along the whole length of the anterior surface of the vertebral bodies from the base of the occiput, anterior tubercle of atlas and body of axis to the first part of the sacrum. It adheres closely to the anterior surface of the discs and to the anterior margins of the vertebral bodies.

Posterior longitudinal—lying within the vertebral canal and extends the whole length of the posterior surface of the vertebral bodies from the body of the axis to the sacrum. It adheres closely to the posterior surface of the discs and vertebral bodies and plays a considerable part in maintaining the positions of the intervertebral discs.

Ligamenta flava—are short ligaments of elastic tissue which connect the laminae of adjacent vertebrae. They are attached above to the lower part of the *anterior* surface of each lamina and below to the upper margin of the *posterior* surface of the lamina of the vertebra below. The ligamenta flava prevent overflexion of the vertebral column and assist the column to return to the vertical position after flexion.

Ligamentum nuchae—a broad, midline fibroelastic membrane which extends from the external occipital protuberance and external occipital crest of the skull to the 7th cervical vertebra, connecting the bifid spinous processes. It represents fused supraspinous and interspinous ligaments.

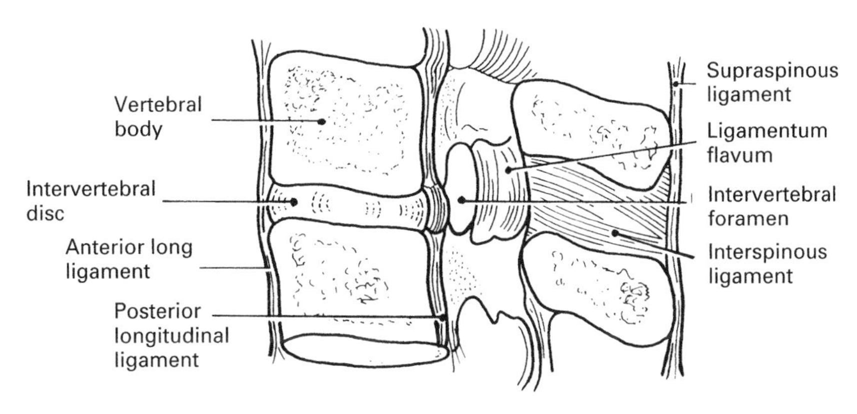

Fig. 3.44 Ligaments of the vertebral column

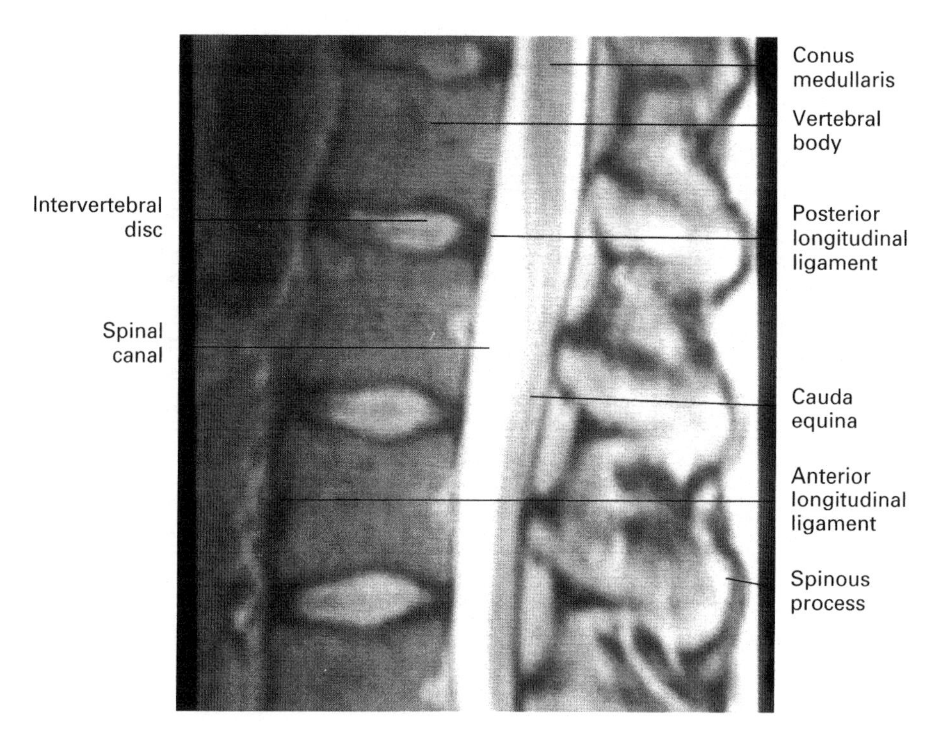

Fig. 3.45 Ligaments of the vertebral column: MR scan

4. The thorax

The skeleton of the thorax is bounded posteriorly by the 12 thoracic vertebrae, anteriorly by the sternum and costal cartilages and laterally by the 12 pairs pf ribs. It is roughly conical in shape—narrower above than below, with a transverse diameter greater than the anteroposterior diameter. On cross section it is kidney-shaped, due to the inward projection of the vertebral bodies and the slight backward curve of the posterior ends of the ribs. The lungs therefore extend behind the plane of the vertebral bodies. All the ribs articulate posteriorly with the vertebral column but anteriorly only the upper seven are attached directly to the sternum by their costal cartilages (Figs 4.1 and 4.2).

The main function of the thorax is in the action of respiration. The protection it affords to the major organs of respiration and circulation is a secondary function. Of the large number of muscles attached to the thorax, not all are involved with the respiratory movement—

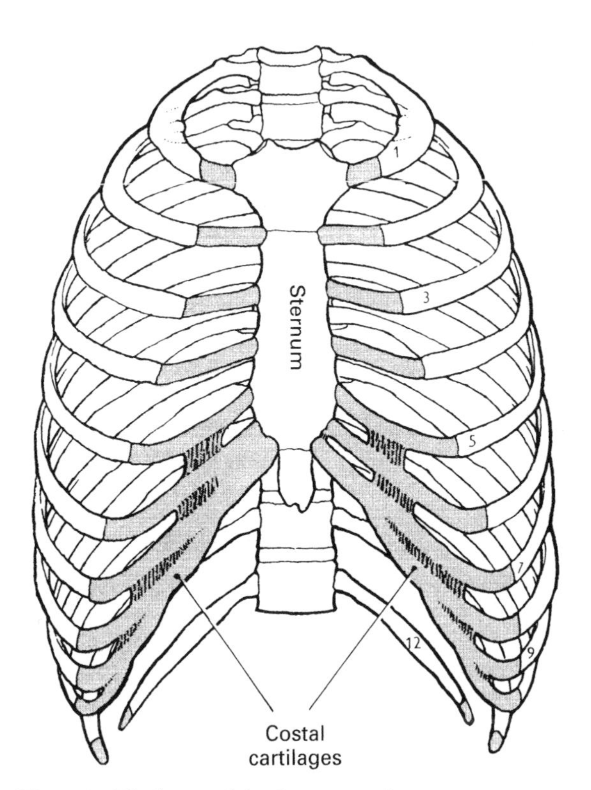

Fig. 4.1 The bones of the thorax: anterior aspect

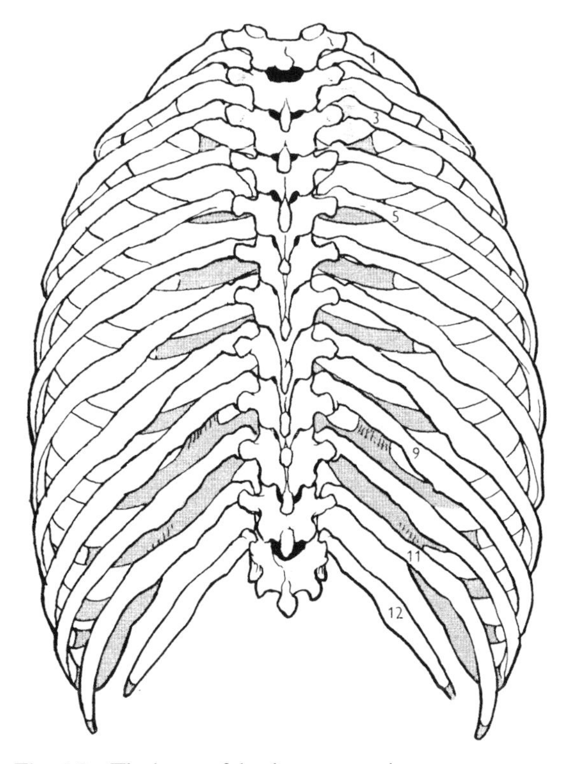

Fig. 4.2 The bones of the thorax: posterior aspect

the muscles of the arm, abdominal wall and vertebral column all have thoracic attachment.

The thoracic inlet is formed posteriorly by the first thoracic vertebra, laterally by the ribs and anteriorly by the manubrium sterni. Through it pass the trachea, the oesophagus, the main vessels to the head and upper limbs, and many nerves, including the sympathetic trunk, the vagus and the phrenic nerves. The apices of the pleural sacs and the lungs project a short distance above the inlet.

The thoracic outlet is larger than the inlet. It is formed posteriorly by the twelfth thoracic vertebra, anteriorly by the costal cartilages and laterally by the ribs. In the living subject, the outlet is closed by the diaphragm (p. 107), which separates the thoracic cavity from the abdominal cavity.

STERNUM

The sternum (Figs 4.3 to 4.6) is a long flat bone lying in the midline in the front of the thorax. It is directed downwards and slightly forwards. Its upper end supports the clavicles at the sterno-clavicular joints (p. 124). Laterally it gives attachment to the costal cartilages of the 7 upper pairs of ribs.

The sternum is composed of highly vascular cancellous bone covered by a layer of compact bone which is thickest in the manubrium sterni, between the clavicular notches. The spaces within the cancellous bone contain red marrow.

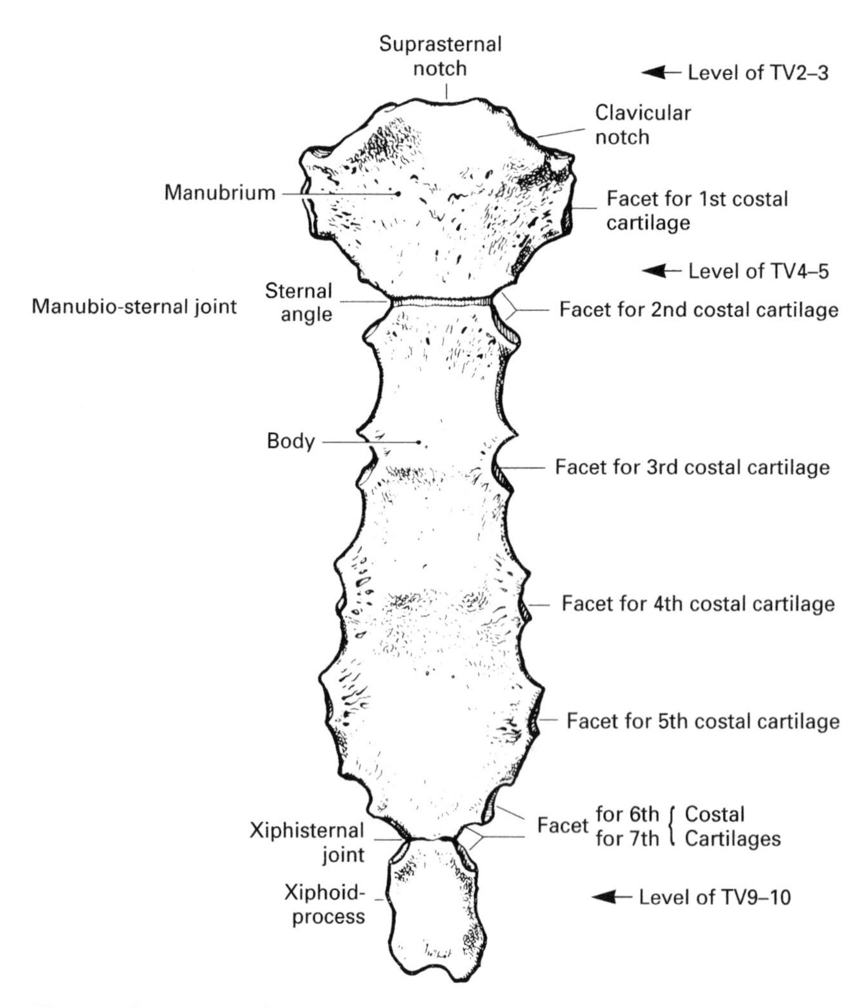

Fig. 4.3 Sternum: anterior aspect

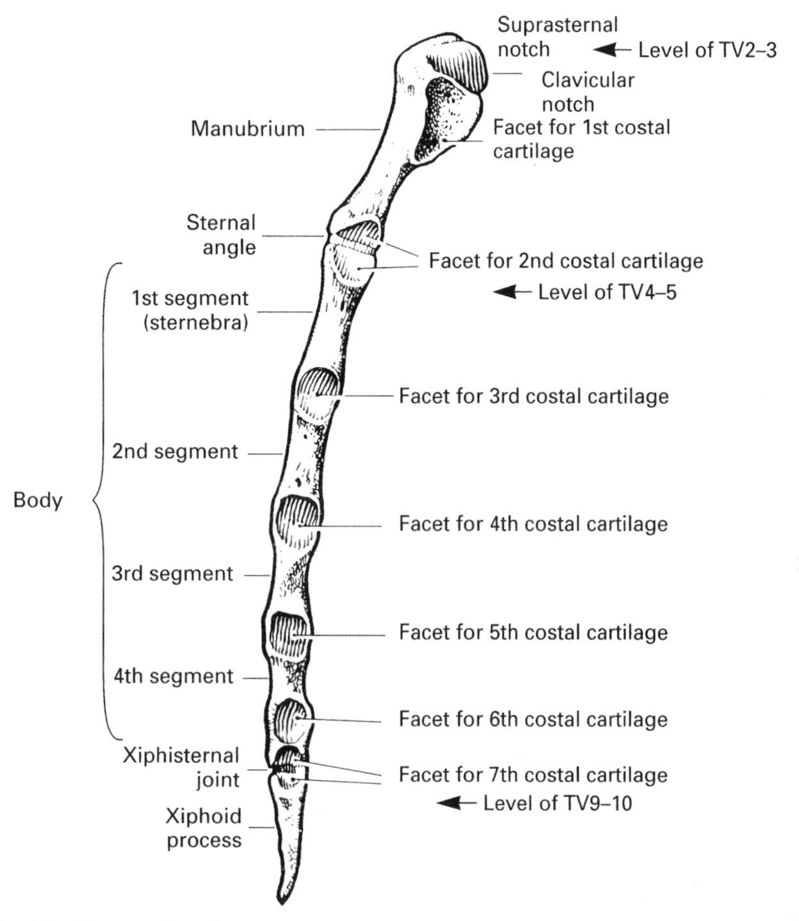

Fig. 4.4 Sternum: lateral aspect

The sternum consists of three parts:

- the manubrium sterni
- the body
- the xiphisternum.

The manubrium sterni is roughly triangular in shape and lies at the level of the 3rd and 4th thoracic vertebrae. Its superior border is thick and in the centre is the suprasternal (jugular) notch. On each side of this notch are articular facets (clavicular notches) for articulation with the medial ends of the clavicles at the sternoclavicular joints (p. 124). At the upper end of each lateral border is a notch for articulation with the costal cartilage of the first rib. A half notch at the lower end, with a similar half notch on the upper end of the body of the sternum, receives the costal cartilage of the 2nd rib.

The inferior border of the manubrium is narrow and articulates with the upper end of the body of the sternum to form the sternal angle—a prominent transverse ridge which can be seen and felt through the skin. The sternal angle lies at the level of the disc space between the 4th and 5th thoracic vertebrae.

The posterior surface of the manubrium forms the anterior border of the superior mediastinum; the lower part is related to the aortic arch, the upper part to the left brachiocephalic vein and to the brachiocephalic artery, left common carotid and left subclavian arteries. The lateral portions of the manubrium are related to the lungs and pleura.

The body of the sternum is long and narrow. It lies opposite the 5th to 9th thoracic vertebrae. The anterior surface is marked by three transverse ridges which show that the bone is formed by the fusion of four segments (sternebrae). Each lateral border has four complete notches for articulation with the 3rd to 6th costal cartilages; at its upper and lower ends are half notches, shared with the manubrium and xiphoid process, and with the 2nd to 7th costal cartilages. The superior border is oval and articulates with the manubrium sterni at the manubriosternal joint (p. 107). The inferior border is narrow and articulates with the xiphisternum.

The posterior surface of the body is related on the right to the right pleura and to the anterior border of the right lung. On the left, the upper two segments of the body are related to the left pleura and lung, and the lower two segments are related to the pericardium.

The xiphisternum (xiphoid process), the smallest part of the sternum, lies in the epigastric region. It articulates with the lower border of the body of sternum at the xiphisternal joint (p. 107) which lies at the level of the 9th to 10th thoracic vertebrae. The xiphisternum is rarely fully ossified in the adult and it is very variable in shape; sometimes it is perforated or bifid or it may be bent to one side. Its superior angle completes the notch for the 7th costal cartilage.

The posterior surface of the xiphisternum is related to the anterior surface of the liver.

Radiographic appearances of the sternum (Figs 4.5 and 4.6)

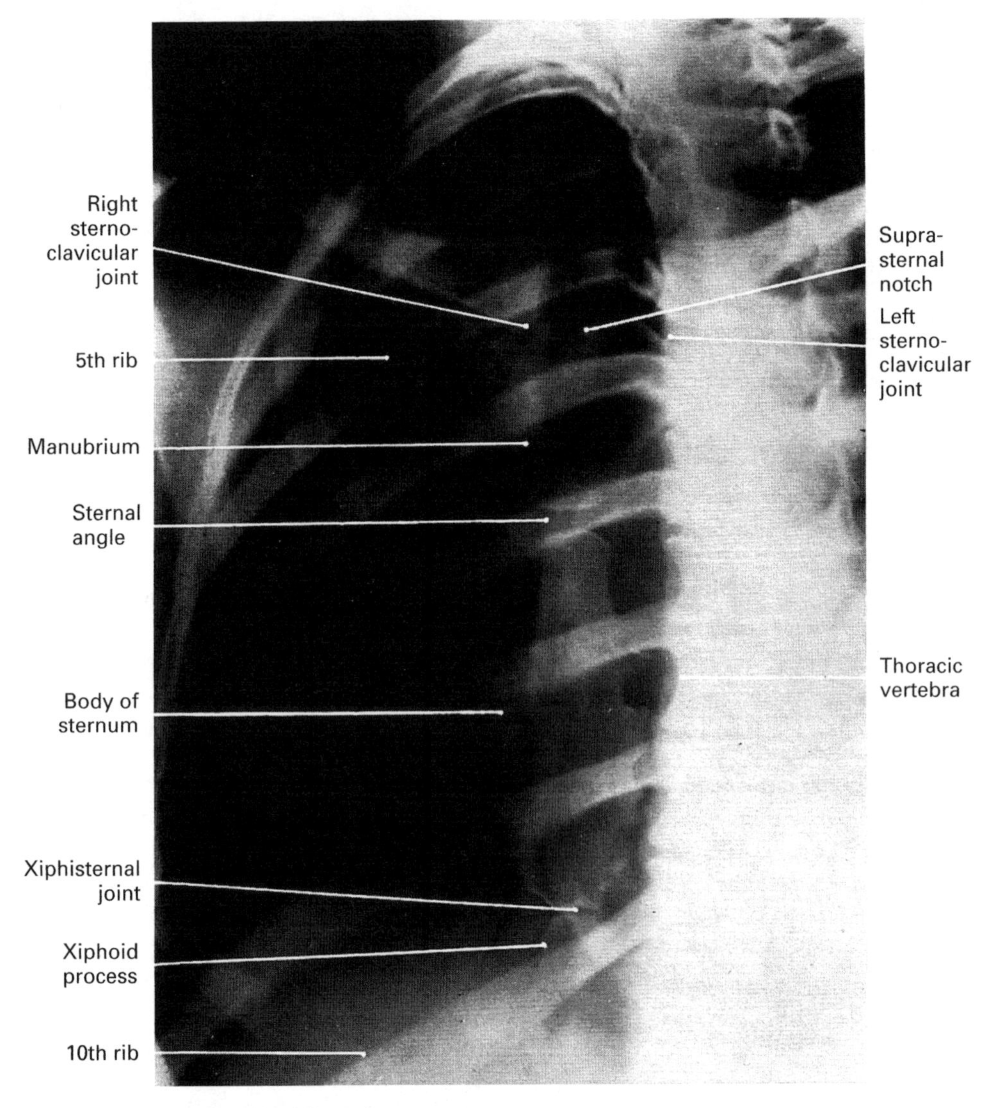

Fig. 4.5 Sternum: anterior oblique view

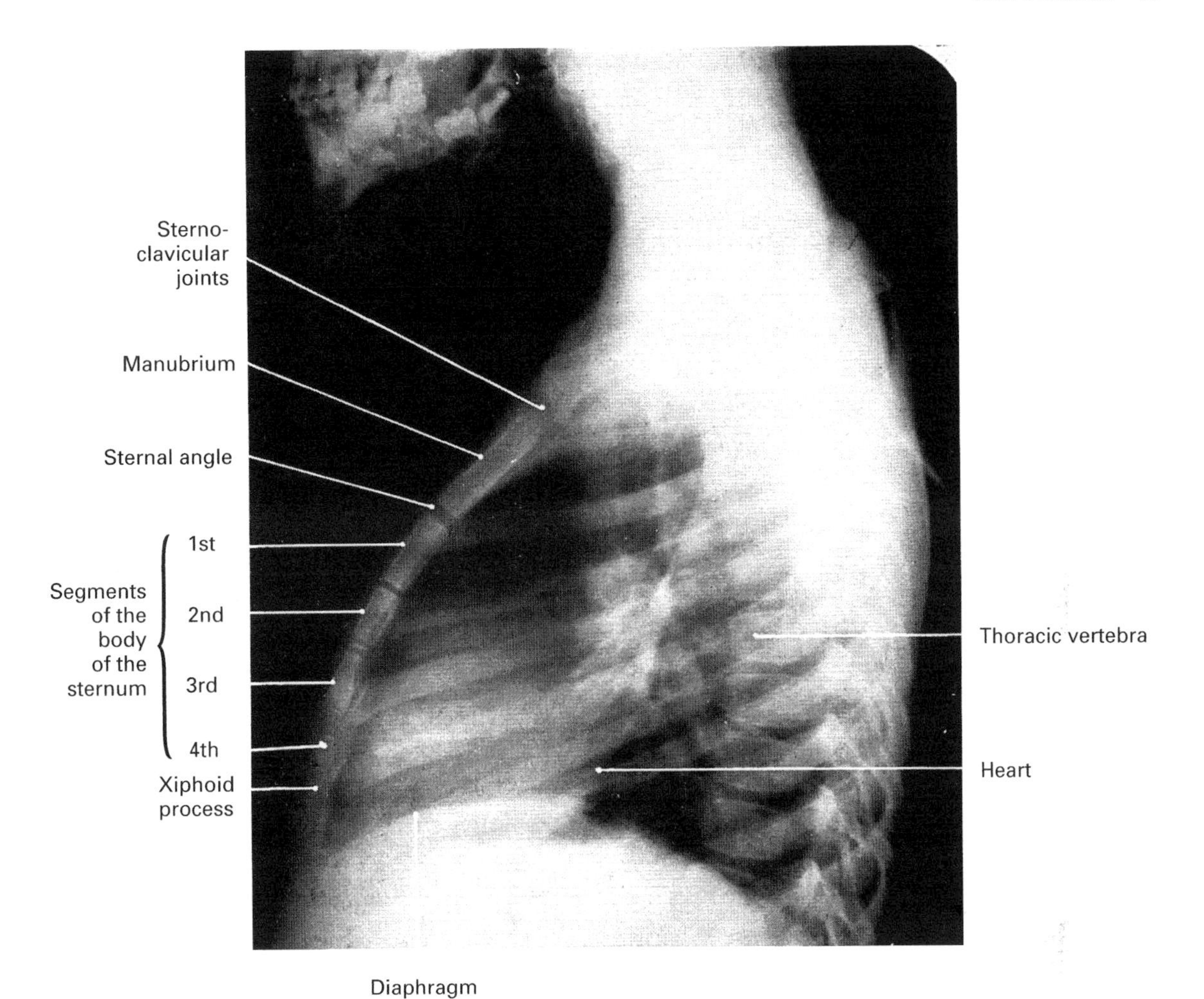

Fig. 4.6 Sternum: lateral view of a child

MUSCLE ATTACHMENTS OF THE STERNUM

Manubrium:

Anterior surface—sternal origins of pectoralis major and sternocleidomastoid.

Posterior surface—sternohyoid.

Body:

Anterior surface—sternal origins of pectoralis major, articular capsules of sternocostal joints.

Posterior surface—transversus thoracis (sternocostalis).

Xiphisternum:

Anterior surface—medial fibres of rectus abdominis, aponeurosis of external and internal oblique muscles, linea alba, aponeurosis of internal oblique and transverse abdominis.

Posterior surface—some of the fibres of the diaphragm.

OSSIFICATION OF THE STERNUM (Fig. 4.7)

Primary centres

- for the manubrium, about the 6th month of intrauterine life.
- for each of the 4 sternebrae, between the 6th and 9th months of intrauterine life.
- for the xiphisternum, about the 3rd year.

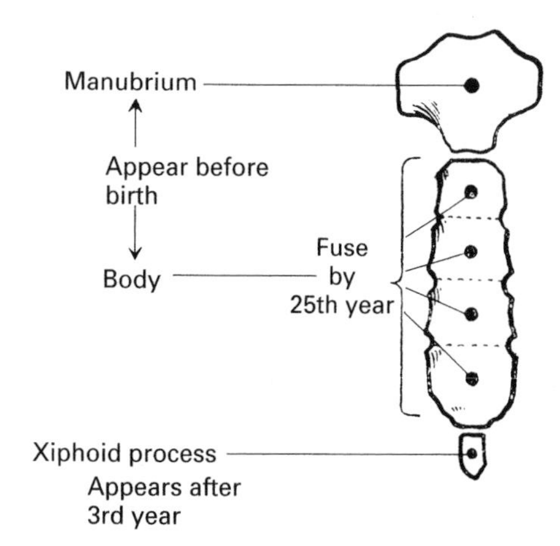

Fig. 4.7 Ossification of the sternum

Fusion takes place from the 15th to the 25th years but the manubrium sterni and the xiphisternum remain as separate bones in adult life.

RIBS

The ribs (Figs 4.1, 4.2, 4.8 to 4.13) are long, slender, elastic arches which form a large part of the skeleton of the thorax. They consist of highly vascular spongy bone enclosed in a thin layer of compact bone. The ribs contain a large proportion of red marrow.

There are normally 12 pairs of ribs but occasionally cervical (p. 102) or lumbar ribs are present. Sometimes the 12th pair of ribs is absent.

The ribs articulate posteriorly with the thoracic vertebrae at the costovertebral joints (p. 105) and they end anteriorly in extensions called costal cartilages.

The 7th upper pairs of ribs are known as true ribs, as their costal cartilages articulate with the sternum. The lower 5 pairs (8th to 12th) are called false ribs since they are not connected directly to the sternum. The costal cartilages of the upper 3 pairs of false ribs (8th to 10th) are each joined to the costal cartilage directly above, whereas the 11th and 12th pairs are unattached, and are known as 'floating ribs'.

The ribs increase in length progressively from the 1st to the 7th and thereafter become shorter. They slope downwards and anteriorly so that the anterior end is lower than the posterior end. They are separated from each other by intercostal spaces, occupied in life by the intercostal muscles, nerves and vessels.

The first two and last three pairs of ribs are slightly different from the others and will be described separately.

A TYPICAL RIB

A typical rib (Fig. 4.8) consists of a long, curved shaft with anterior and posterior ends.

The anterior end is slightly concave to receive the costal cartilage of the rib.

The posterior end consists of

- the head
- the neck
- the tubercle.

The *head* is covered with articular cartilage and is divided into upper and lower halves by a transverse crest. The facets articulate with demifacets on adjacent vertebrae to form costovertebral joints (p. 105). The crest is attached by a ligament to the intervertebral disc, the rib taking the number of the lower vertebra.

The *neck* is a short, flattened portion of the rib between the head and the tubercle. It lies anterior and slightly superior to the transverse process of the corresponding vertebra to which it is attached by ligaments.

The *tubercle* is a bony protuberance on the posterior surface of the rib, at the junction of the neck and the shaft. It is divided into a lateral nonarticular part and a medial articular facet which articulates with a facet on the transverse process of the corresponding vertebra to form the costotransverse joint (p. 106).

The shaft is long and is flattened from side to side, so that it has external and internal surfaces. The superior border is rounded. On the inner side of the inferior border is the costal groove along which the intercostal vessels and nerves pass.

The posterior end of the rib slopes obliquely downwards and backwards as far as the angle, where the direction of the shaft changes to that of a gentle forward and inward curve. The shaft

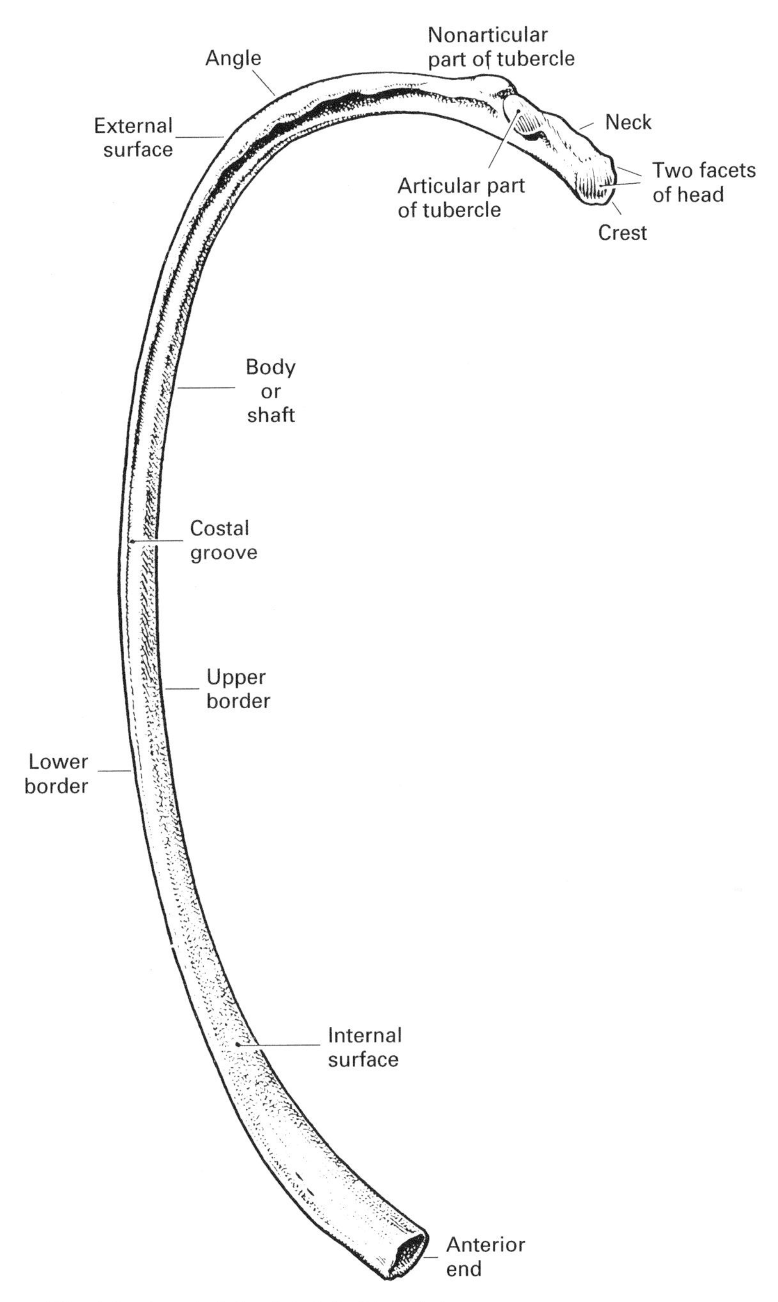

Fig. 4.8 A typical rib of the left side: inferior aspect

twists slightly so that the anterior end of the inner surface faces slightly downwards.

THE FIRST RIB (Fig. 4.9)

The 1st rib is short and broad and it slopes very obliquely downwards and forwards. In contrast with a typical rib, it is flattened above and below and so it has upper and lower surfaces and an outer and an inner border. The head bears a single articular facet, because it articulates only with the body of the first thoracic vertebra. The tubercle is prominent and at this point the direction of the slope changes and therefore the tubercle and the

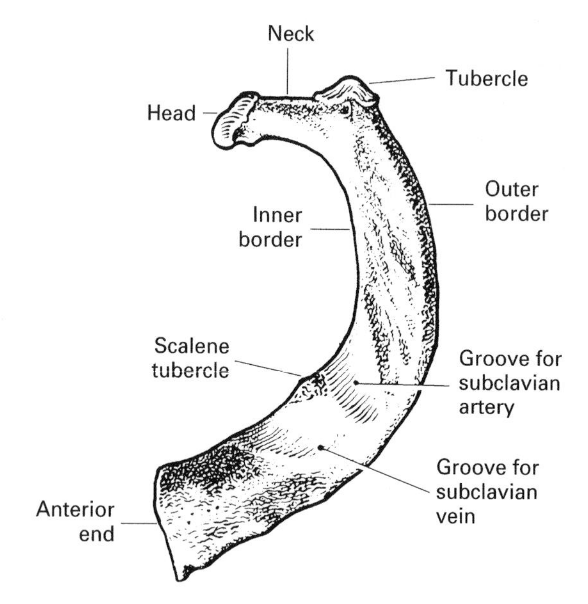

Fig. 4.9 First rib of left side: superior aspect

angle coincide. Two wide, shallow grooves lie on the middle of the upper surface, separated by a small projection—the scalene tubercle—on the inner border. The anterior groove is for the subclavian vein, the posterior groove for the subclavian artery. The 1st rib has no costal groove and its costal cartilage passes under the clavicle to articulate with the manubrium sterni.

THE SECOND RIB (Fig. 4.10)

The 2nd rib is almost twice as long as the 1st rib. The direction of its surfaces lies between that of the 1st rib and that of a typical rib. The external surface faces upwards and outwards. This rib is not twisted so that it will rest evenly on a flat surface, which enables it to be identified easily. Near the middle of its outer surface is a roughened area for attachment of part of the serratus anterior muscle.

THE TENTH, ELEVENTH AND TWELFTH RIBS

The 10th, 11th and 12th ribs each has a single facet on its head. The 11th and 12th ribs do not possess a tubercle or a neck, and their anterior ends are pointed and tipped with cartilage.

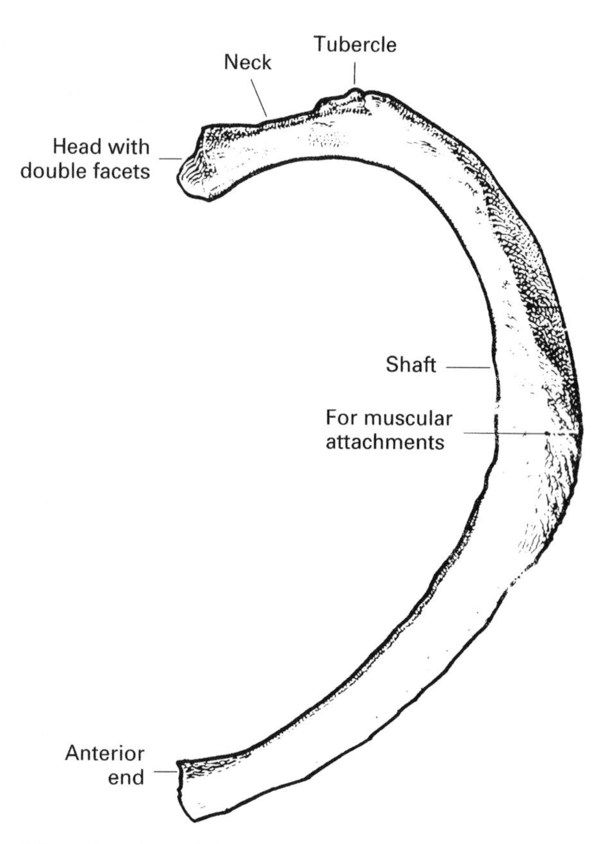

Fig. 4.10 Second rib of left side: superior aspect

The costal cartilages are flattened bars of hyaline cartilage which are joined to the anterior end of the ribs. The 7 upper pairs of cartilages connect the anterior ends of their respective ribs to the sternum. The medial ends of the 8th, 9th and 10th cartilages articulate only with the lower border of the cartilage above and form the lower costal margins between which lies the infrasternal (subcostal) angle.

CERVICAL RIBS

Occasionally the costal element of the 7th cervical vertebra develops as a cervical rib. The condition may be unilateral or, more commonly, bilateral. The anterior extremity may be cartilaginous or it may be connected to the sternum by a band of fibrous tissue. Thus the true length of the costal element may not always be demonstrable radiographically.

Radiographic appearances of the ribs (Figs 4.11 to 4.13)

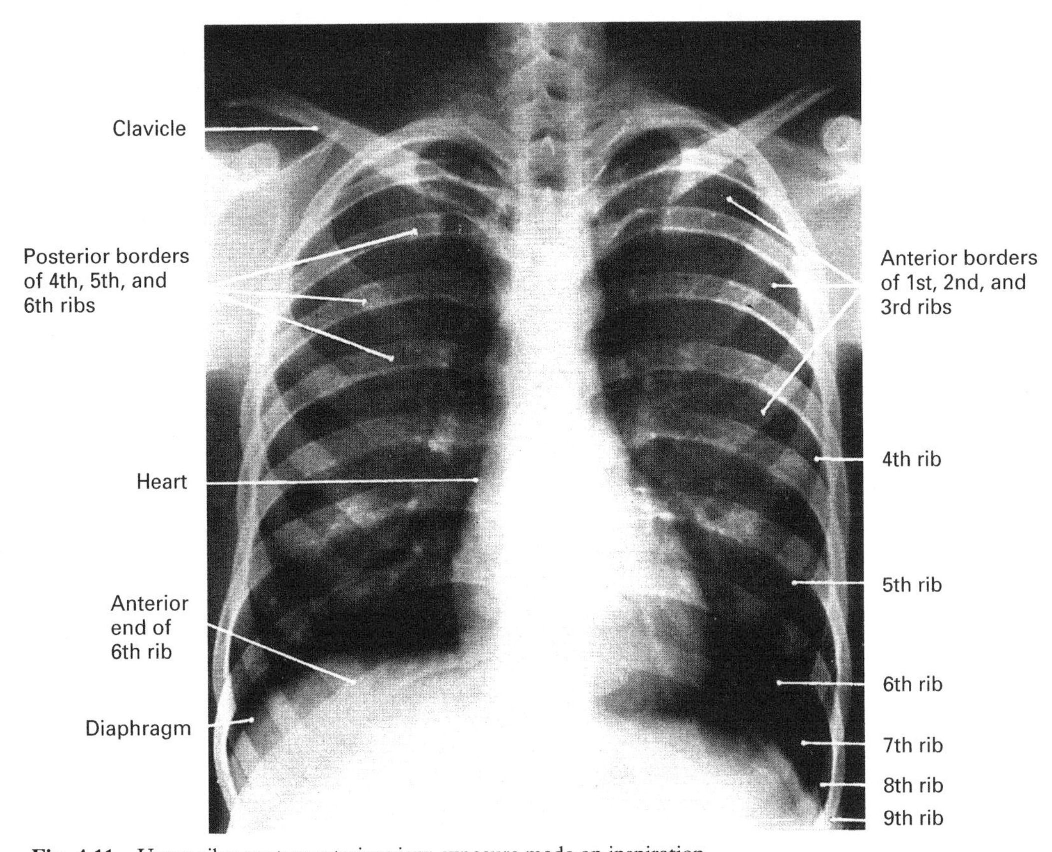

Fig. 4.11 Upper ribs: posteroanterior view: exposure made on inspiration

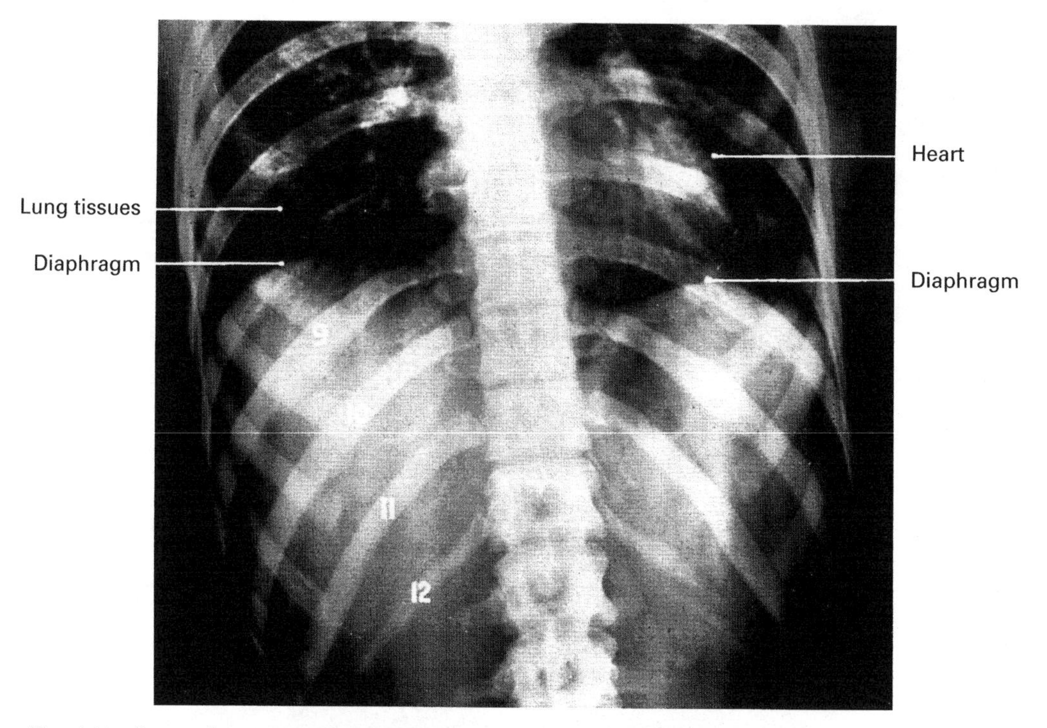

Fig. 4.12 Lower ribs: anteroposterior view: exposure made on expiration

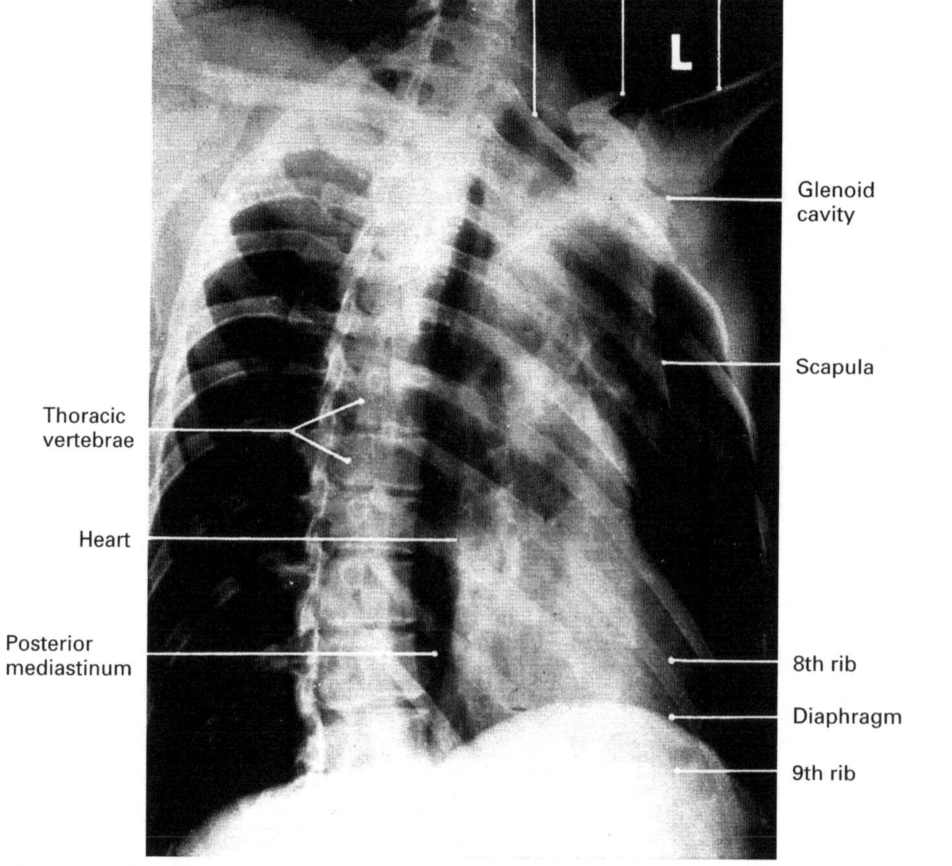

Fig. 4.13 Ribs: oblique view

OSSIFICATION OF THE RIBS (Fig. 4.14)

Primary centre

Appears in the shaft of each rib about week 8 of intrauterine life.

Secondary centres

These appear at puberty.

1st rib

- for the head
- for the tubercle.

2nd–10th ribs

- for the head
- for the nonarticular part of the tubercle
- for the articular part of the tubercle.

11th and 12th ribs

- for the head of the rib. (These ribs do not possess tubercles.)

The epiphyses unite with the shafts in about the 20th year.

JOINTS OF THE THORAX

Each rib articulates with the vertebral column at two joints, one between the head of the rib and the body of the vertebra (**costovertebral joint**) and the other between the tubercle of the rib and the transverse process of the vertebra (**costo-transverse joint**).

COSTOVERTEBRAL JOINTS

Type: Synovial, plane.

Articular surfaces: Head of ribs 2–9 articulate with a facet on upper lateral surface of vertebra from which it takes its number, also with facet on lower lateral surface of vertebra above, and with intervertebral disc between them. The heads of ribs 1, 10, 11 and 12 articulate only with their own vertebra.

Capsule: Fibrous, surrounds head of rib and articular cavity between its vertebra and intervertebral disc. Capsule is thickened anteriorly to form radiate ligament.

Ligaments:

Radiate—joins head of each rib to sides of corresponding vertebral bodies and intervertebral disc.

Intra-articular—lies within joint and is attached to crest between two facets of the head and to the intervertebral disc.

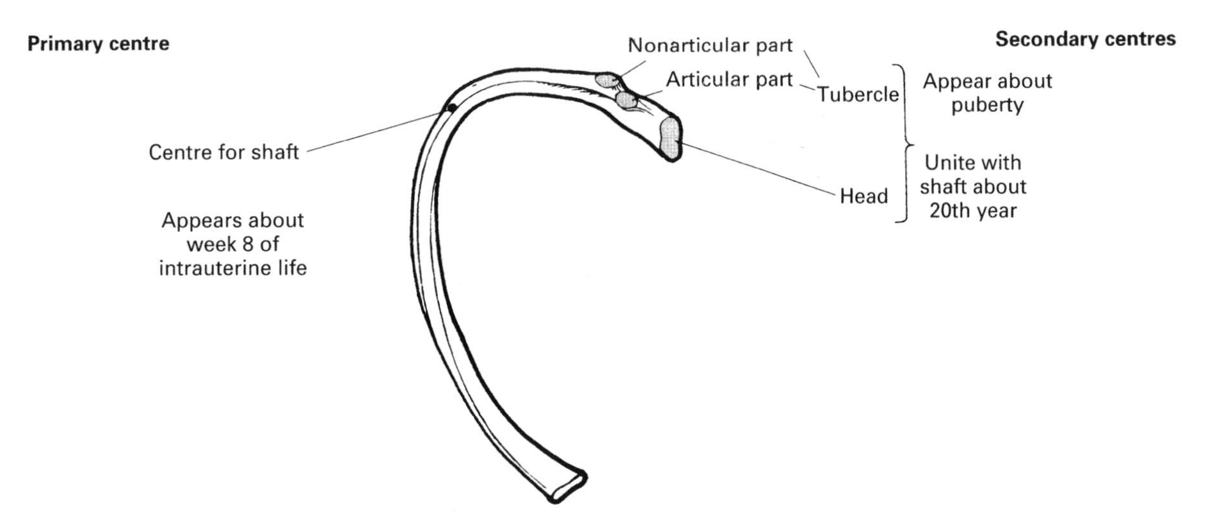

Fig. 4.14 Ossification of a typical rib

COSTOTRANSVERSE JOINTS (Fig. 4.15)

Type: Synovial, plane.

Articular surfaces: Articular part of tubercle of upper ten ribs with transverse process of their own vertebrae.

Capsule: Fibrous, lined with synovial membrane. Attached to periphery of articular surfaces.

Ligaments:

Superior costotransverse—joins neck of each rib to transverse process of vertebra above it. 1st rib lacks this ligament.

Costotransverse—very short. Joins posterior of neck of each rib to anterior surface of transverse process of its own vertebra.

Lateral costotransverse—joins tubercle of each rib to apex of transverse process of its own vertebra.

Movements: Slight gliding during respiration.

STERNOCOSTAL JOINTS

Type: 1st = synchondrosis, 2nd–9th = synovial, plane.

Articular surfaces: Costal cartilages and concave depressions on lateral border of sternum.

Capsule: Fibrous, surrounds each joint. Present at 2nd to 7th joints.

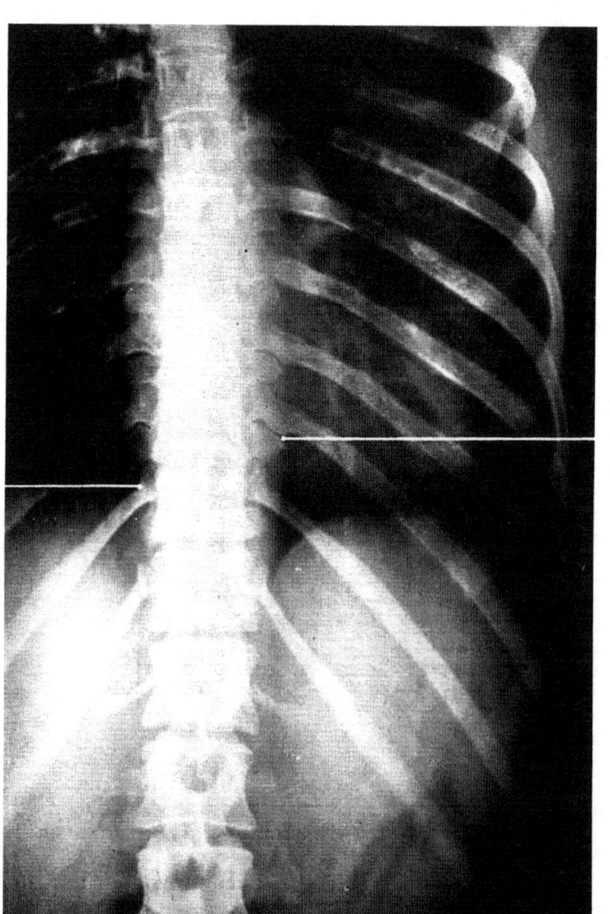

Fig. 4.15 Ribs: antero-posterior view showing costotransverse joints

Ligaments:

Radiate sternocostal—thin but wide. Joins sternal ends of costal cartilages to anterior and posterior surfaces of sternum.

Costoxiphoid—joints 7th costal cartilage (sometimes the 6th also) to the xiphoid process by anterior and posterior fibres.

Intra-articular—divides joint into two parts by passing between sternum and costal cartilage. Most constant at 2nd joint.

INTERCHONDRAL JOINTS

Type: Synovial, plane.

Articular surfaces: Adjacent borders of lower costal cartilages articulate with each other by elongated synovial joints.

Capsule: Surrounds joint.

Ligaments:

Interchondral—pass between costal cartilages.

COSTOCHONDRAL JOINTS

The medial end of each rib articulates with its costal cartilage forming a primary cartilaginous joint with periosteum continuous with the perichondrium.

MANUBRIOSTERNAL JOINT

This is a symphysis between the manubrium sterni and the body of the sternum. The surfaces are covered with hyaline cartilage and are connected by a fibrocartilaginous disc which may become ossified.

XIPHISTERNAL JOINT

This is a symphysis between the xiphoid process and the body of the sternum. Usually it becomes a synostosis by about the 40th year but sometimes it remains ununited.

STERNAL JOINTS

In the newborn the sternum is composed of four sternebrae, united by primary cartilaginous joints. Between puberty and the mid-20s synostoses occur from below upwards.

THE DIAPHRAGM

The diaphragm is a large, dome-shaped, musculofibrous sheet separating the thoracic from the abdominal cavity. It is composed of a trilobed central tendon from the edges of which muscle fibres radiate out to the periphery of the thoracic outlet—the vertebral column, the sternum and the lower ribs. The muscle fibres are in three groups:

Sternal—anterior, arising from the posterior surface of the xiphisternum.

Costal—lateral, arising from the lower six ribs and costal cartilages.

Lumbar—posterior, arising from the lumbar vertebrae by two crura and two arcuate ligaments. The right crus, which is broader and longer than the left, arises from the upper three lumbar vertebrae; the left crus arises from the upper two lumbar vertebrae. The arcuate ligaments extend from the transverse processes of the 1st lumbar vertebra medially to the tips of the 12th ribs laterally.

These muscular fibres are inserted into the central tendon of the diaphragm. When the diaphragm is relaxed, the central tendon is at the level of the 8th thoracic vertebra (Fig. 4.12). When the diaphragm contracts, the muscles shorten, the central tendon is pulled downwards and the thoracic cavity is increased in length (Fig. 4.11). Simultaneously, the intercostal muscles (p. 108) contract so that the thorax, as well as being increased in length, is also increased in size from front to back (upper ribs) and from side to side (lower ribs).

The nerve supply to the diaphragm is from the phrenic nerves.

RELATIONS OF THE DIAPHRAGM

Superior—the upper surface of the diaphragm is covered by the pleura and by the pericardium, separating the diaphragm from the lungs and heart.

Inferior—most of the inferior surface is covered by peritoneum. The liver, the right kidney and right suprarenal gland are in close association with the right side of the diaphragm. The left kidney, the left suprarenal gland and the spleen are in close association with the left side.

APERTURES OF THE DIAGRAM

The aortic aperture is the lowest and most posterior of the apertures. It lies between the diaphragm and the vertebral column so is in fact *behind* the diaphragm. It lies at the level of the 12th thoracic vertebra.

The aortic aperture transmits the aorta, the thoracic duct and the azygos vein.

The oesophageal aperture lies at the level of the 10th thoracic vertebra, above and a little in front of the aortic aperture. It transmits the oesophagus, the vagal trunk and the oesophageal branches of the left gastric vessels.

The vena caval aperture lies at the level of the 8th thoracic vertebra. It transmits the vena cava and branches of the right phrenic nerve.

MUSCLES OF RESPIRATION

The diaphragm (p. 107) is the major muscle of inspiration

INTERCOSTALS

There are three groups of intercostal muscles and there are 11 pairs of muscles in each group. They are all incomplete, i.e. they do not encompass the entire hemithorax.

External intercostals:
The fibres run downwards and forwards.
- origin—lower inner border of rib, attached from tubercle to costal cartilage. External intercostal membrane unites costal end and sternum.
- insertion—upper border of rib below.
- function—elevation of ribs during inspiration.

Internal intercostals:
The fibres run obliquely downwards and backwards, at right angles to external intercostal muscles.
- origin—lower inner border of rib, from floor of costal groove.
- insertion—upper border of rib below. Internal intercostal membrane unites portion between rib angle and vertebrae.
- function—depression of ribs during forced expiration.*

Inner intercostals:
The fibres extend downwards and backwards.
- origin—inner surface of rib.
- insertion—upper surface of rib below.
- function—depression of ribs (with internal intercostals).
- nerve supply (all intercostals)—adjacent intercostal nerves.

ACCESSORY MUSCLES OF RESPIRATION

These muscles are used in addition to the diaphragm and the intercostal muscles during forced respiration.

Muscles	Action
Scalene (p. 92) Sternocleidomastoid (p. 92)	to elevate 1st and 2nd ribs, sternum and clavicle.
Pectoralis major (p. 119) Pectoralis minor (p. 119) Serratus minor (p. 119) Trapezius (p. 125)	to elevate ribs when scapula is fixed.
Levator scapulae (p. 125) Rhomboid major (p. 125) Rhomboid minor (p. 125)	to elevate and fix scapula.
Quadratus lumborum (p. 92) Subcostals (p. 109) Transversus thoracis (p. 109)	to act with internal intercostals in depressing ribs in forced expiration.

*In quiet respiration, expiration is passive.

Subcostals:
origin—lower inner border of rib
insertion—2nd or 3rd rib below
nerve supply—intercostal nerves

Transversus thoracis:
origin—posterior surface of body and xiphoid process of sternum, costal cartilages 4 to 7.
insertion—postero-inferior borders of costal cartilages 2 to 6 by muscle slips which pass obliquely upwards and laterally.
nerve supply—intercostal nerves.

5. The shoulder girdle

The shoulder girdle consists of:

- the scapula
- the clavicle.

From the shoulder girdle is suspended the upper limb. Anteriorly the shoulder girdle is attached to the sternum but posteriorly there is no bony connection with the trunk.

The scapula is situated on the upper postero-lateral aspect of the thorax to which it is attached by muscles only. It articulates with the head of the humerus at the shoulder joint (p. 118).

The clavicle lies almost horizontally, high up on the anterior wall of the thorax and it is subcutaneous throughout its length. The lateral end of the clavicle articulates with the acromion process of the scapula at the acromioclavicular joint (p. 125). The medial end of the clavicle articulates with the sternum at the sterno-clavicular joint (p. 124). The clavicle acts as a prop to support the shoulder clear of the trunk and allow the upper limb freedom of movement. To a large extent it transmits the weight of the upper limb to the trunk.

SCAPULA (Figs 5.1 to 5.7)

The scapula is a flat bone, triangular in shape. It has a costal (anterior) surface and a dorsal (posterior) surface and three borders; superior, medial and lateral. At the junction of these borders are the superior, inferior and lateral angles. The lateral angle is flattened and massive and it is often referred to as the 'head' of the scapula. On its lateral aspect is the glenoid cavity for articulation with the head of the humerus at the shoulder joint.

The costal surface (Fig. 5.1) faces forwards and slightly medially when the arm is by the side. It is slightly concave and conforms roughly to the curvature of the chest wall. It is crossed by a number of ridges for attachment of the sub-scapularis muscle. A thickened ridge of bone runs parallel with the lateral border and is most prominent at its upper end, immediately below the glenoid cavity.

The dorsal surface (Fig. 5.2) faces backwards and outwards. It is crossed by the spine of the scapula which divides it into the supraspinous fossa above and the infraspinous fossa below. The spine ends just short of the head of the scapula and the angle so formed between the lateral edge of the spine and the neck of the scap-ula is the spinoglenoid notch. The two fossae can communicate with each other only through this notch.

The superior border is the shortest of the three borders. At its lateral end is a small indentation—the suprascapular notch. Beyond this the beak-like coracoid process projects forwards and outwards.

The medial (vertebral) border lies parallel with the vertebral column when the arm is by the side. It is easily palpable for most of its length.

The lateral (axillary) border extends from the inferior angle to the glenoid cavity. It is so well covered by muscles that it can only be felt near the inferior angle.

The superior angle is at the junction of the superior and medial borders. It lies almost level

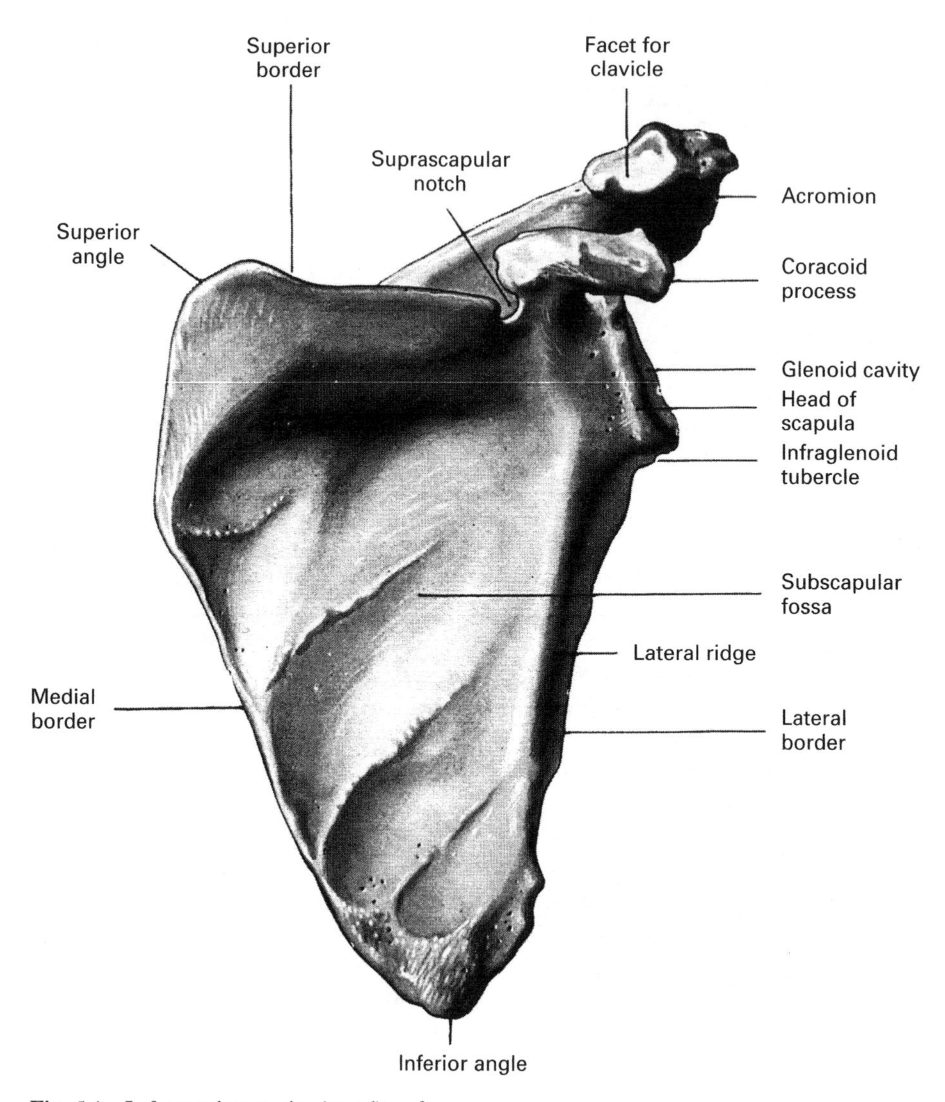

Fig. 5.1 Left scapula: anterior (costal) surface

with the midpoint of the clavicle when the arm is by the side. It is easily palpable.

The inferior angle is at the junction of the medial and lateral borders and is more acute than the superior angle. It lies over the posterior end of the seventh rib and is easily palpable.

The lateral angle is broad and bears the glenoid cavity. Immediately behind the head is a slight constriction—the neck of the scapula.

The glenoid cavity is a pear-shaped, articular surface which is slightly concave (Fig. 5.3). It faces laterally with a slight upward and forward tilt. The glenoid labrum is a fibrocartilaginous rim attached to the periphery of the glenoid cavity; it deepens the cavity of the joint. The shallow socket and the comparatively large humeral head permit great freedom of movement, but render this ball and socket joint unstable. The joint capsule is thin and hangs loosely inferiorly. The ligaments are also weak. The stability of the joint depends mainly on surrounding muscles and it is a common site of dislocation. The supraglenoid and infraglenoid tubercles are small areas of roughened bone immediately above and below the glenoid cavity.

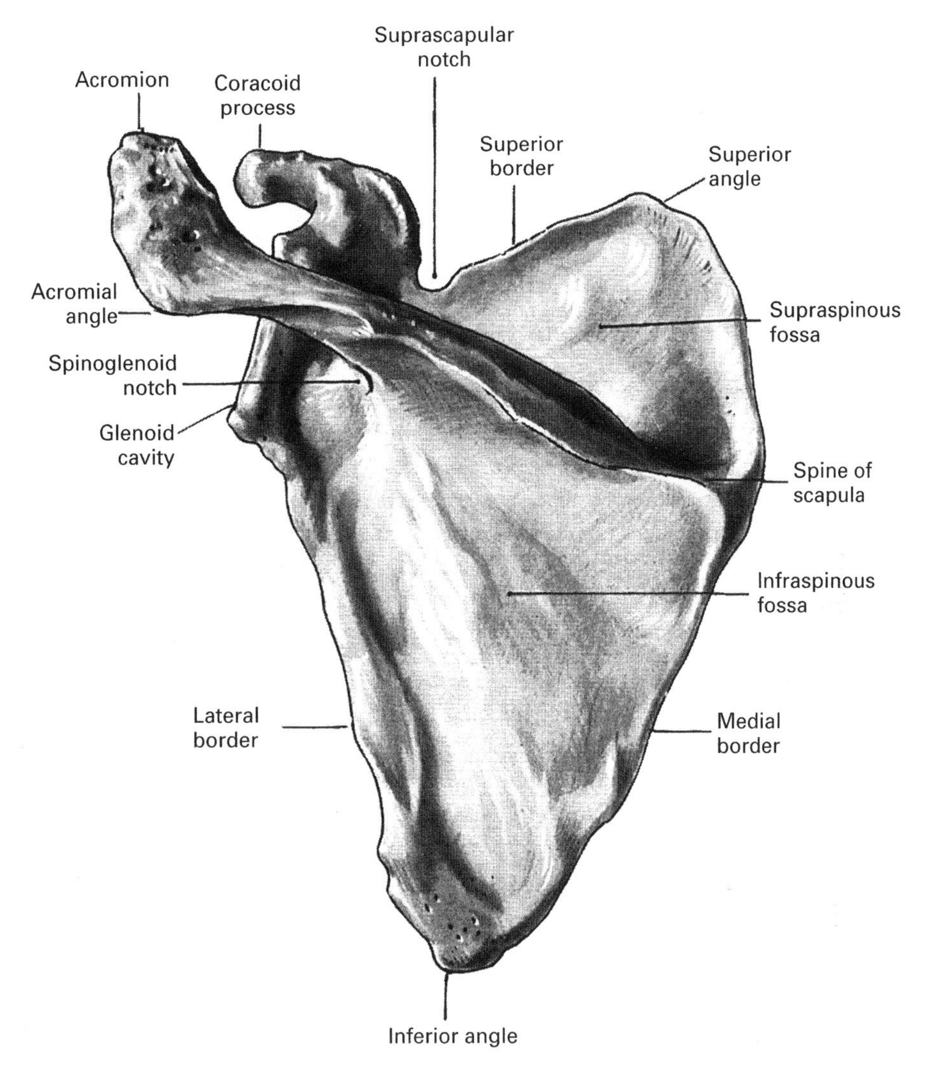

Fig. 5.2 Left scapula: posterior (dorsal) surface

They are the sites of attachment for the long heads of the biceps and of the triceps muscles respectively.

The coracoid process arises from the lateral end of the superior border of the scapula, adjacent to the superior margin of the head. It projects upwards and forwards and then turns sharply laterally over the superior aspect of the shoulder joint. The tip of this process can be felt on deep pressure in a depression—the infraclavicular fossa—which lies 25 mm below the outer third of the clavicle under the anterior border of the deltoid muscle. The coracoid process gives attachment to the short head of the biceps, the coracobrachialis muscle and to ligaments which stabilize the acromioclavicular joint.

The spine of the scapula is a thin shelf of bone on the posterior surface. It runs outwards and slightly upwards from the vertebral border to end by becoming continuous with the acromion and it separates the supraspinous and infraspinous fossae. The crest of the spine is roughened and gives attachment to the trapezius and deltoid muscles. It is easily palpable for the whole of its length.

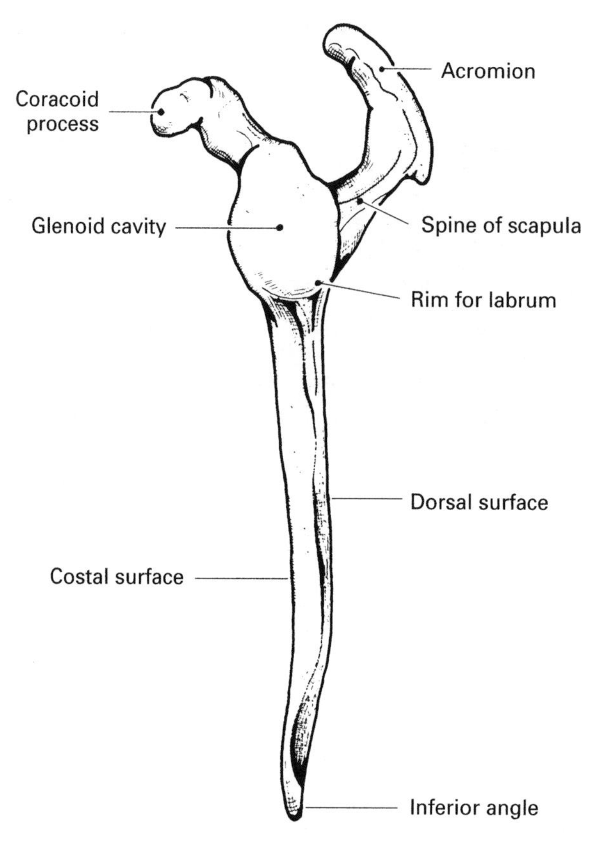

Fig. 5.3 Left scapula: lateral aspect

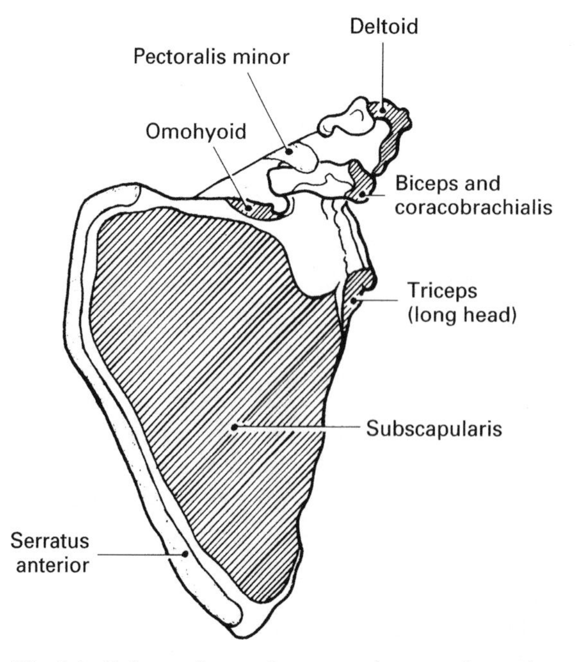

Fig. 5.4 Left scapula: costal aspect to show muscle attachments

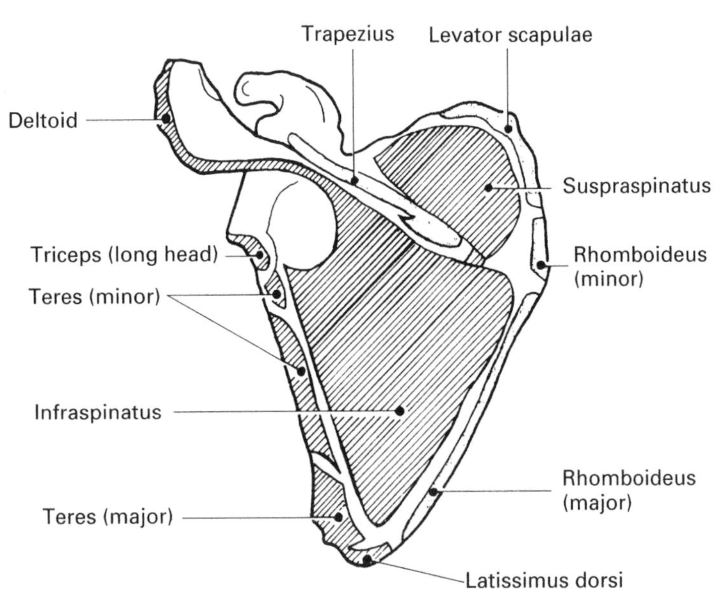

Fig. 5.5 Left scapula: dorsal aspect to show muscle attachments

The acromion is the flattened lateral projection of the crest of the spine. Its plane is roughly at right angles to the spine and it extends laterally to overhang the shoulder joint. Its lateral border is continuous with the inferior border of the crest of the spine and at their junction is the acromial angle. The medial border of the acromion is continuous with the superior border of the crest of the spine and at its anterior end there is a small oval articular facet which articulates with the clavicle to form the acromioclavicular joint. These features can be felt through the skin. The coracoacromial ligament is attached between the acromion and the coracoid process and forms a protective shelf above the shoulder joint.

Radiographic appearances of the scapula (Figs 5.6 and 5.7)

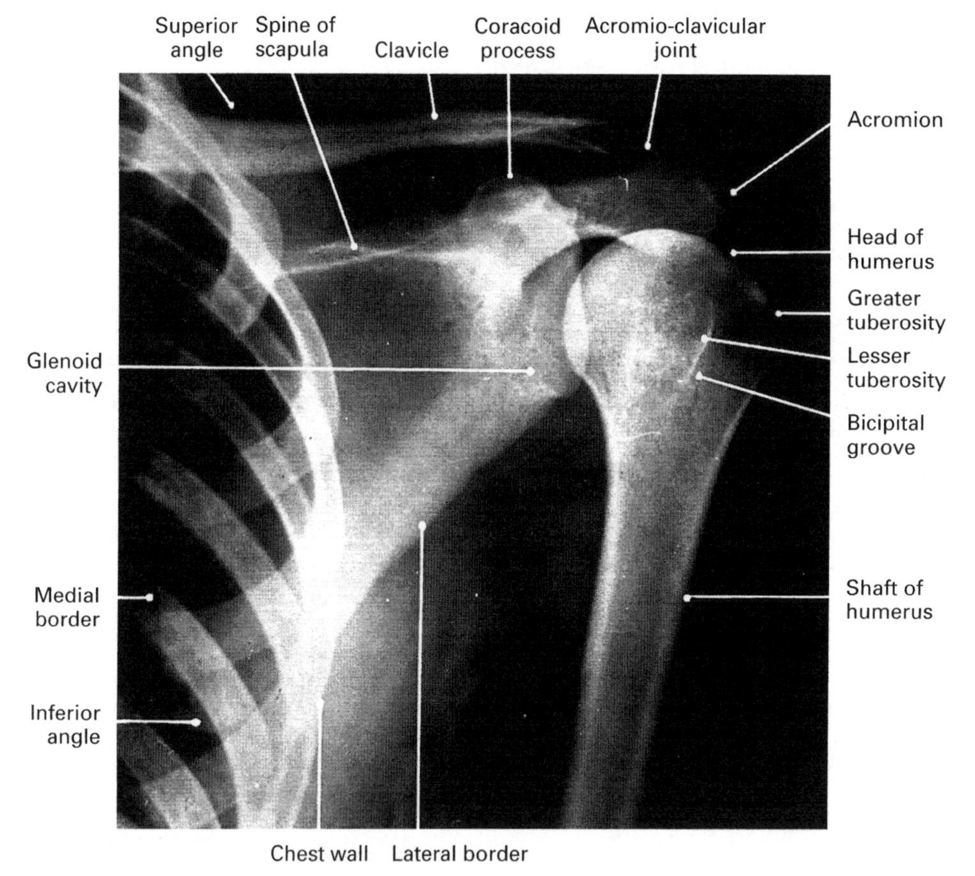

Fig. 5.6 Left shoulder joint and scapula: anteroposterior view

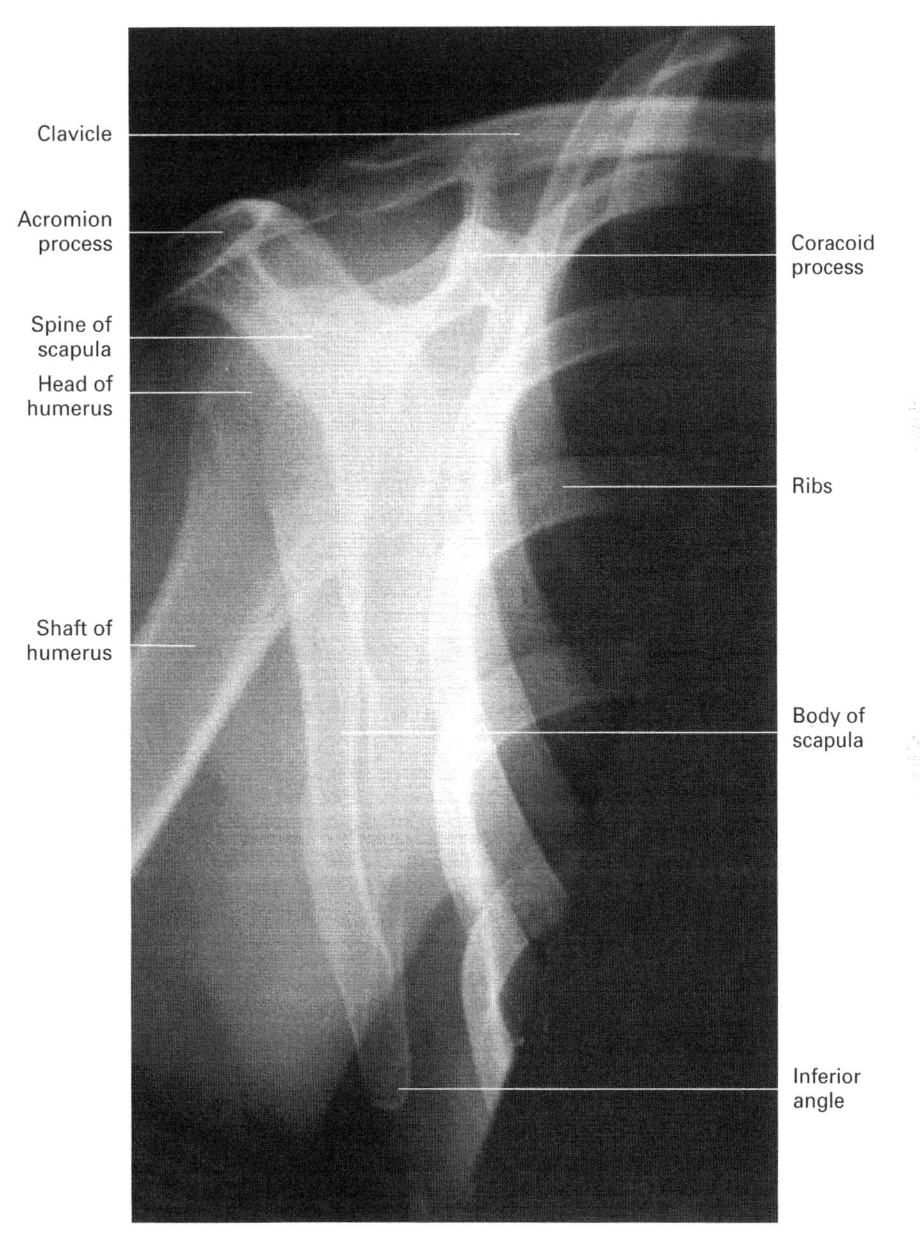

Fig. 5.7 Left scapula: lateral view

SHOULDER JOINT (Figs 5.8 to 5.11)

Type: Synovial, ball and socket, multiaxial.

Articular surfaces: Glenoid fossa of scapula and hemispherical head of humerus. The shallow glenoid fossa is deepened by a fibrocartilaginous rim—the glenoid labrum.

Synovial membrane: Lines capsule and extends over anatomical neck of humerus. Tendon of long head of biceps passes from superior glenoid tubercle and lies within a sheath of synovial membrane.

Ligaments:

Capsular—envelops joint. Is loose inferiorly (to allow free movement of joint). Attached to outer surface of glenoid labrum. Supported by very strong muscles, subscapularis, supraspinatus, infraspinatus and teres minor ('rotator cuff' muscles), the tendons of which insert into the capsular ligament.

Transverse humeral—attached to greater and lesser tuberosities, holding the tendon of long head of biceps within the bicipital groove. Also provides attachment for capsular ligament.

Coracohumeral—attached between coracoid process and greater tuberosity of humerus. Strengthens upper part of capsule.

Glenohumeral—three bands of minor importance, attached medially to upper and medial parts of glenoid labrum, lesser tuberosity and anatomical neck of humerus.

Coracoacromial—thick band between entire lateral surface of coracoid and anterior part of acromion.

Blood supply: From branches of axillary and subclavian arteries.

Nerve supply: From circumflex humeral, subscapular and suprascapular nerves.

Bursae:

Subacromial—between capsule and deltoid, extending below acromion. (Does not communicate with joint.)

Subscapular—between subscapularis tendon and joint capsule.

Others lie above acromion; between coracoid and capsule; between teres major and long head of triceps; between infraspinatus and capsule.

Movements and muscles:

Flexion—deltoid (anterior fibres), pectoralis major (clavicular head), coracobrachialis, biceps.

Extension—deltoid (posterior fibres), pectoralis major (sternodorsal head), teres major, latissimus dorsi.

Abduction—supraspinatus. (Rotator cuff muscles provide stability during this movement.)

Adduction—pectoralis major, latissimus dorsi, subscapularis, teres major and minor, coracobrachialis, triceps (long head).

Medial rotation—pectoralis major, latissimus dorsi, teres major, deltoid (anterior fibres), subscapularis.

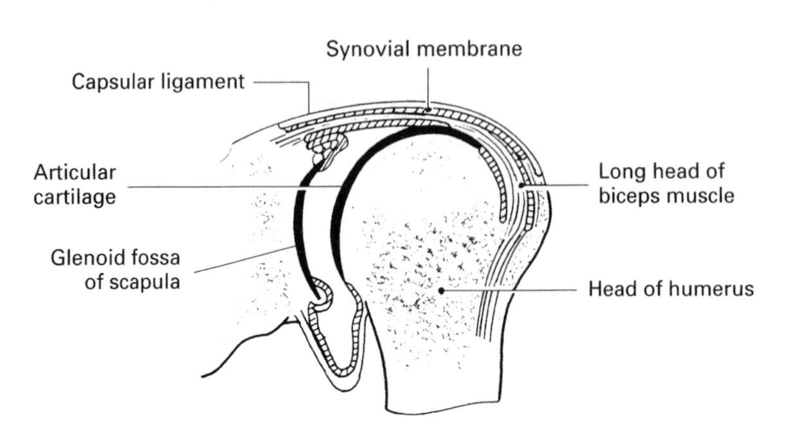

Fig. 5.8 Schematic diagram of the shoulder joint

Lateral rotation—deltoid (posterior fibres), infraspinatus, teres minor.
Circumduction—combination of all these movements.

Subscapularis:
origin—subscapular fossa of scapula
insertion—lesser tuberosity of humerus, via a tendon.
function—medial rotation of arm, stabilization of humeral head.
nerve supply—subscapular nerves (upper and lower).

Supraspinatus:
origin—supraspinous fossa of scapula.
insertion—greater tuberosity of humerus.
function—abduction, stabilization of humeral head.
nerve supply—suprascapular nerve.

Infraspinatus:
origin—infraspinous fossa of scapula.
insertion—greater tuberosity of humerus via a tendon.
function—lateral rotation of arm.
nerve supply—suprascapular nerve.

Teres minor:
origin—inferior angle of scapula laterally.
insertion—medial lip of bicipital groove, via a tendon.
function—abduction and medial rotation of arm.

(The four muscles listed above are the 'rotator cuff' muscles.)

Deltoid:
origin—(extensive), clavicle, acromion process and spine of scapula.
insertion—deltoid tuberosity of humerus.
function—extension (posterior fibres), flexion (anterior fibres) abduction.
nerve supply—axillary nerve.

Pectoralis major:
origin—front of medial half of clavicle, anterior surface of sternum and 2nd–6th costal cartilages.
insertion—lateral lip of bicipital groove.
function—adduction and medial rotation of arm, flexion (clavicular head), extension (sternocostal head).
nerve supply—lateral and medial pectoral nerves.

Teres major:
origin—inferior angle of scapula laterally.
insertion—medial lip of bicipital groove, via a tendon.
function—abduction and medial rotation of arm.
nerve supply—lower subscapular nerve.

Latissimus dorsi:
origin—spines of lower 6 thoracic vertebrae, spines of lumbar and sacral vertebrae (via thoracolumbar fascia), posterior part of iliac crest.
insertion—floor of bicipital groove.
function—adduction, extension, medial rotation of arm.
nerve supply—thoracodorsal nerve.

Coracobrachialis:
origin—coracoid process of scapula.
insertion—middle of medial border of humerus.
function—flexion and adduction of arm.
nerve supply—musculocutaneous nerve.

Serratus anterior:
origin—outer surface of 1st–8th ribs.
insertion—costal surface of entire medial border of scapula.
function—pulls scapula forwards around thoracic wall, stabilization of scapula.
nerve supply—long thoracic nerve.

Pectoralis minor:
origin—3rd, 4th and 5th ribs.
insertion—coracoid process of scapula.
function—depression of shoulder. Aids serratus anterior in pulling scapula around thoracic wall.
nerve supply—medial and lateral pectoral nerves.

Radiographic appearances of the shoulder joint (Figs 5.9 to 5.11)

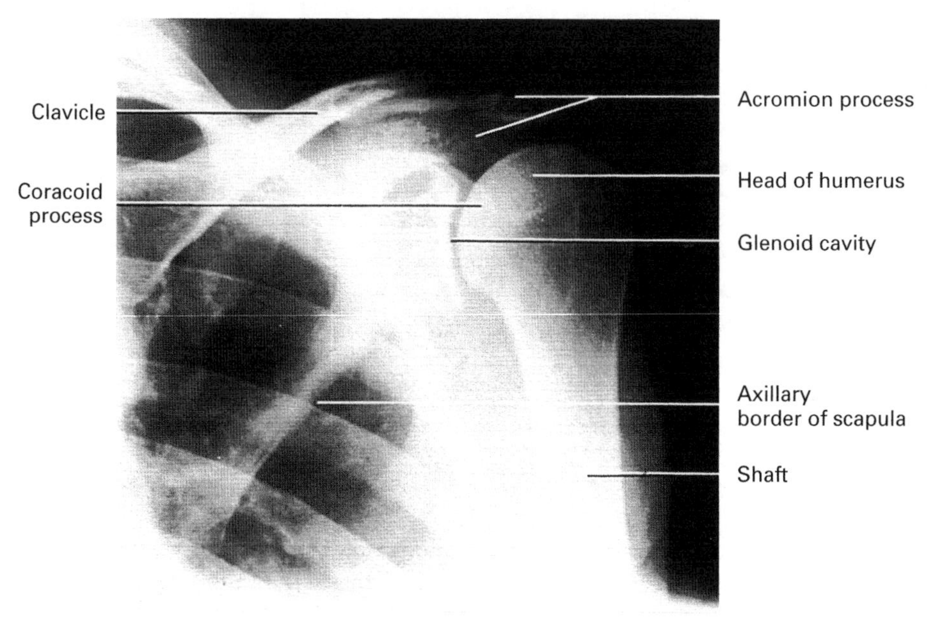

Fig. 5.9 Left shoulder: anteroposterior view

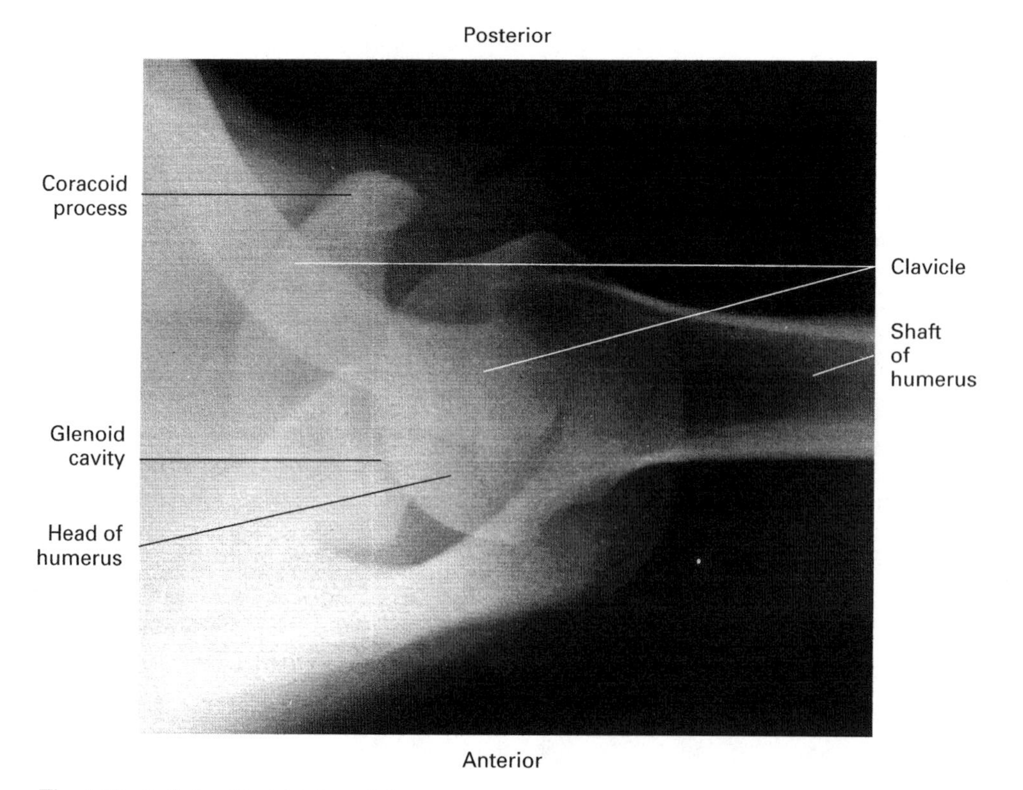

Fig. 5.10 Left shoulder joint: lateral view

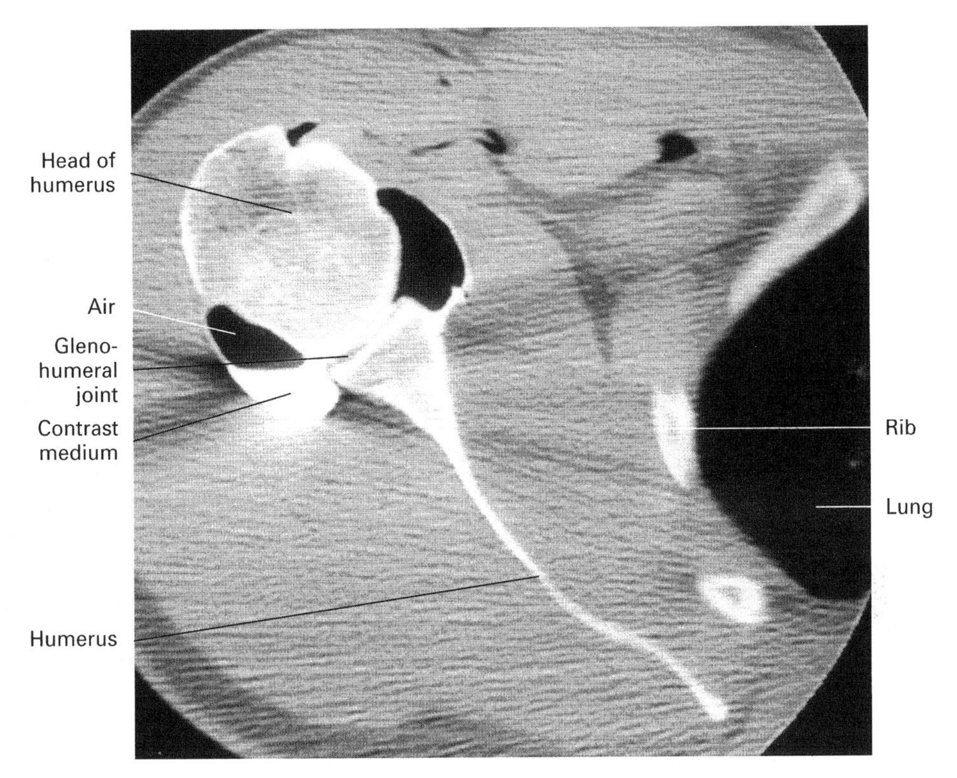

Fig. 5.11 Shoulder: double contrast CT arthrogram

CLAVICLE (Figs 5.12 to 5.17)

The clavicle (Figs 5.12 and 5.13) differs from other long bones in that it has no medullary cavity. The clavicle is composed of cancellous bone covered with a layer of compact bone. It consists of:

- the sternal (medial) end
- the shaft
- the acromial (lateral) end.

The sternal end is enlarged and articulates with the clavicular notch of the manubrium sterni at the sternoclavicular joint (p. 124).

The acromial end is flattened and articulates with the medial side of the acromion at the acromioclavicular joint (p. 125). The under surface of the acromial end has two small bony prominences.

The shaft is gently curved. The medial part (two-thirds) is convex forwards and the lateral third is concave forwards. The medial part is cylindrical and on its inferior surface, near the sternal end, is a roughened, slightly depressed oval area for the insertion of the costoclavicular ligament which joins the clavicle and 1st rib to the first costal cartilage.

The lateral part of the shaft is flattened. Its anterior border is concave and gives attachment to the deltoid muscle. Its posterior border is convex and part of the trapezius muscle is inserted into it. On the inferior surface is a small prominence—the conoid tubercle—from which a ridge extends laterally. This is the trapezoid ridge to which is attached the trapezoid ligament. The coracoclavicular ligament, which transmits the weight of the upper limb, is attached to the conoid tubercle and to the trapezoid ridge.

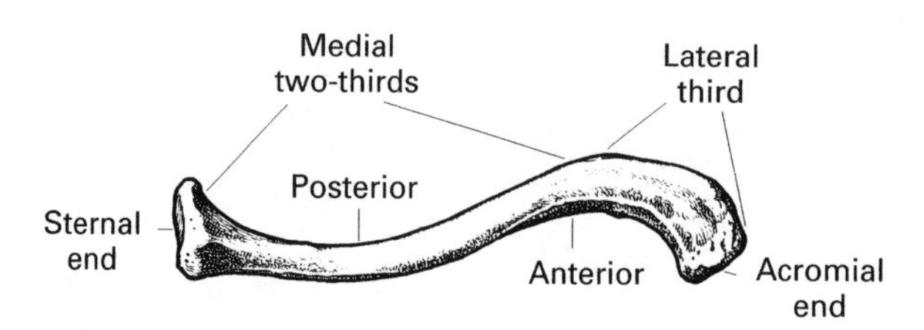

Fig. 5.12 Left clavicle: upper surface

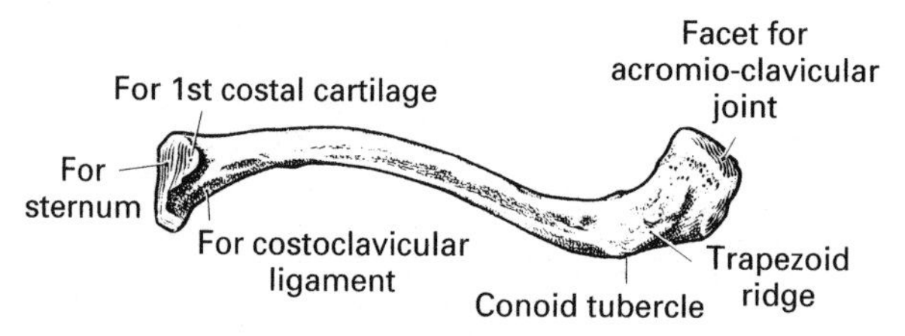

Fig. 5.13 Left clavicle: lower surface

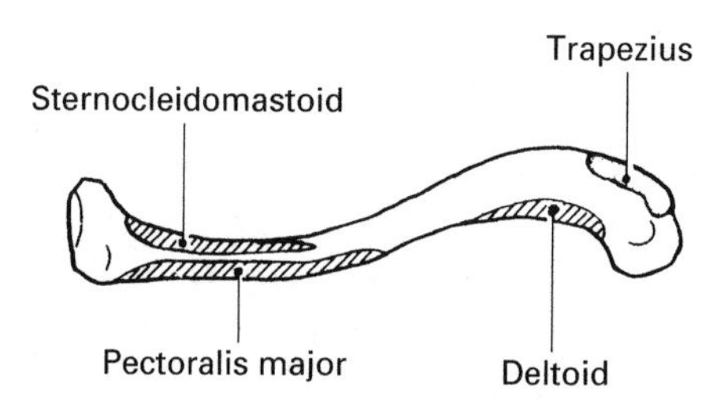

Fig. 5.14 Left clavicle: upper surface to show muscle attachments

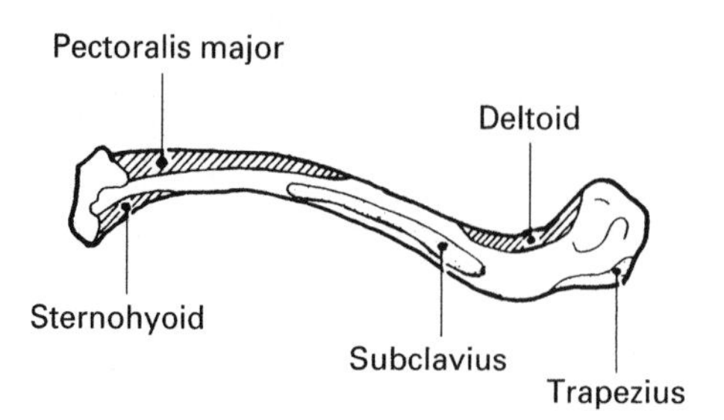

Fig. 5.15 Left clavicle: lower surface to show muscle attachments

Radiographic appearances of the clavicle (Figs 5.16 and 5.17)

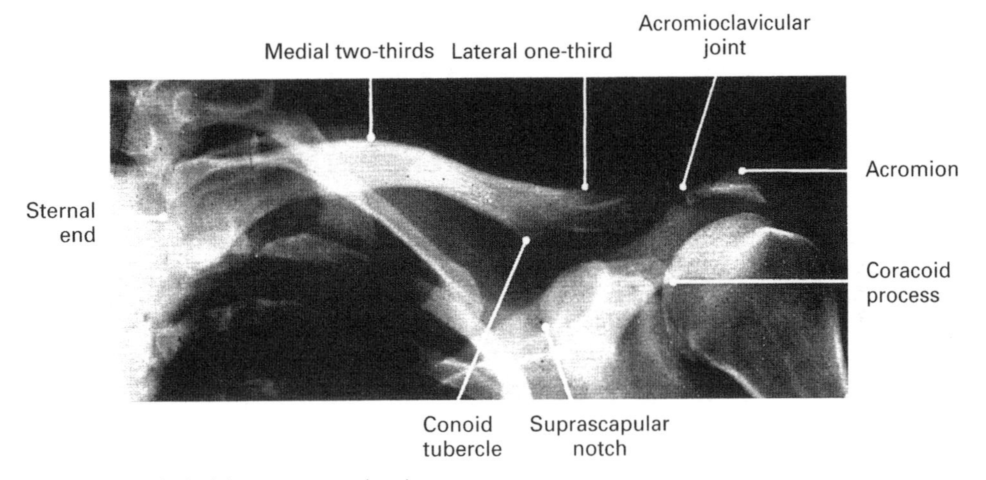

Fig. 5.16 Left clavicle: anteroposterior view

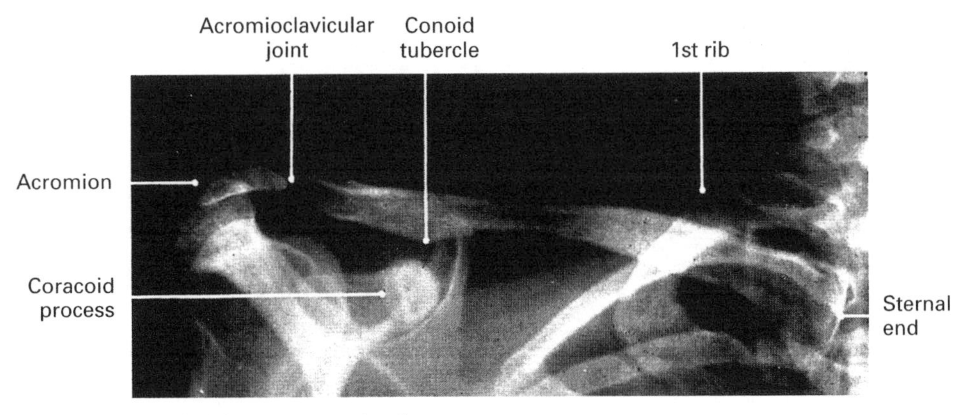

Fig. 5.17 Left clavicle: posteroanterior view

STERNOCLAVICULAR JOINT (MANUBRIOSTERNAL JOINT) (Fig. 5.18)

Type: Synovial, saddle.

Articular surfaces: Sternal end of clavicle with the clavicular notch of manubrium sterni and upper surface of cartilage of 1st rib. Articular surface of clavicle is much larger than that of the sternum. An articular disc divides joint into two parts.

Ligaments:

Capsular—fibrous, surrounds articular surfaces. Thickened and strengthened by ligaments anteriorly and posteriorly.

Sternoclavicular—(anterior and posterior)—attached to sternal end of clavicle and to anterior manubrium and 1st costal cartilage.

Interclavicular—extends between sternal ends of clavicle and superior surface of manubrium.

Costoclavicular—extends between upper surfaces of 1st rib and under surface of medial end of each clavicle.

Articular disc: Flat, nearly circular. Attached at its circumference to joint capsule. Divides joint into two parts. Separates manubrium and clavicle superiorly, but 1st costal cartilage and clavicle articulate directly. Plays a large part in the strength of the joint, particularly in the movement of depression.

Blood supply: From branches of internal thoracic and suprascapular arteries.

Nerve supply: From branches of medial supraclavicular nerve.

Movements and muscles:

Elevation—trapezius, levator scapulae (the clavicle moves with the scapula).

Depression—pectoralis minor, serratus anterior.

Protraction (anterior)—pectoralis minor, serratus anterior.

Retraction (posterior)—trapezius, rhomboid major and minor.

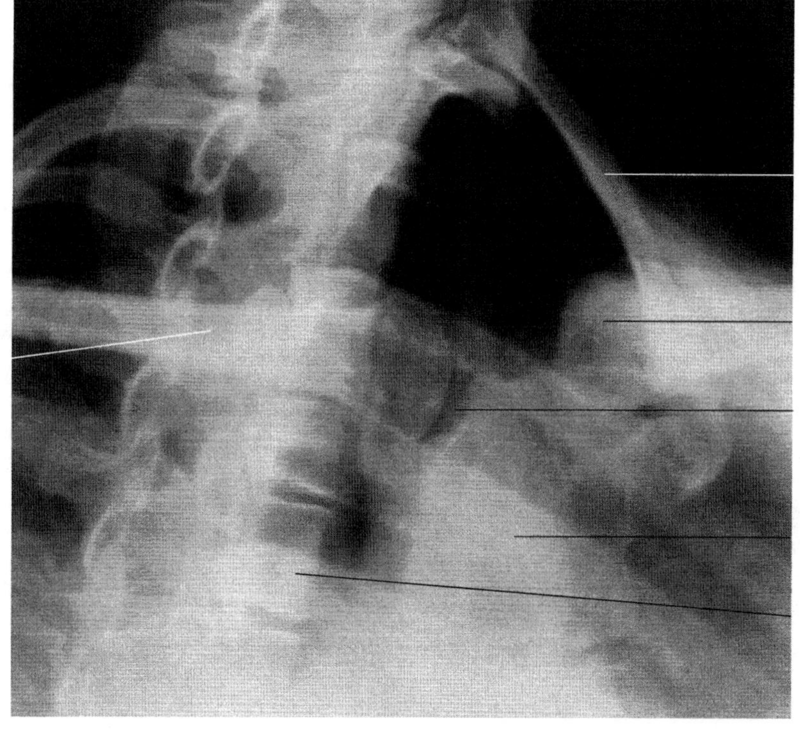

Fig. 5.18 Sternoclavicular joint: anterior oblique view

Rotation—trapezius, serratus anterior (along long axis of clavicle).

Trapezius:
- origin— superior nuchal line of occipital bone, spinous processes of 7th cervical and all thoracic vertebrae.
- insertion—lateral third of clavicle, acromion process and spine of scapula.
- function—stabilizes position of clavicle.
- nerve supply—accessory nerve.

Levator scapulae:
- origin—posterior tubercles of transverse processes of 1st–4th cervical vertebrae.
- insertion—superomedial border of scapula.
- function—elevation and rotation of scapula.
- nerve supply—dorsal scapular nerve.

Pectoralis minor, Serratus anterior } see page 119.

Rhomboid major:
- origin—spinous processes of 2nd–5th thoracic vertebrae.
- insertion—medial border of scapula.
- function—stabilizes position of scapula.
- nerve supply—dorsal scapular nerve.

Rhomboid minor:
- origin—spinous processes of 7th cervical and 1st thoracic vertebrae.
- insertion—medial side of spine of scapula.
- function, nerve supply } —as for rhomboid major.

ACROMIOCLAVICULAR JOINT (Fig. 5.19)

Type: Synovial, plane.

Articular surfaces: Acromial end of clavicle and medial end of acromion process of scapula. An articular disc is sometimes present.

Ligaments:

Capsular—fibrous, attached to articular margins. Surrounds joint.

Acromioclavicular—thickening of capsular ligament. Consists of:
- *Conoid* (medial) part—conical in shape. Attached to conoid tubercle of calvicle and to posterior part of coracoid process. Prevents posterior movement of clavicle.
- *Trapezoid* (lateral) part—attached laterally to conoid ligament and anteriorly to coracoid process. Prevents anterior movement of clavicle.

Movements: These are in association with those of the scapula. Rotational movement occurs also along the long axis of the clavicle.

OSSIFICATION OF THE SCAPULA AND CLAVICLE (Fig. 5.20)

Primary centres

Scapula—for the body and spine, appears near

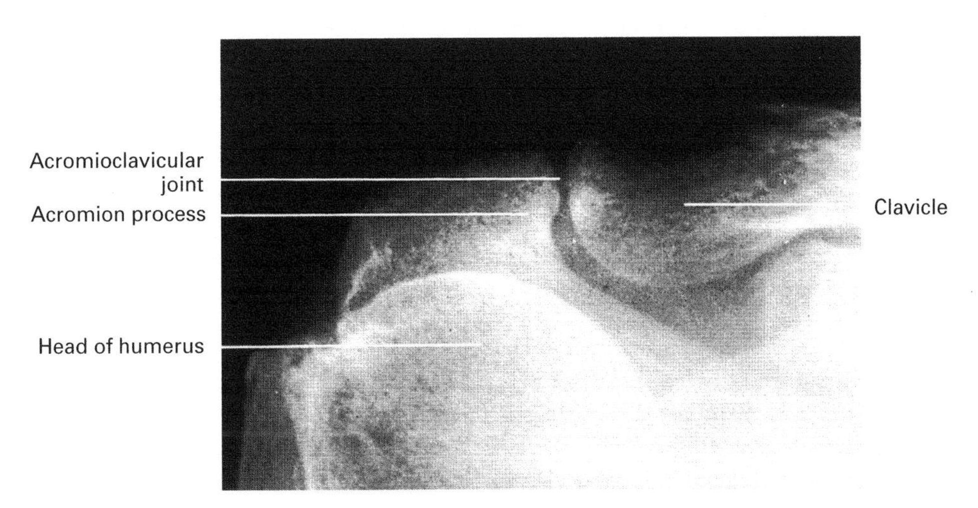

Fig. 5.19 Right acromioclavicular joint: anteroposterior view

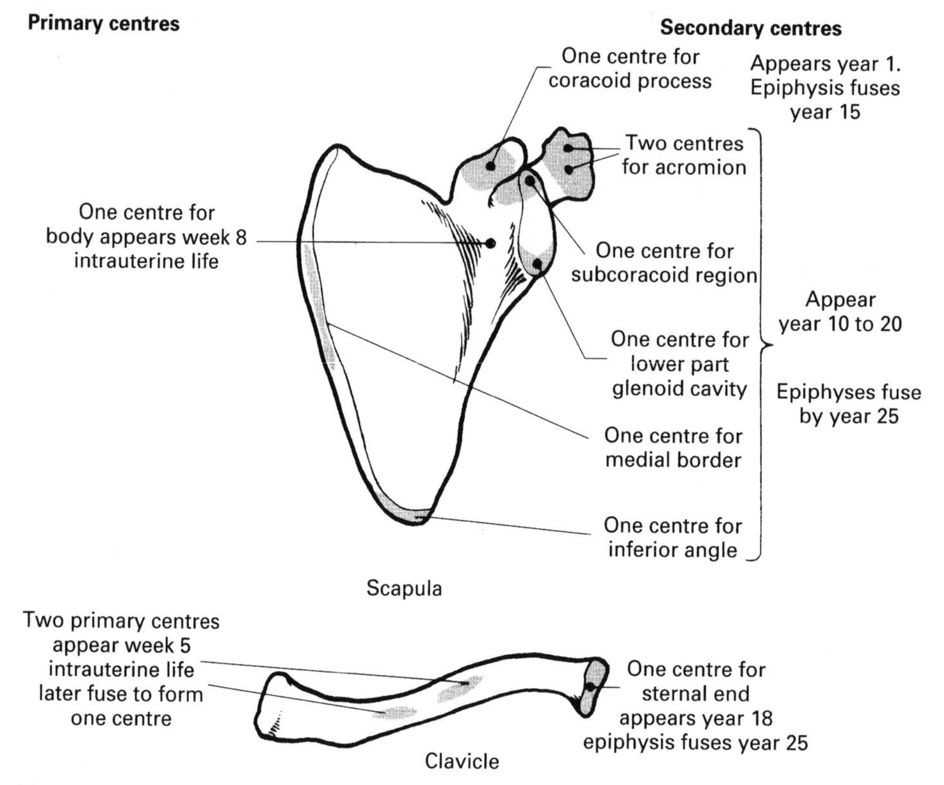

Fig. 5.20 Ossification of the scapula and clavicle

the glenoid cavity at about week 8 of intrauterine life.

Clavicle—for the shaft about week 5 of intrauterine life. It is the first bone in the body to ossify and the only long bone to ossify primarily in membrane.

Secondary centres

Scapula

For the coracoid process, appears during 1st year.

- for the subcoracoid region
- for the acromion (2 centres)
- for the inferior angle
- for the medial border
- for the lower part of the glenoid cavity

} appear at puberty.

The epiphysis for the coracoid unites with the body at about the 15th year. The other six epiphyses unite at about the 25th year.

Clavicle—in the sternal end, appears between the 18th and 25th years. This epiphysis unites with the shaft at about 25 years.

6. The upper limb

The bones of the upper limb are those of the arm, the forearm and the hand.

HUMERUS (Figs 6.1 to 6.5)

The humerus, the bone of the arm, is the largest and longest bone in the upper limb. The humerus consists of:

- the (expanded) upper end
- the shaft
- the lower end.

The upper end articulates with the glenoid cavity of the scapula to form the shoulder joint (p. 118). The lower end articulates with the radius and ulna to form the elbow joint (p. 136).

The upper end consists of:

- the head
- the greater tuberosity
- the lesser tuberosity.

The head is smooth, rather less than hemispherical and is covered with hyaline cartilage. When the arm is in the anatomical position, the head is directed upwards, inwards and slightly backwards. It is separated from the shaft by the anatomical neck which gives attachment to the capsular ligament of the shoulder joint.

The greater tuberosity is situated on the lateral side of the upper end of the humerus. It projects laterally beyond the acromion process of the scapula and causes the normal rounded appearance of the shoulder. It bears three areas for attachment of muscles which arise from the scapula. The upper facet is for the supraspinatus muscle, the middle one is for the infraspinatus and the lowest one is for teres minor.

The lesser tuberosity is on the anterior aspect of the humerus just below the anatomical neck. It gives attachment to the subscapularis muscle.

The surgical neck is situated at the upper end of the shaft, just below the tuberosities. It is a frequent site of fracture.

The bicipital groove lies between the greater and lesser tuberosities. It is a vertical channel in which lies the tendon of the long head of the biceps. The pectoralis muscle is attached to the lateral lip of the groove, the latissimus dorsi and teres major muscles are inserted into its medial lip.

The shaft is cylindrical in shape but becomes flattened and widened at its distal end. The shaft has three borders:

- anterior
- medial
- lateral.

The anterior border begins as the lateral margin of the bicipital groove and becomes ill-defined and rounded towards the lower end.

The medial border begins as the medial margin of the bicipital groove and at its lower end becomes continuous with the medial supracondylar ridge.

The lateral border is ill-defined superiorly but inferiorly becomes continuous with the lateral

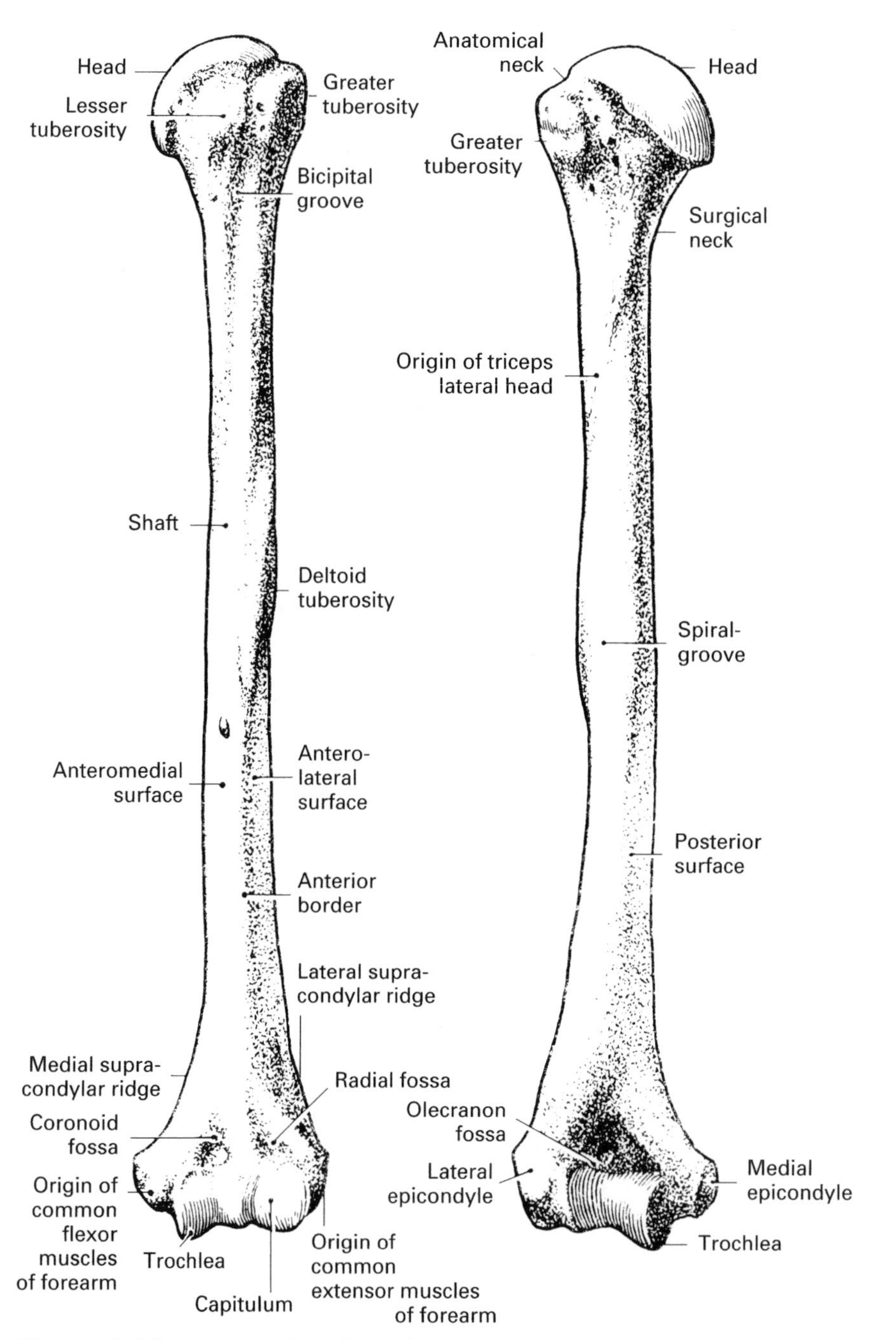

Fig. 6.1 Left humerus: anterior and posterior aspects

supracondylar ridge. In the centre of this border is a depression which runs obliquely forwards and downwards. This is the spiral groove along which passes the radial nerve.

On the anterolateral surface is the deltoid tuberosity to which is attached the deltoid muscle. At the same level on the anteromedial surface there is a nutrient foramen.

The lower end of the humerus is expanded from side to side and has both articular and non-articular parts. The articular part consists of:

- the capitulum
- the trochlea.

The non-articular part consists of:

- the medial epicondyle
- the lateral epicondyle.

The capitulum is a rounded surface on the lateral aspect. It extends anteriorly and inferiorly and it articulates with the head of the radius.

The trochlea is larger than the capitulum and is pulley-shaped. It is on the medial aspect of the humerus and extends from the posterior to the anterior surfaces. The two margins of the trochlea are of unequal size, the larger being the medial, thus influencing the 'carrying angle' i.e. the angle between the long axis of the humerus and that of the supinated forearm. This angle is approximately 160° on the lateral side, less in the female.

The medial epicondyle is an easily palpable projection on the medial side of the condyle. The ulnar nerve crosses its smooth posterior surface and here it can be pressed against the bone to produce the sensation of 'pins and needles' (paraesthesia) in the medial hand. The superficial flexor muscles of the forearm arise from the anterior surface of the medial condyle.

The lateral epicondyle is smaller and less prominent than the medial. The superficial extensor muscles of the forearm arise from its anterior surface.

The coronoid and radial fossae are two depressions on the anterior aspect of the humerus, above the trochlea and the capitulum respectively. On flexion of the forearm, these depressions accommodate the coronoid process and the head of the radius respectively.

The olecranon fossa is a deep depression on the posterior aspect of the humerus, just above the trochlea. On extension of the forearm, this fossa accommodates the olecranon.

Supraspinatus
Subscapularis
Pectoralis major
Latissimus dorsi
Teres major
Deltoid
Coracobrachialis
Brachialis
Brachioradialis
Extensor carpi radialis longus
Common extensor origin
Common flexor origin

Fig. 6.2 Left humerus: anterior aspect to show muscle attachments

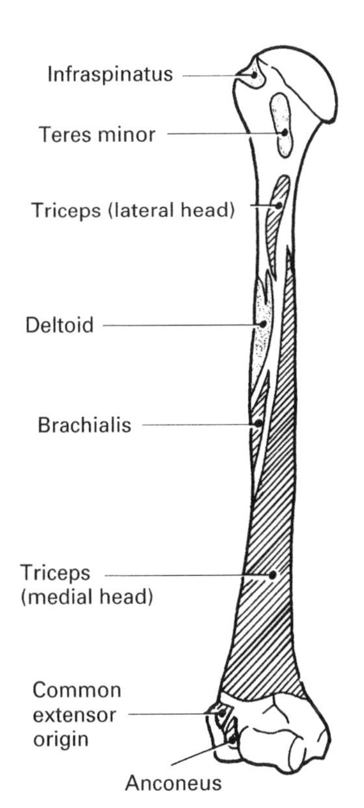

Fig. 6.3 Left humerus: posterior aspect to show muscle attachments

Radiographic appearances of the humerus (Figs 6.4 and 6.5)

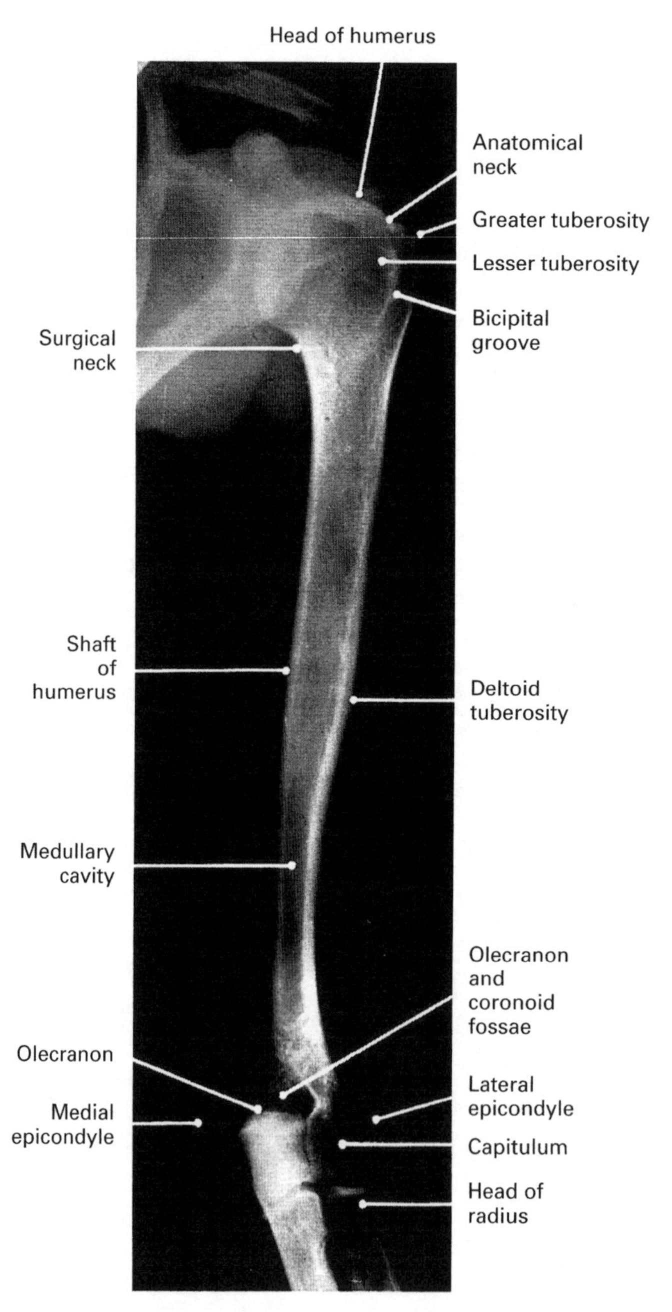

Fig. 6.4 Left humerus: anteroposterior view

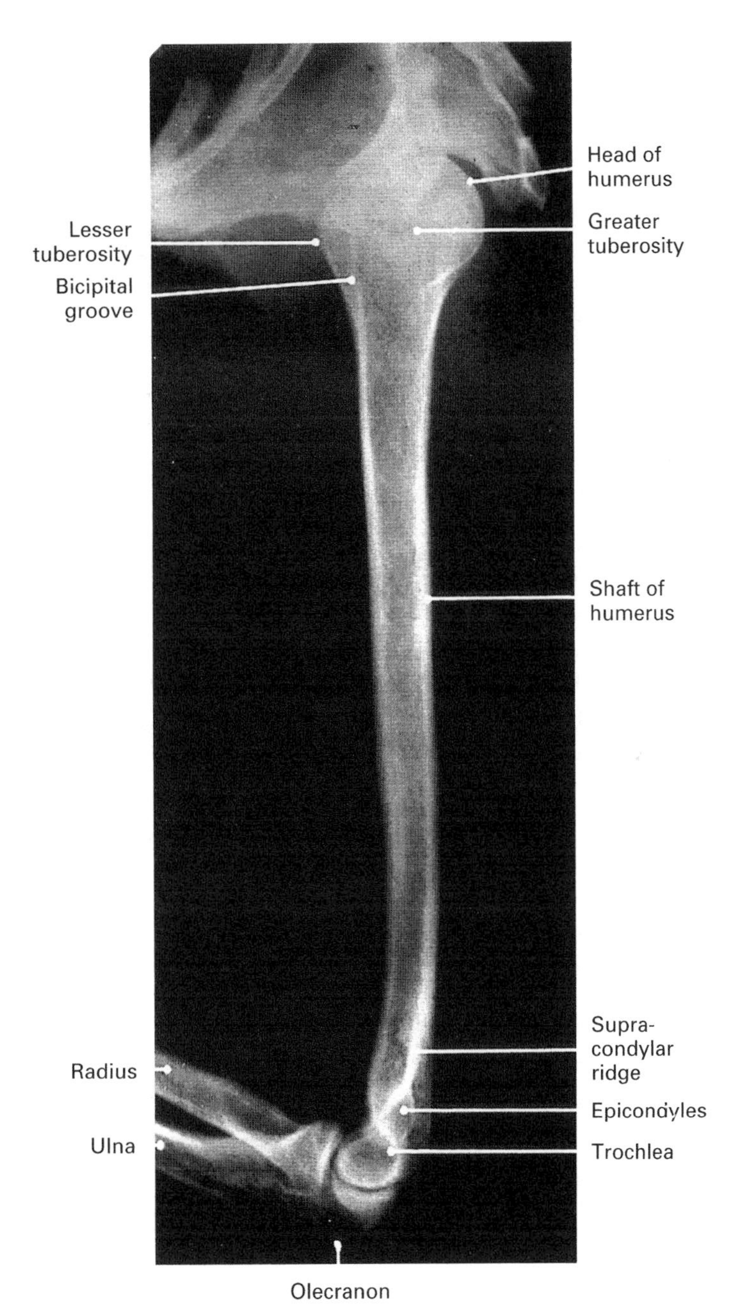

Fig. 6.5 Left humerus: lateral view

OSSIFICATION OF THE HUMERUS
(Figs 6.6–6.12)

Primary centre

- for the shaft, appears about week 8 of intra-uterine life.

Secondary centres

Proximal end

- for the head appears about 1st year
- for the greater tuberosity, about 2nd year
- for the lesser tuberosity, about 5th year

Distal end

- for the capitulum, about 2nd year
- for the medial epicondyle, about 5th year
- for the trochlea, about 11th year
- for the lateral epicondyle, about 12th year

The centres for the proximal end join to form one large epiphysis about the 6th year. It fuses with the shaft between the 18th and 20th years.

The centres for the distal end appear between the 2nd and 12th years and, except for that of the medial epicondyle, join together to form one epiphysis which fuses with shaft between the 16th and 18th years. The epiphysis for the medial epicondyle develops separately and fuses with the shaft at about the 20th year.

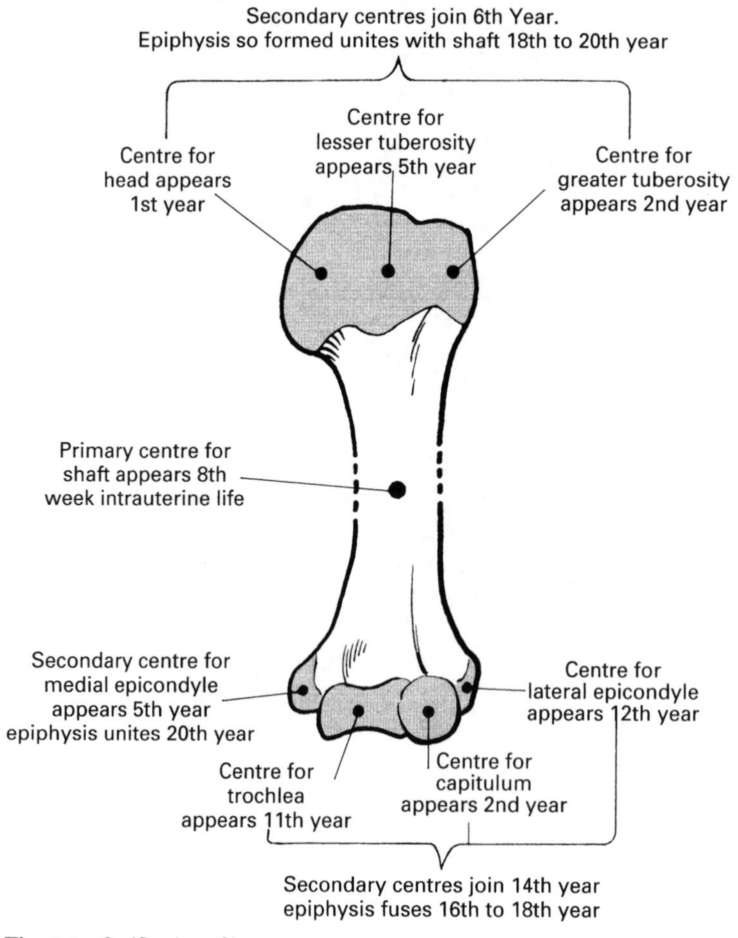

Fig. 6.6 Ossification of humerus

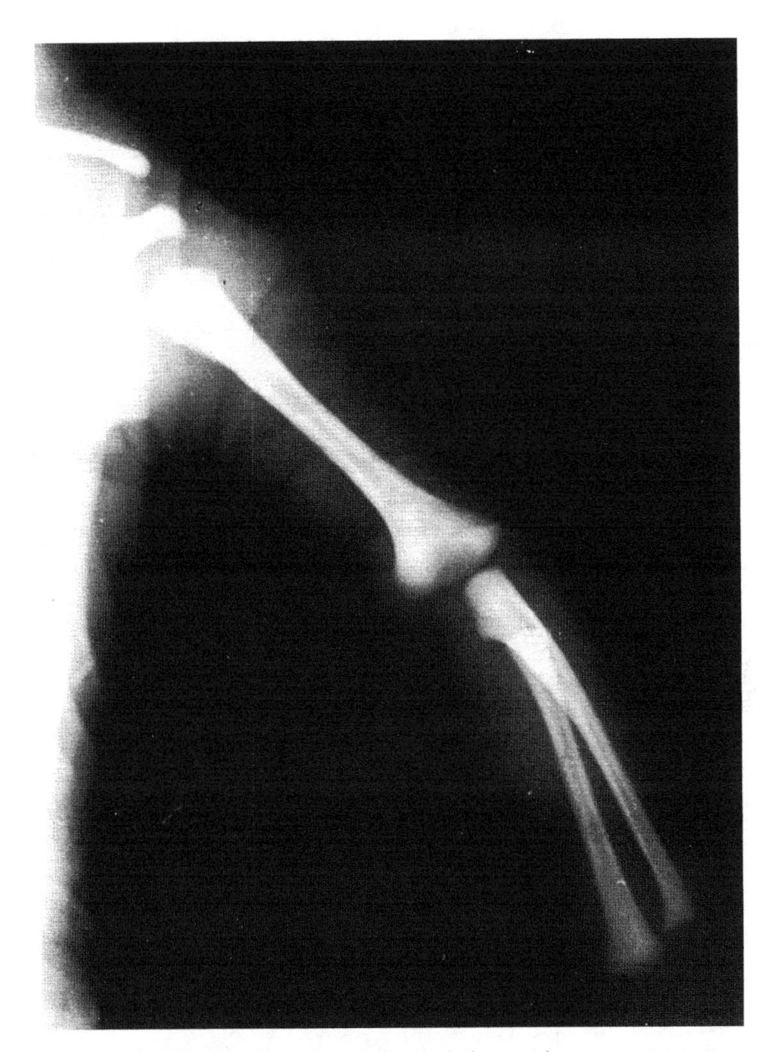

Fig. 6.7 Ossification of humerus, radius and ulna: newborn

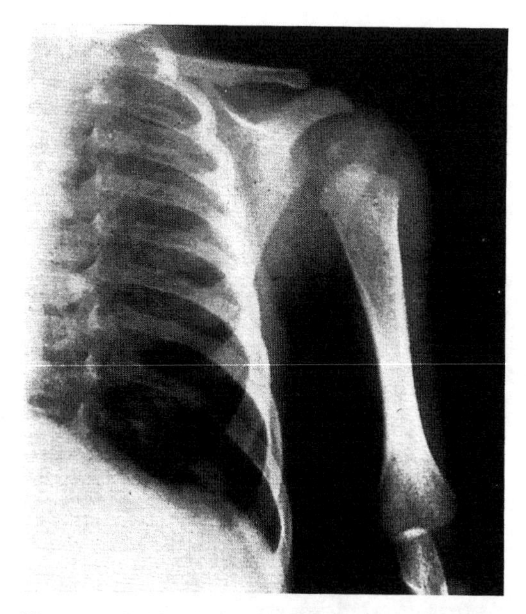

Fig. 6.8 Ossification at the shoulder joint: 6 months

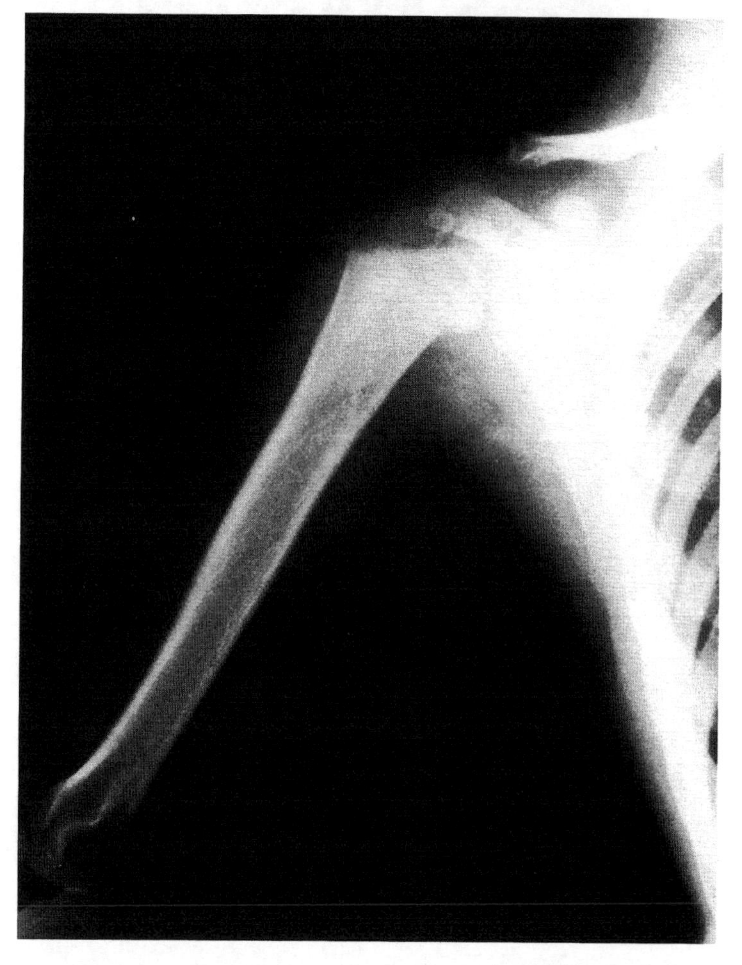

Fig. 6.9 Ossification at the shoulder joint: $2\frac{1}{2}$ years

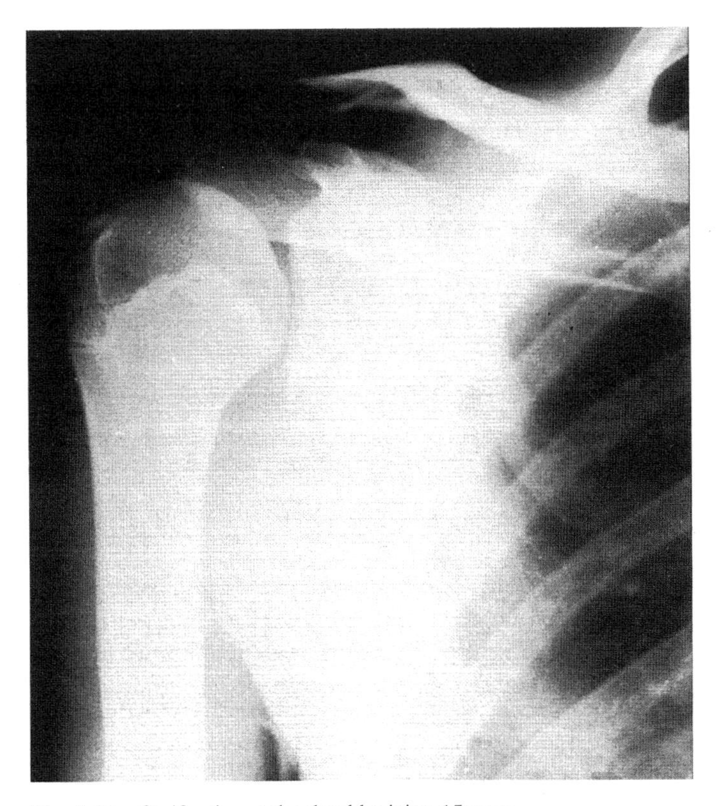

Fig. 6.10 Ossification at the shoulder joint: 15 years

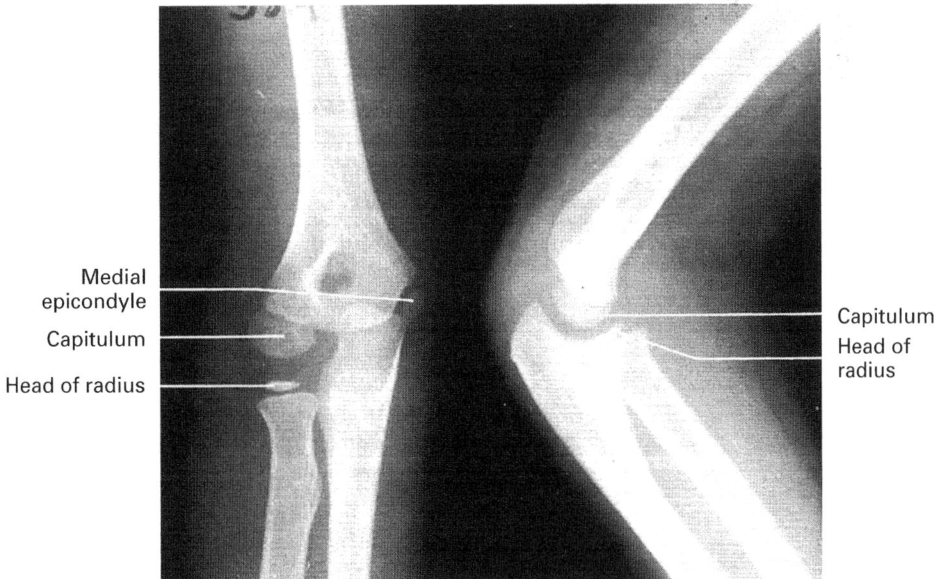

Fig. 6.11 Ossification of the elbow: 6 years

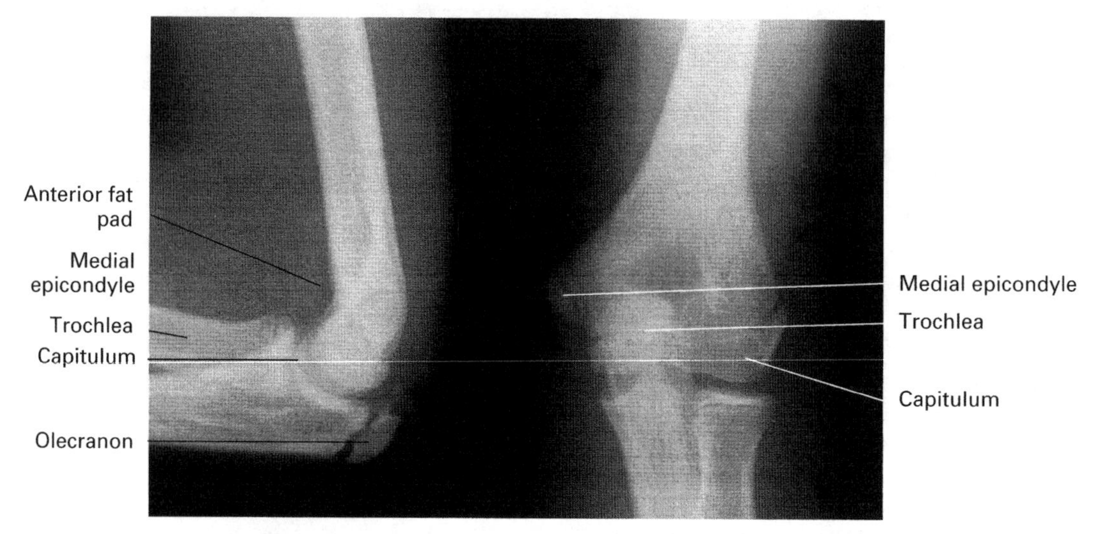

Fig. 6.12 Ossification of the elbow: 13 years

ELBOW JOINT (Figs 6.13 to 6.18)

Type: Synovial, hinge.

Articular surfaces: Trochlea of humerus with the trochlear notch of ulna, and capitulum of humerus with head of radius.

Synovial membrane: Lines capsule. Attached to margins of articular surfaces of humerus, covers fossae (olecranon, radial and coronoid). Posteriorly it folds between the radius and ulna and partly lines the annular ligament. Fat pads lie between the capsule and synovial membrane at the fossae and are clearly visible when the joint is swollen.

Ligaments:

Capsular—broad, fibrous layer attached to medial epicondyle, upper margins of the radial and coronoid fossae, ulnar and radial

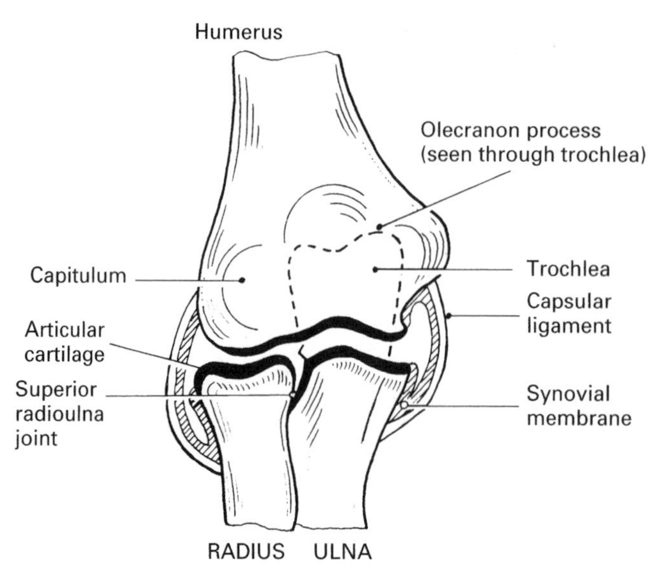

Fig. 6.13 Schematic diagram of the elbow joint

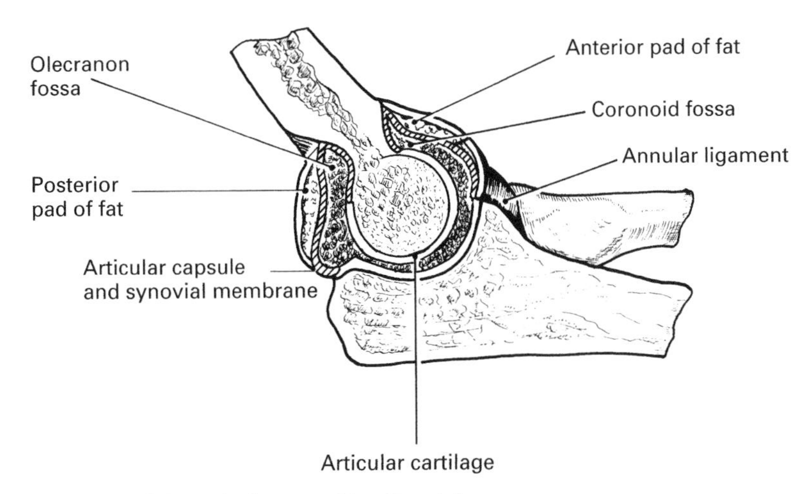

Fig. 6.14 Schematic diagram of the elbow joint

collateral ligaments, coronoid process of ulna and the annular ligament. Posteriorly attached to margin of olecranon fossa, capitulum, trochlea and olecranon process.

Radial collateral—attached to lateral epicondyle and to annular ligament.

Ulnar collateral—triangular, apex attached to medial epicondyle and to medial side of olecranon and coronoid processes.

Blood supply: From branches of the radial, brachial and ulnar arteries which form an anastomosis around the joint.

Nerve supply: From the musculocutaneous and radial nerves and to a lesser extent the ulnar and median nerves.

Bursae:

Olecranon—between joint capsule and triceps tendon.

Biceps—between rough portion of radial tuberosity and biceps tendon.

Anconeus—between anconeus muscle and the skin.

Movements and muscles:

Flexion—biceps, brachialis, brachioradialis.

Extension—triceps, anconeus.

Biceps brachii:

origin

- long head—supraglenoid tubercle of scapula
- short head—coracoid process of scapula.

insertion—radial tuberosity.

function—flexion of elbow, supination of forearm, stabilization of shoulder joint.

nerve supply—musculocutaneous nerve.

Triceps:

origin

- long head—infraglenoid tubercle of scapula
- lateral head—posterolateral surface of humerus, above radial groove, by means of a flat tendon
- short head—posteromedial surface of humerus below radial groove, by large triangular surface of origin.

insertion—olecranon process of ulna.

function—extension of elbow joint.

nerve supply—radial nerve.

Brachialis:

origin—lower anterior humerus.

insertion—tuberosity of ulna and coracoid process of ulna.

function—flexion of elbow (assists biceps).

nerve supply—musculocutaneous nerve.

Anconeus (small triangular muscle):

origin—posterior surface of lateral epicondyle.

insertion—lateral aspect of olecranon process and small area of posterior proximal ulnar shaft.

function—extension of elbow (acts with triceps).

nerve supply—radial nerve.

Brachioradialis:
origin—lateral supracondylar ridge of humerus.
insertion—lateral distal radius proximal to styloid process.
function—flexion of elbow, assists supination.
nerve supply—radial nerve.

Radiographic appearances of the elbow joint (Figs 6.15 to 6.18)

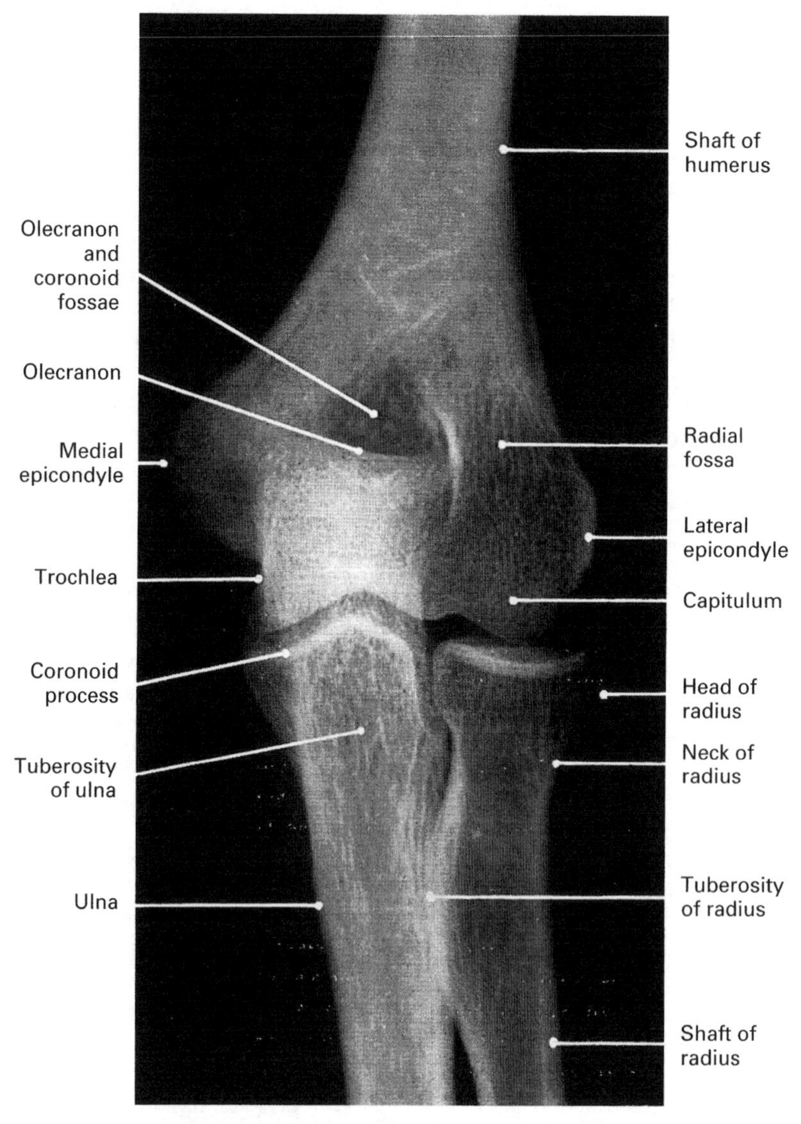

Fig. 6.15 Left elbow joint: anteroposterior view

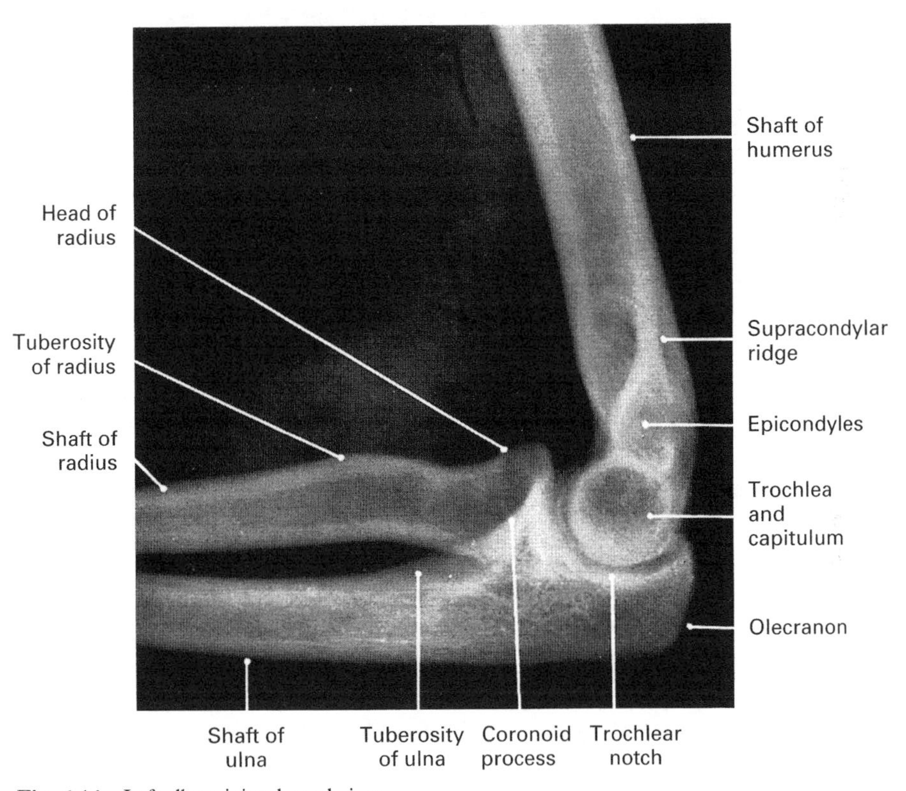

Fig. 6.16 Left elbow joint: lateral view

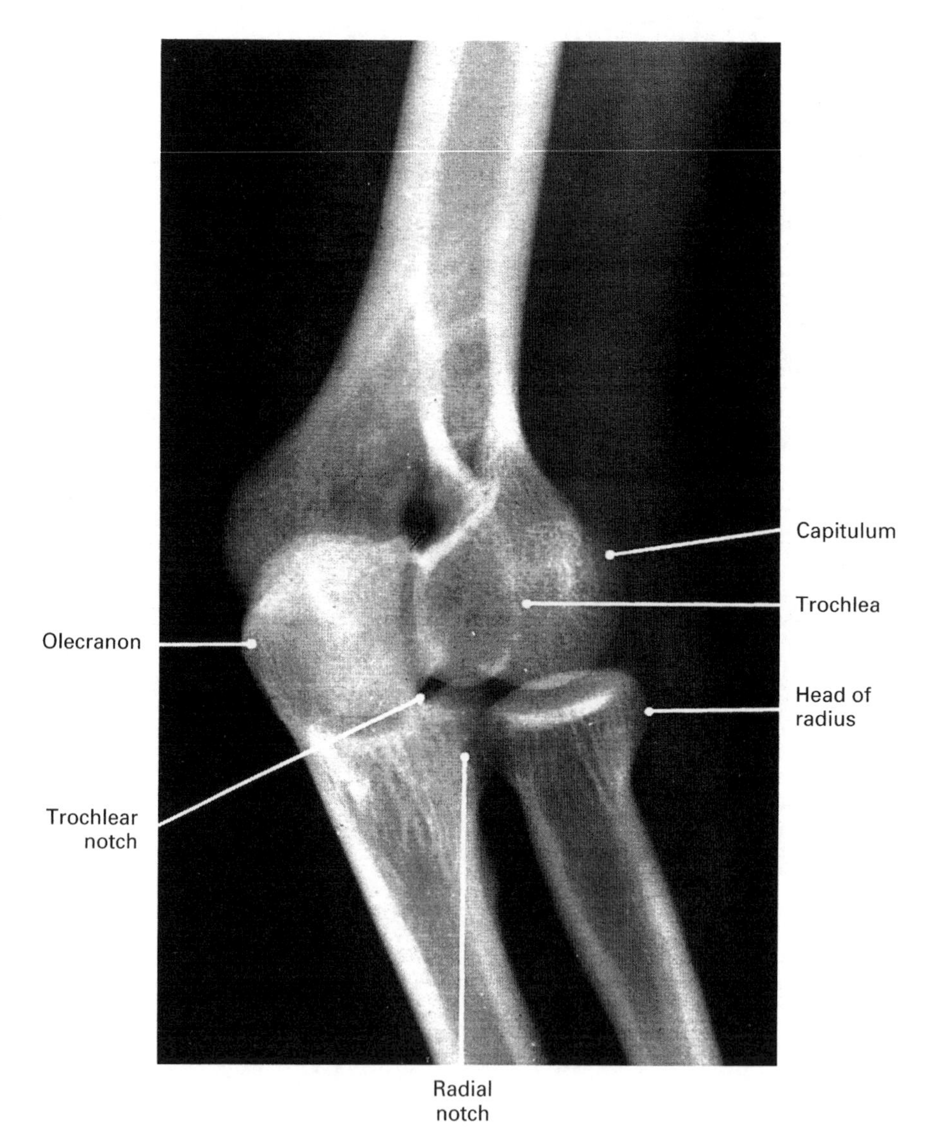

Fig. 6.17 Left elbow joint: oblique view

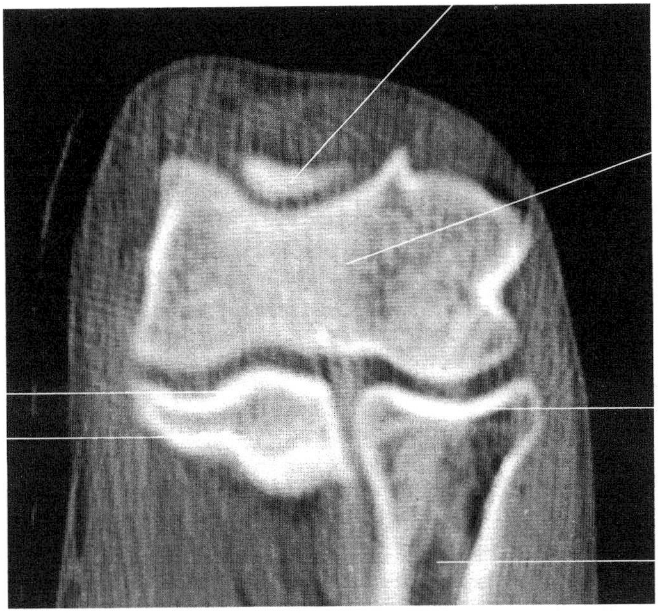

Fig. 6.18 Elbow: CT scan, coronal section

RADIOULNAR JOINTS

The radius and ulna are joined at the proximal and distal ends by the superior and inferior radio-ulnar joints. The bone shafts are joined by the interosseous membrane forming the radioulnar syndesmosis.

Superior radioulnar joint

Type: Synovial, pivot.

Articular surfaces: Head of radius and radial notch of ulna.

Capsule: That of the elbow joint.

Ligaments:

Annular—strong band which encircles the head of radius. Attached to anterior and posterior margins of radial notch of ulna.

Quadrate—small, passes between radial neck and radial notch of ulna and annular ligament.

Inferior radioulnar joint

Type: Synovial, pivot.

Articular surfaces: Head (lower end) of ulna and ulnar notch of radius. Articular surfaces are enclosed in an articular capsule and held together by a triangular articular disc.

Ligaments:

Capsular—between anterior and posterior margins of ulnar notch of radius and anterior and posterior margins of ulnar head.

Interosseous membrane (radioulnar syndesmosis)

The interosseus membrane is a broad, thin sheet of fibres passing downwards and medially from the interosseous border of the radius to that of the ulna. This membrane increases the surface for attachment of the muscles of the forearm and it joins the two bones also. An independent interosseous band—the oblique cord—extends from the tuberosity of the radius to the coronoid process. The posterior interosseous vessels pass between the oblique cord and the interosseous membrane, whereas the anterior interosseous vessels pass through an aperture in the membrane near its lower end.

Movements and muscles:

Pronation and supination. These are rolling movements of the radius around the fixed ulna.

Pronation—pronator quadratus, pronator teres.

Supination—supinator, biceps, brachioradialis.

Pronator quadratus:
- origin—inferior shaft of anterior ulna.
- insertion—distal anterior and lateral shaft of radius.
- function—pronation, supports distal radioulnar joint.
- nerve supply—anterior interosseous branch of medial nerve.

Pronator teres:
- origin—above medial epicondyle of humerus, coronoid process of ulna.
- insertion—lateral border of mid-radius.
- function—pronation, flexion of elbow.
- nerve supply—median nerve.

Supinator:
- origin, two heads
  - lateral epicondyle of humerus.
  - supinator crest of ulna and medial ulna distal to radial notch.
- insertion—lateral neck and oblique line of radius.
- function—supination.
- nerve supply—posterior interosseous branch of radial nerve.

Biceps brachii
see elbow joint, page 137.

Brachioradialis
see elbow joint, page 138.

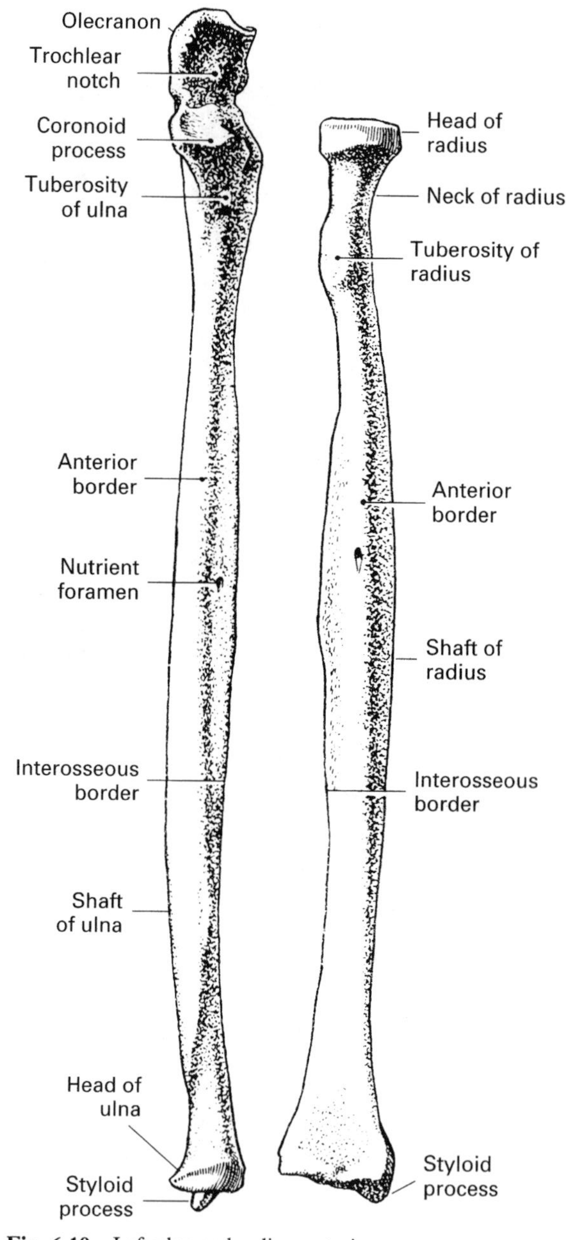

Fig. 6.19 Left ulna and radius: anterior aspect

FOREARM (Figs 6.19 to 6.24)

There are two long bones in the forearm: (Figs 6.19 and 6.20).

- the radius
- the ulna.

When the hand is supinated the radius and ulna lie side by side, the radius on the lateral and the ulna on the medial side.

Proximally the radius and ulna articulate with each other at the superior radioulnar joint (p. 141) and with the humerus at the elbow joint (p. 136). The lower end of the radius and the articular disc at the lower end of the ulna articulate with the proximal row of carpal bones to form the wrist joint (p. 156). Distally the radius and ulna articulate with each other at the inferior radioulnar joint (p. 141).

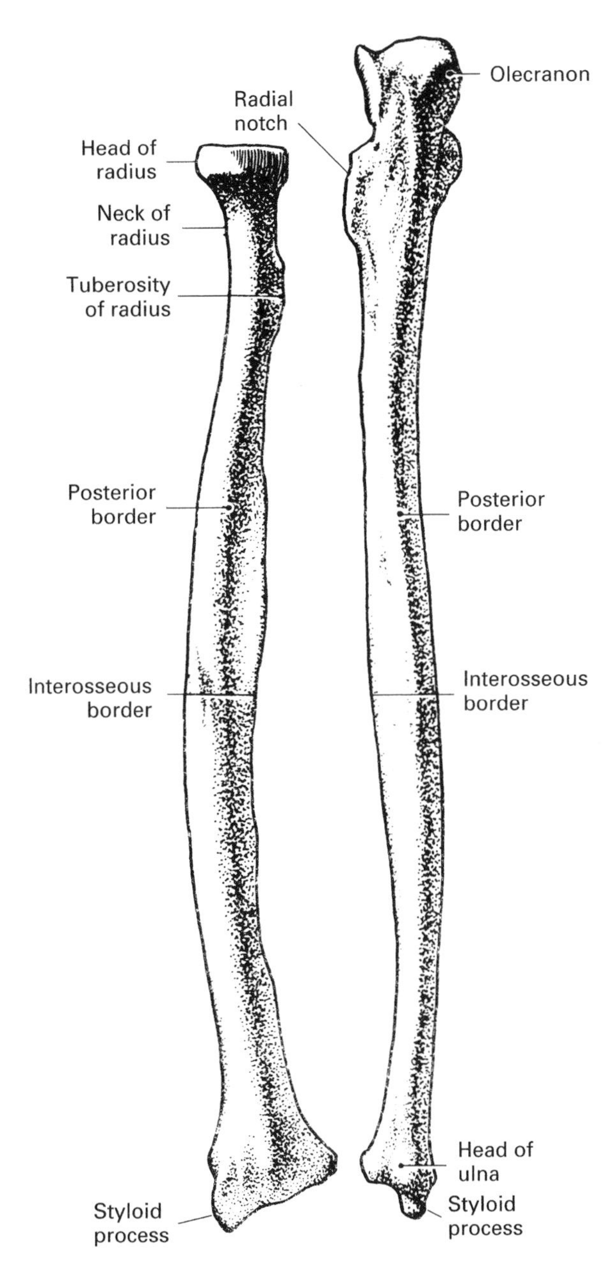

Fig. 6.20 Left radius and ulna: posterior aspect

RADIUS

The radius consists of expanded upper and lower ends and a shaft. The lower end is much larger than the upper.

The upper end consists of:

- the head
- the neck
- the tuberosity.

The head is disc-shaped. Its superior surface is concave for articulation with the capitulum of the humerus on the lateral side of the elbow joint. The circumference of the head articulates medially with the corresponding radial notch of the ulnar to form the superior radioulnar joint. The annular ligament arises from the margins of the radial notch of the ulna to convert the notch into a complete articular ring in which the head of the radius can revolve freely.

The neck is a slightly narrowed area immediately below the head.

The tuberosity is on the medial side of the bone, below the neck. Into its posterior part is inserted the tendon of the biceps brachii muscle.

The shaft is curved, being convex towards its lateral aspect. It is roughly triangular in cross-section and presents three borders—medial, anterior and posterior.

The medial (interosseous) border is a sharp ridge giving attachment to the interosseous membrane (p. 141) which connects the radius and ulna for three-quarters of their length.

The posterior border is only clearly defined in the middle third of the shaft.

The anterior border begins as an oblique line below the radial tuberosity; it is indistinct at the middle of the shaft but becomes clearly defined on the lateral side of the lower end.

The middle of the anterior surface is pierced by a nutrient artery which transmits blood vessels to the interior of the bone. The lateral and posterior surfaces do not present any notable bony features.

The lower end is the widest part of the bone and is roughly quadrilateral in cross-section, thus presenting four surfaces. The lateral surface is slightly roughened. It projects further distally than the rest of the bone thus forming the styloid process of the radius which is easily palpable. On the medial surface is the ulnar notch where the head of the ulna articulates to form the inferior radioulnar joint. The anterior surface is at first smooth and concave. It ends distally in a prominent ridge 250 mm proximal to the base of the thenar eminence and is easily palpable. The posterior surface is marked by a series of grooves and ridges for the extensor tendons passing from the forearm to the hand.

The distal articular surface of the radius is divided by a faint ridge into a quadrilateral medial facet and a triangular lateral facet. The medial facet articulates with the lunate. The triangular lateral facet articulates with the scaphoid.

ULNA

The ulna consists of:

- the large upper end
- the shaft
- the smaller lower end.

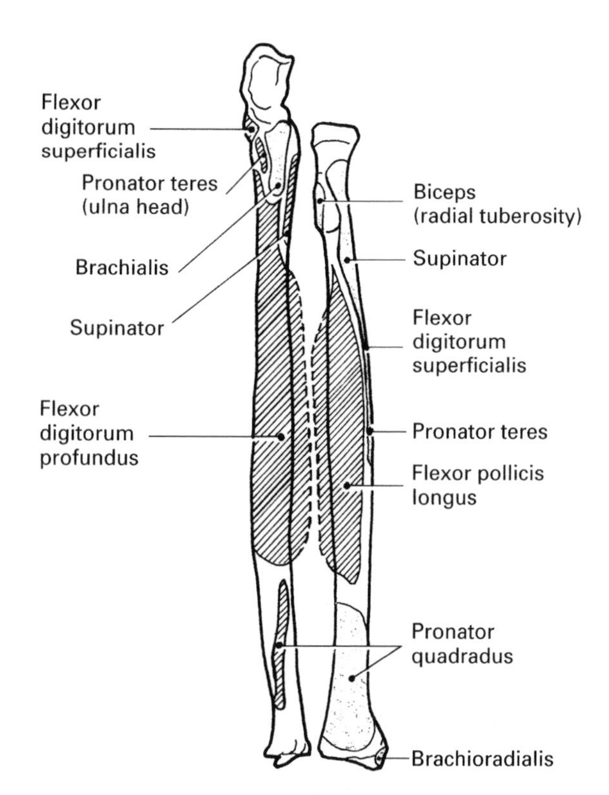

Fig. 6.21 Left radius and ulna: anterior aspect to show muscle attachments

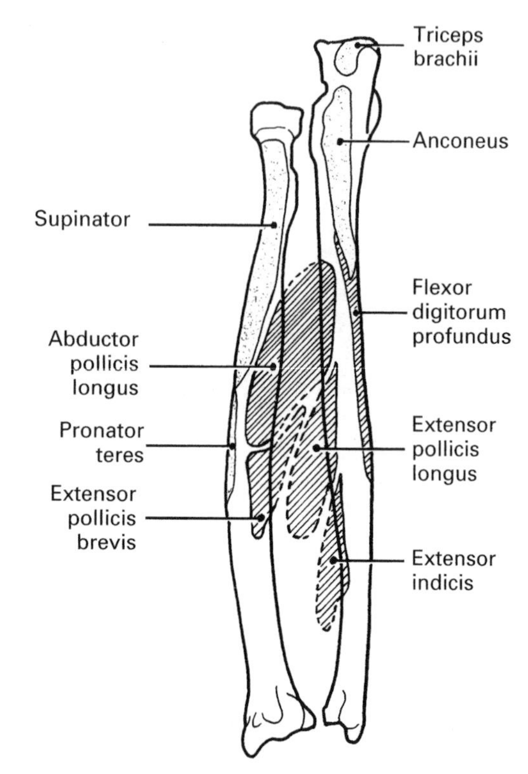

Fig. 6.22 Left radius and ulna: posterior aspect to show muscle attachments

The upper end consists of two large bony prominences:

- the olecranon
- the coronoid process

and two articular areas

- the trochlear notch
- the radial notch

The olecranon has a prominent 'beak' at the upper edge which projects into the olecranon fossa of the humerus when the forearm in extended. The posterior margin of the olecranon is roughly triangular and can be felt easily as the subcutaneous tip of the elbow.

The coronoid process is smaller than the olecranon and projects forward from the anterior surface of the bone immediately below the olecranon. Its upper surface forms the lower part of the trochlear notch and is smooth and articular. On its anterior surface is a roughened tuberosity into which the tendon of the brachialis muscle is inserted.

The trochlear notch is a semilunar-shaped articular surface formed by the anterior surface of the olecranon and the superior surface of the coronoid process. It is divided by a vertical ridge into a larger medial and a smaller lateral surface which conform with the shape of the trochlea of the humerus with which the notch articulates.

The radial notch is a shallow articular surface on the lateral aspect of the coronoid process. It articulates with the head of the radius. To its anterior and posterior margins is attached the annular ligament which embraces the head of the radius.

The shaft has a slight convexity towards the medial side in its lower half—separation of the radius and ulna being widest at that level therefore when the hand and forearm are supinated. The shaft is triangular on cross-section, similar to that of the radius. It presents three borders:

- anterior
- posterior
- interosseous.

The anterior border is thick and rounded. It extends from just below the coronoid process to the styloid process.

The posterior border begins at the posterior end of the olecranon and is indistinct in the lower third of the shaft.

The interosseous border is sharply defined and faces the medial border of the radius.

The lower end consists of:

- the head
- the styloid process.

The head is small and on its lateral side is a small surface for articulation with the radius at the inferior radioulnar joint. An articular disc (triangular cartilage) is interposed between the inferior surface of the head and the adjacent carpal bone, the triquetrum.

The styloid process is small and projects distally from the medial side of the head. It is palpable and lies 1 cm proximal to the radial styloid process.

Radiographic appearances of the radius and ulna (Figs 6.23 and 6.24)

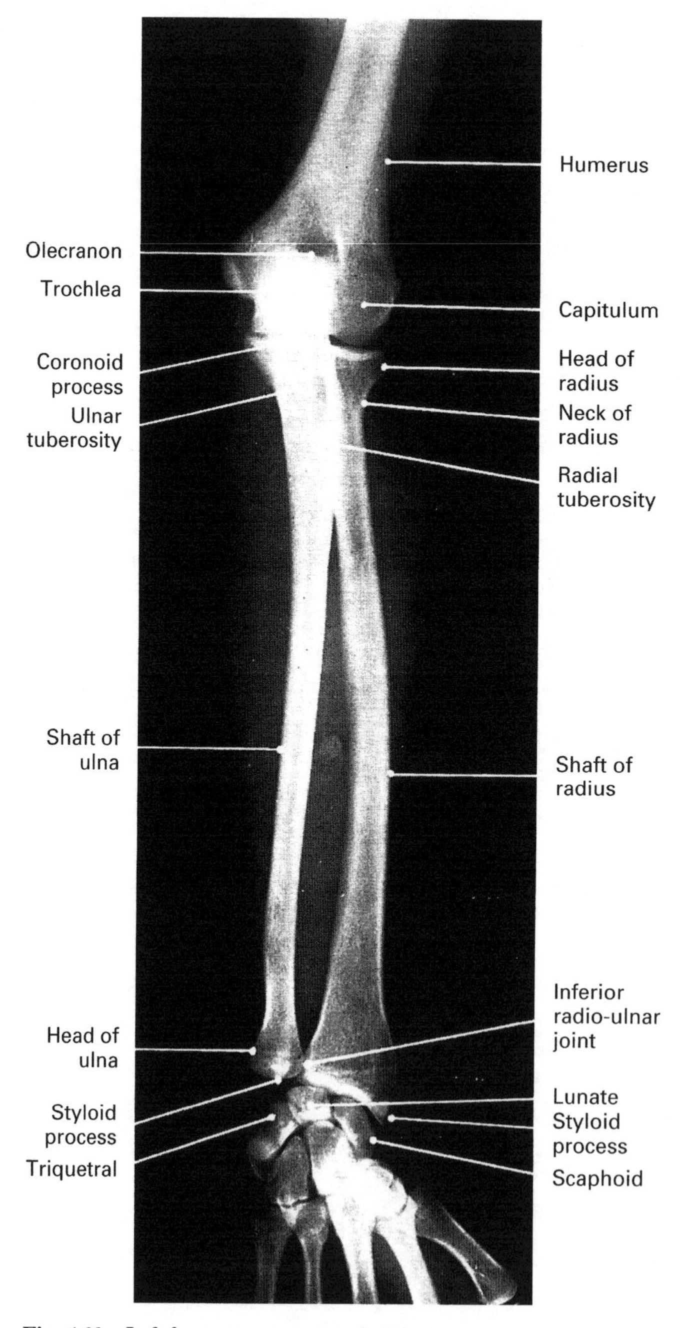

Fig. 6.23 Left forearm: anteroposterior view

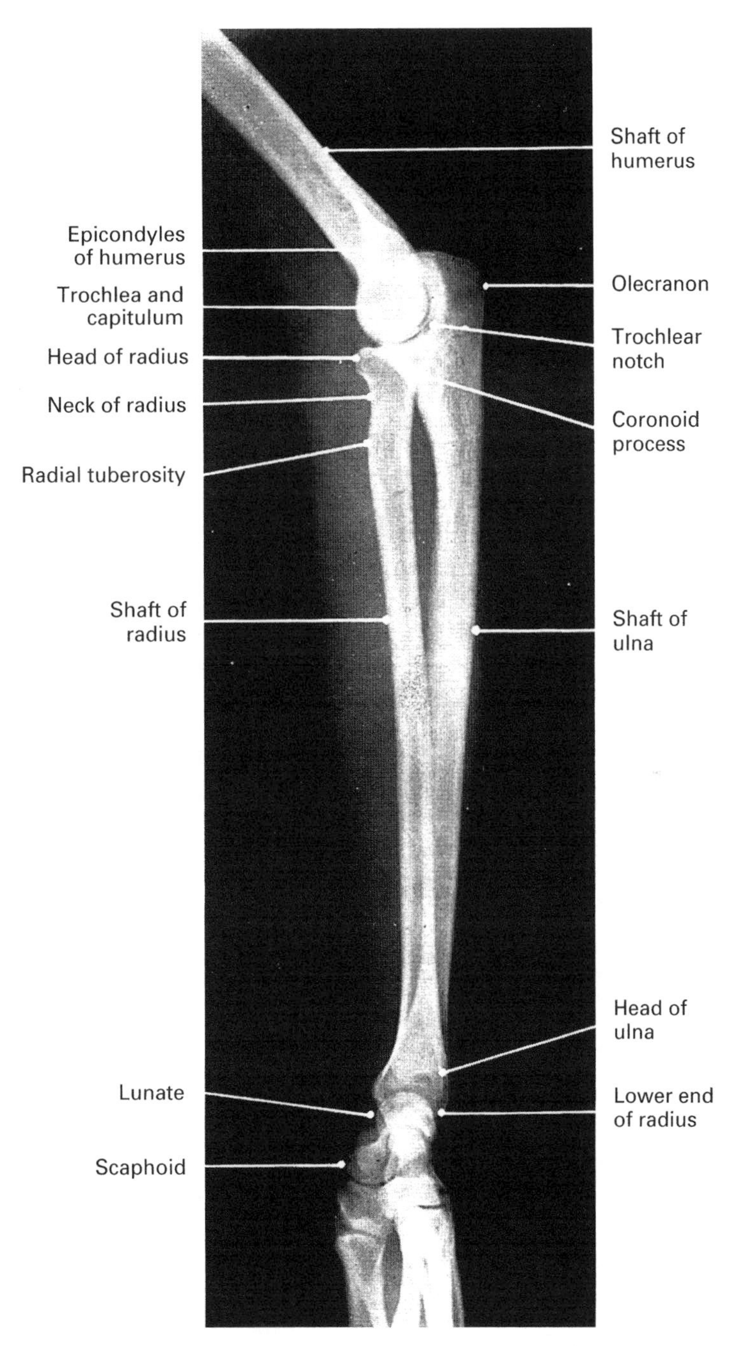

Fig. 6.24 Left forearm: lateral view

OSSIFICATION OF THE RADIUS AND ULNA (Fig. 6.25)

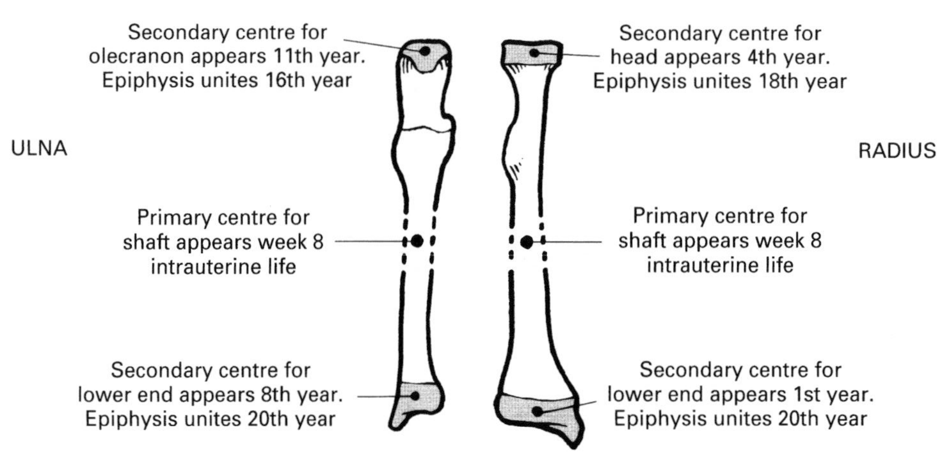

Fig. 6.25 Ossification of radius and ulna

Primary centres

- appear in the shafts about week 8 of intrauterine life.

Secondary centres

- for the head of radius appears about 4th year
- for the olecranon, about 11th year
- for the lower end of radius, at end of 1st year
- for the lower end of ulna, about 8th year.

The epiphyses for the proximal end unite with the shafts between the 16th and 18th years. Those for the distal end unite with the shafts at the 20th year.

HAND (Figs 6.26 to 6.36)

There are three groups of bones in the hand:

- the carpus
- the metacarpus
- the phalanges.

CARPUS

The carpus consists of eight small, irregularly shaped bones arranged in two rows. The bones of the proximal and distal rows from lateral to medial are as shown opposite →

The bones of the distal row articulate with the bases of the matacarpal bones. The bones of the proximal row, except the pisiform, articulate both with radius and with the articular disc (triangular cartilage) at the distal end of the ulna.

The dorsal and palmar surfaces are roughened for the attachment of ligaments. (It should be noted that when referring to the hand it is usual to employ the terms 'dorsal' and 'palmar' in place of posterior and anterior, but the terms 'proximal' and 'distal' are employed in the usual sense, the terminal phalanges being the distal bones of the hand.)

Proximal row

The scaphoid is the largest bone in the proximal row. It is roughly boatshaped, with a narrow waist—a common site for fracture. It lies with its long axis directed laterally and slightly forward. It

LATERAL SIDE			**MEDIAL SIDE**
	Distal row		
Trapezium	Trapezoid	Capitate	Hamate
	Proximal row		
Scaphoid	Lunate	Triquetrum	Pisiform

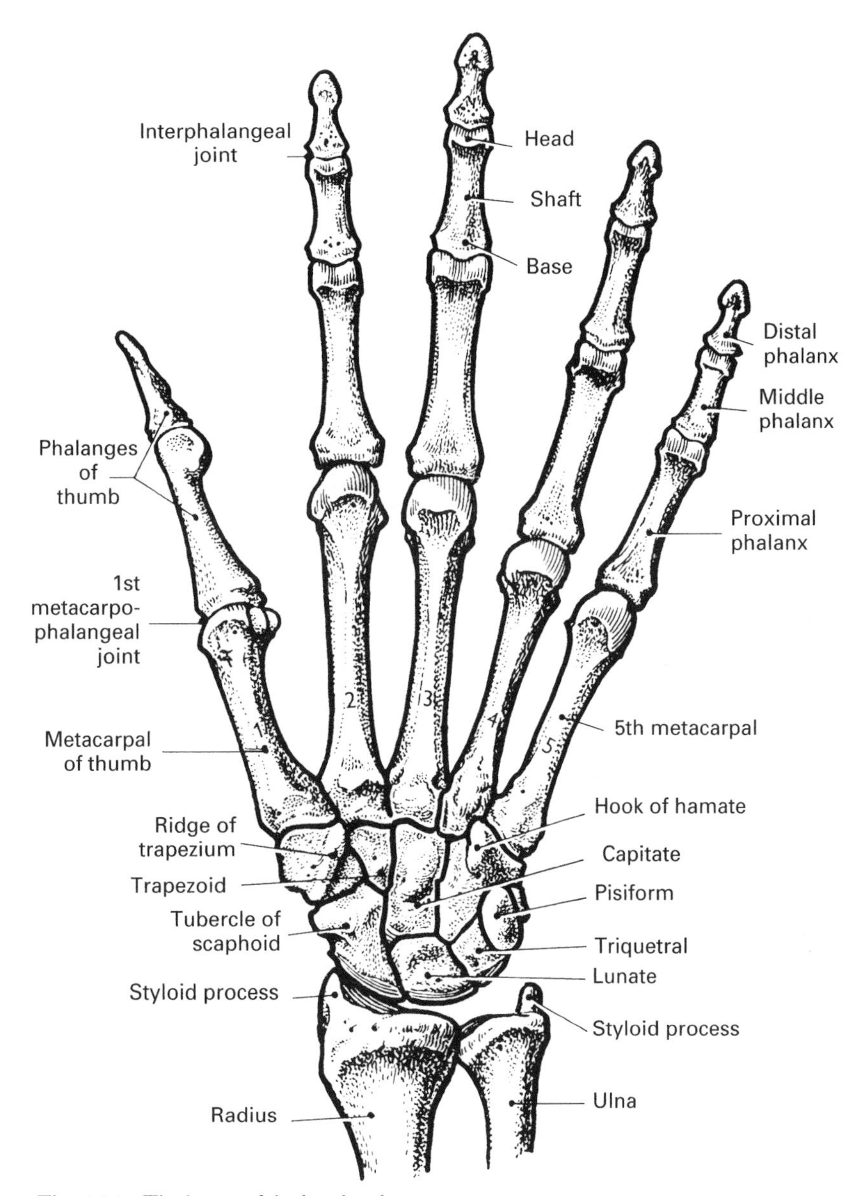

Fig. 6.26 The bones of the hand: palmar aspect

has a rounded tubercle on the anterior margin of the distal end which gives attachment to the flexor retinaculum muscle and is easily palpable. The scaphoid articulates with the trapezius, trapezoid, capitate, lubate and radius.

The lunate is crescent-shaped. It articulates with the scaphoid, capitate, hamate, triquetrum and radius.

The triquetrum is pyramidal in shape. It articulates with the hamate and on its palmar surface it articulates with the pisiform also.

The pisiform is really a sesamoid bone within the flexor carpi ulnaris tendon. It is pea-shaped and it articulates only with the triquetrum.

Distal row

The trapezium is saddle-shaped. It bears a groove on its palmar surface for the flexor carpi

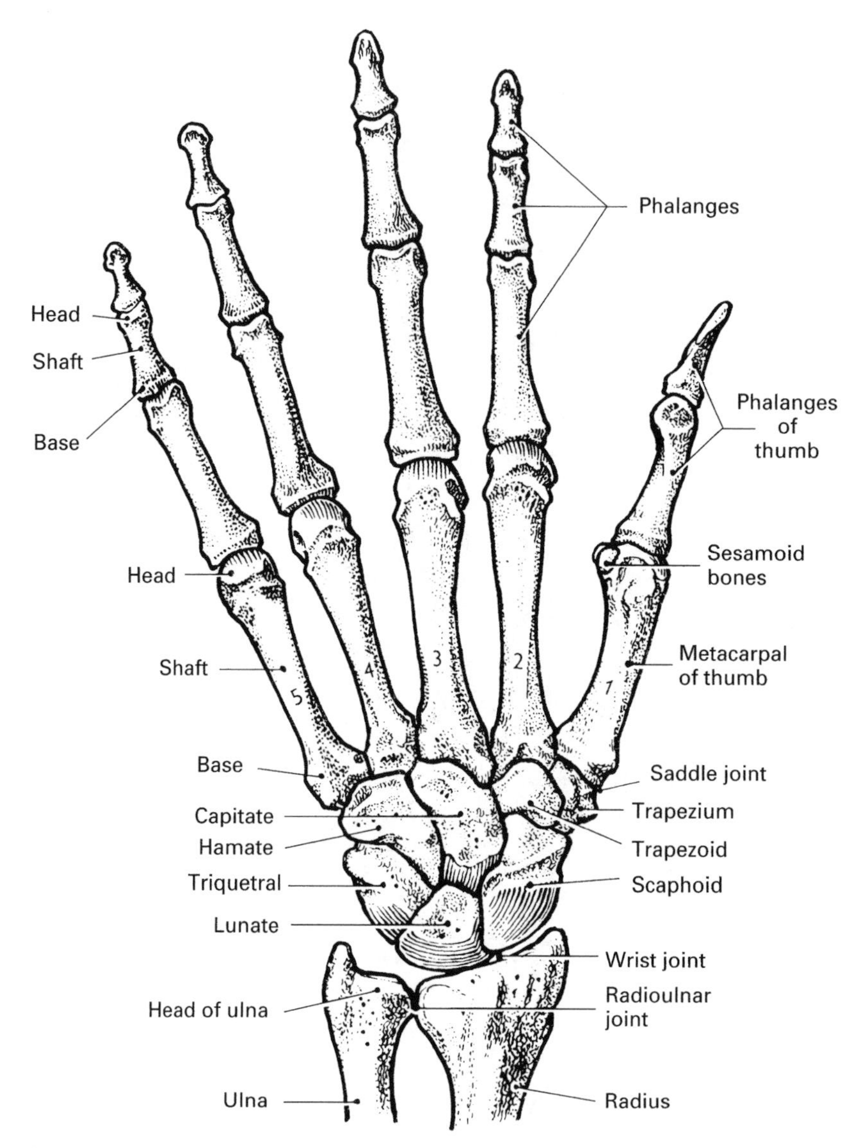

Fig. 6.27 The bones of the hand: dorsal aspect

radialis tendon. On its palmar surface there is also a tubercle to which are attached the muscles of the thenar eminence (the 'ball' of the thumb). The trapezium articulates with the 1st and 2nd metacarpal bones and with the trapezoid and scaphoid.

The trapezoid is small and irregular in shape. It articulates with the 2nd metacarpal bone and with the trapezium, capitate and scaphoid.

The capitate is the largest of the carpal bones. It articulates with the 2nd, 3rd and 4th metacarpal bones and with the trapezoid, scaphoid, lunate and hamate.

The hamate is wedge-shaped and has a hook-like process—the hamulus—on the palmar aspect. The hamulus forms part of the carpal tunnel and to its tip is attached the flexor retinaculum ligament.

Carpal tunnel (Fig. 6.28)

The carpus forms an arch, concave on its palmar aspect. This concavity is accentuated by the presence of four bony prominences—the pisiform, the hamulus of the hamate, the tubercle of the scaphoid and the tubercle of the trapezium. The concavity is converted into an osseofibrous tunnel by the flexor retinaculum ligament which is attached to these bony prominences. Through the tunnel are transmitted the flexor tendons, median nerve and branches of the radial and ulnar arteries.

METACARPUS

There are five metacarpal bones and they form the bony framework of the palm of the hand. They are miniature long bones. From the lateral to the medial side they are numbered 1 to 5. The heads of the metacarpal bones articulate with the proximal phalanges at the metacarpophalangeal joints (p. 166). The 'knuckles' are formed by the metacarpal heads. The bases of the metacarpal bones articulate with the distal row of carpal bones at the carpometacarpal joints (p. 165). In addition, the bases of metacarpal bones 2 to 5 articulate with each other.

The 1st metacarpal bone is shorter and thicker than the others and the head is less convex (i.e. flatter) and broader. Its base forms a saddle joint with the trapezius of the carpus. This particular form of joint brings the thumb into an oblique position in relation to the rest of the hand, enabling it to pass diagonally across the palm to the finger tips, so increasing the grasping power of the hand. On the palmar aspect of the first metacarpophalangeal joint there are two sesamoid bones.

The 2nd metacarpal bone is the largest and, like the remaining metacarpal bones, it has a rounded head which overlaps the palmar aspect of the shaft. Its base articulates with three carpal bones—the trapezium, trapezoid and capitate. On the medial side of the base there is a facet for articulation with the 3rd metacarpal bone.

The 3rd metacarpal bone is distinguished by a small projection—the styloid process—which projects proximally from the dorsal surface. The base articulates with the capitate of the carpus and also medially and laterally with the bases of the 2nd and 4th metacarpal bones.

The 4th metacarpal bone is shorter and thinner. Its base articulates with the hamate and with the 3rd and 5th metacarpal bones.

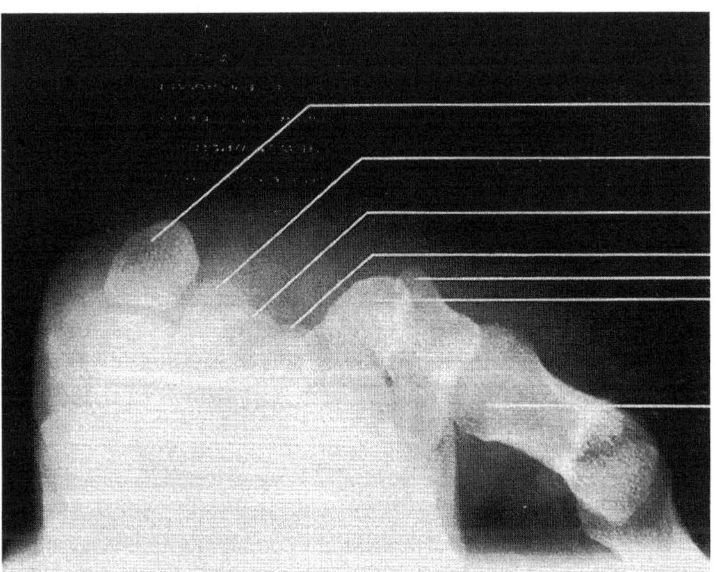

Fig. 6.28 Carpal tunnel: axial view

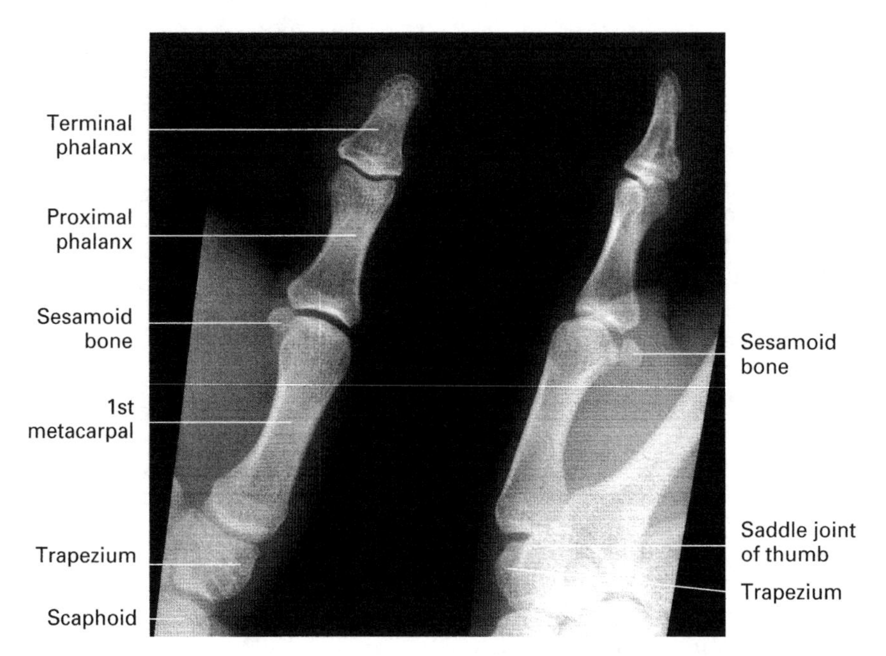

Fig. 6.29A Right thumb: anteroposterior view

Fig. 6.29B Right thumb: lateral view

The 5th metacarpal bone is short and thin and its base articulates with the hamate also. On the lateral aspect of the base there is a facet for articulation with the 4th metacarpal bone.

The heads of the metacarpal bones are wider on the palmar than on the dorsal aspect. The collateral ligaments of the metacarpophalangeal joints are attached to the dorsal surface. Thus the fingers are locked in flexion but can move sideways in extension.

PHALANGES

There are 14 phalanges—three in each finger (proximal, middle and distal) and two in the thumb (proximal and distal). Each phalanx consists of a head (distal end), a shaft and a base (proximal end). The shaft is narrow and flattened on its palmar aspect. The heads of the middle and proximal phalanges are pulley-shaped. The bases of the distal and middle phalanges have two concave facets, separated by a smooth ridge, to accomodate the pulley shape. These interphalangeal joints (p. 167) are hinge joints and allow flexion and extension only. The bases of the proximal phalanges have concave, oval facets for articulation with the metacarpal bones at the metacarpophalangeal joints (p. 166). The head of each distal phalanx is expanded to support the pulp of the finger tip.

Radiographic appearances of the hand (Figs 6.30 to 6.32)

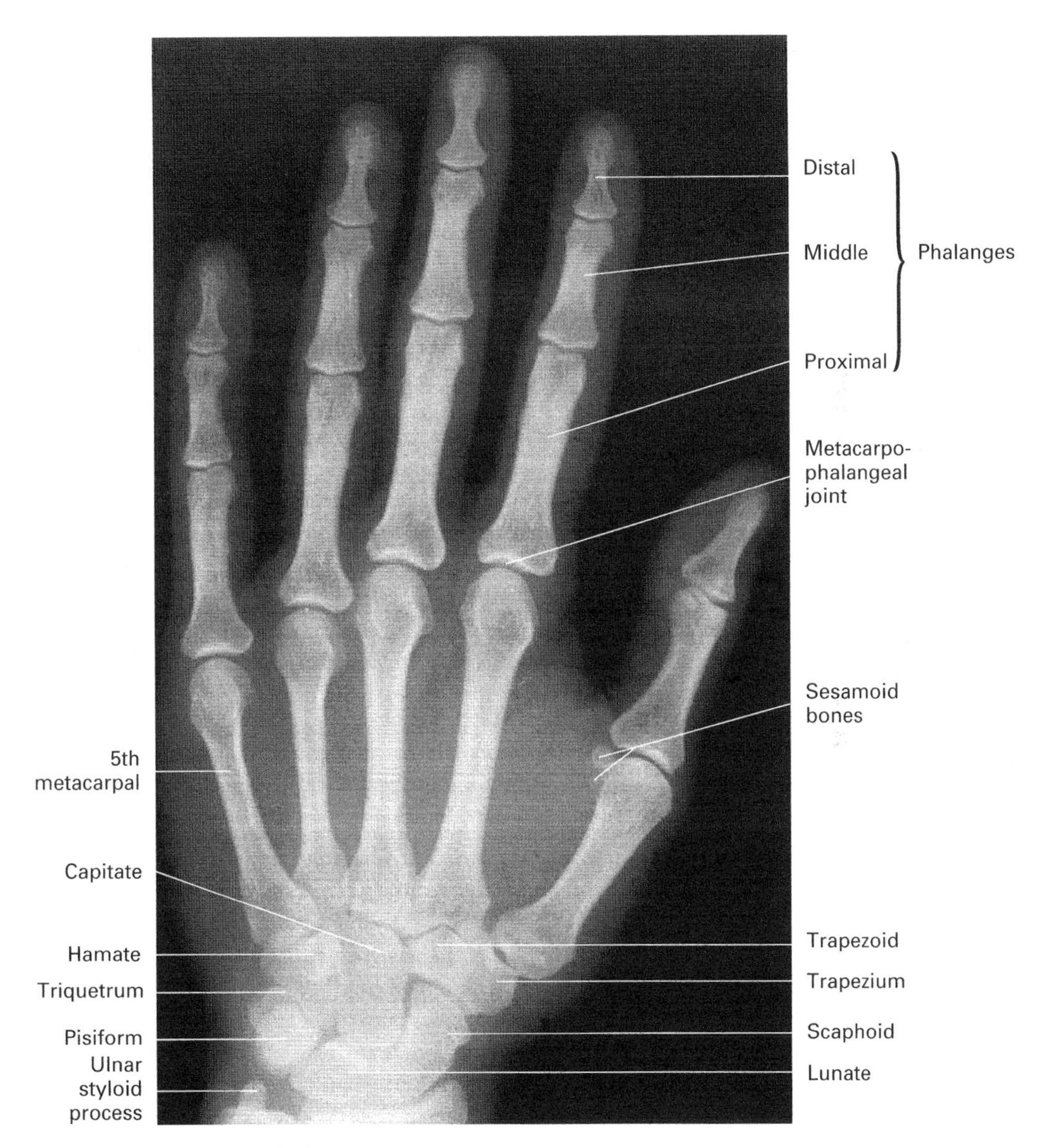

Fig. 6.30 Left hand: dorsipalmar view

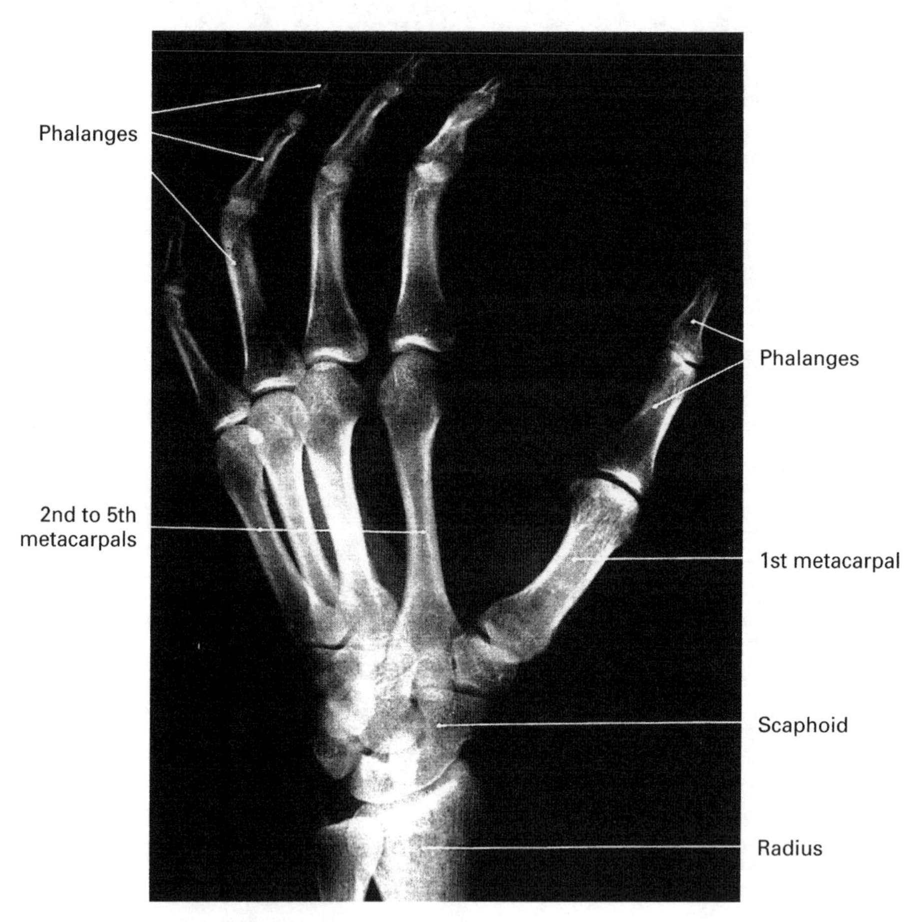

Fig. 6.31 Left hand: dorsipalmar oblique view

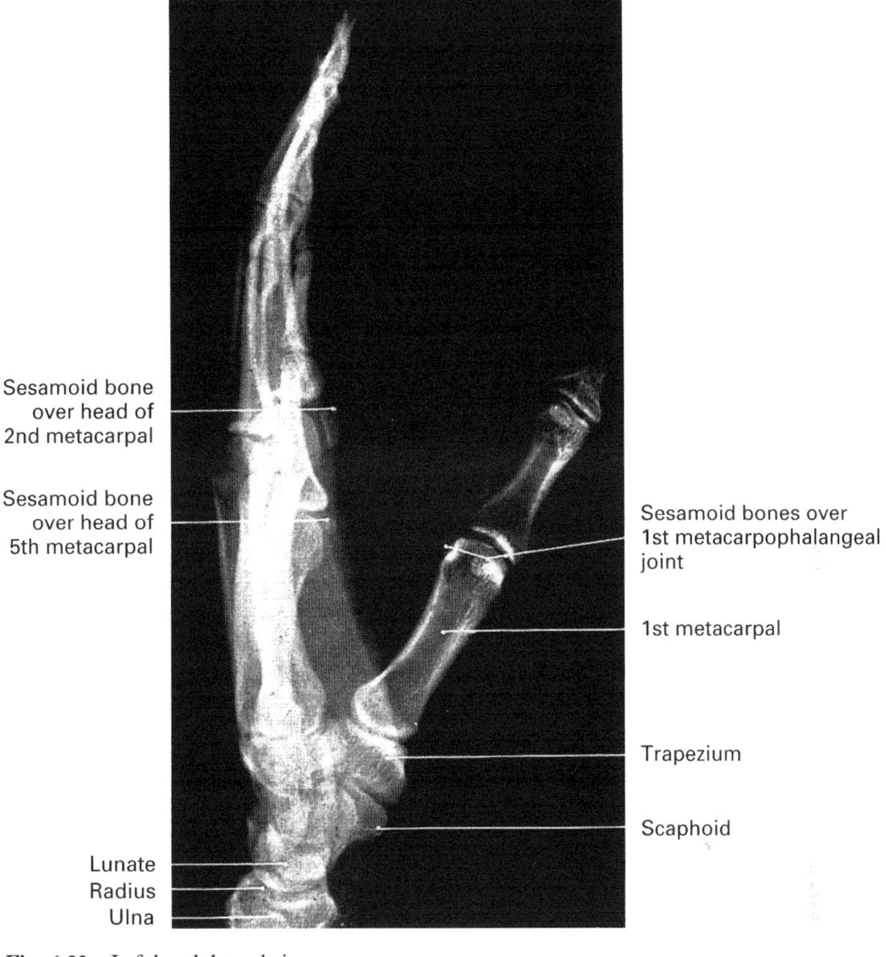

Fig. 6.32 Left hand: lateral view

OSSIFICATION OF THE BONES OF THE HAND (Figs 6.33–6.36)

Primary centres

Carpus—Each carpal bone ossifies from one primary centre only. None of these is present at birth (cf. tarsal bones, p. 234). The carpal bones ossify in a 'circular' anticlockwise order.

- Capitate 1st year
- Hamate 2nd year
- Triquetrum 3rd year
- Lunate 4th year
- Scaphoid 5th year
- Trapezium 6th year
- Trapezoid 7th year
- Pisiform 10th year.

Metacarpus—Each metacarpal bone ossifies from one primary centre in the shaft between weeks 8–12 of intrauterine life.

Phalanges. Each phalanx develops from one primary centre in the shaft between weeks 8 and 12 of intrauterine life.

Secondary centres

Metacarpus—There is one secondary centre for each metacarpal bone. Those for metacarpals 2 to 5 appear at the distal ends of the bones but that for the 1st metacarpal develops at the proximal end. These secondary centres develop between the 2nd and 3rd years and the epiphyses fuse with the shafts about the 18th year.

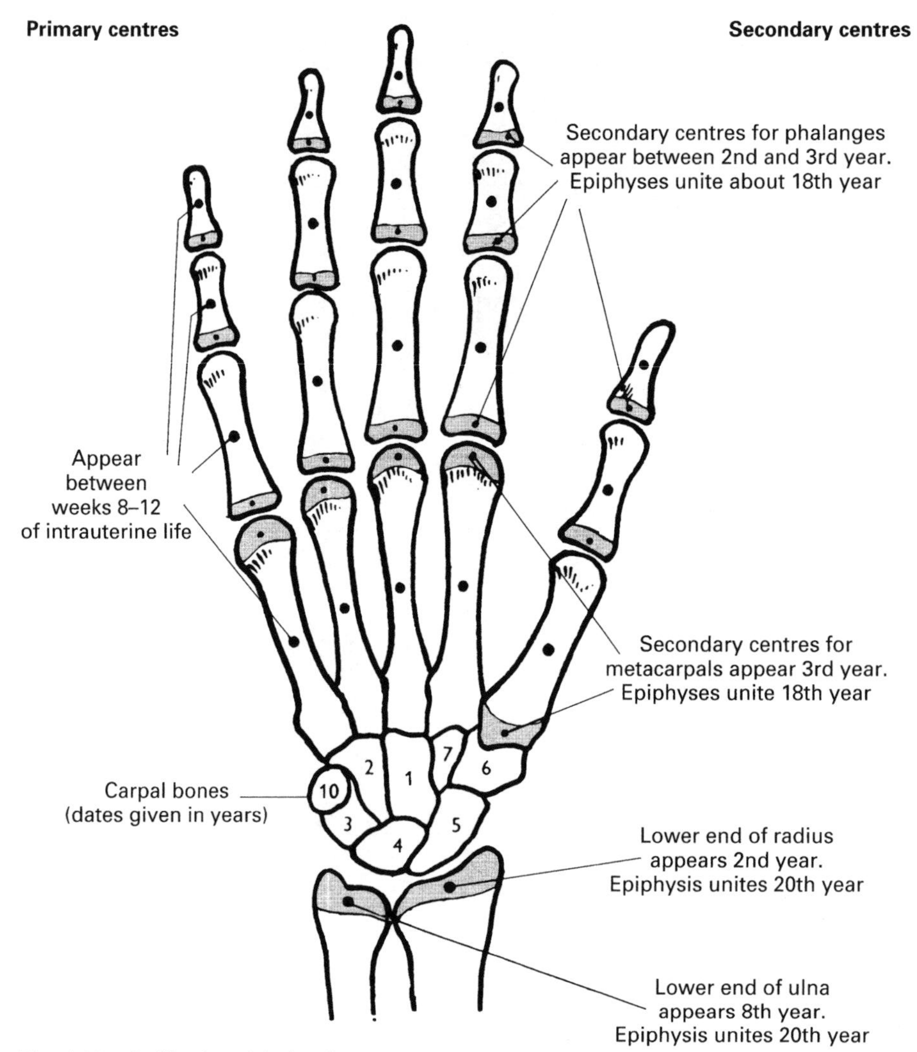

Fig. 6.33 Ossification of the hand

Phalanges—There is one secondary centre for each phalanx. Their position and time of development and fusion are similar to those of the metacarpal bones.

WRIST JOINT (RADIOCARPAL JOINT)

(Figs 6.37–6.45)

Type: Synovial, ellipsoid.

Articular surfaces: Distal end of radius and proximal articular surfaces of scaphoid, lunate and triquetrum. Head of ulna is separated from carpal bones by triangular fibrocartilaginous articular disc which articulates with the carpal bones and separates inferior radioulnar joint from wrist joint.

Synovial membrane: Forms separate joint space from that of inferior radioulnar joint and intercarpal joints.

Ligaments:

Capsular—surrounds the joint.

Radiocarpal

- *palmar*—attached to distal end of radius and anterior surfaces of scaphoid, lunate and triquetrum
- *dorsal*—attached to distal end of radius and to posterior surfaces of scaphoid, lunate and triquetrum.

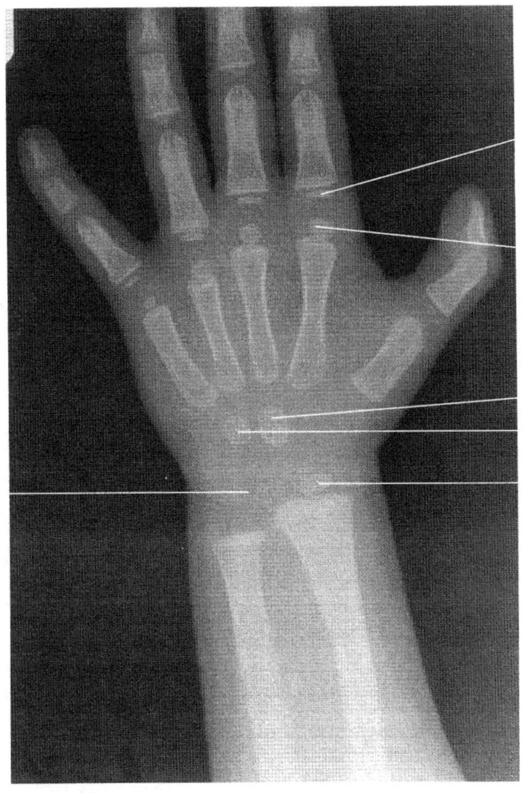

Fig. 6.34 Ossification of the hand: $2\frac{1}{2}$ years

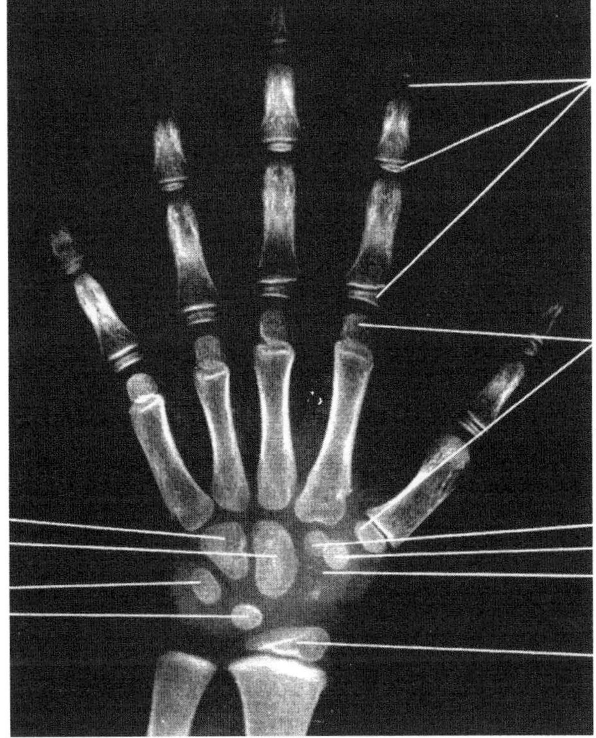

Fig. 6.35 Ossification of the hand: 7 years

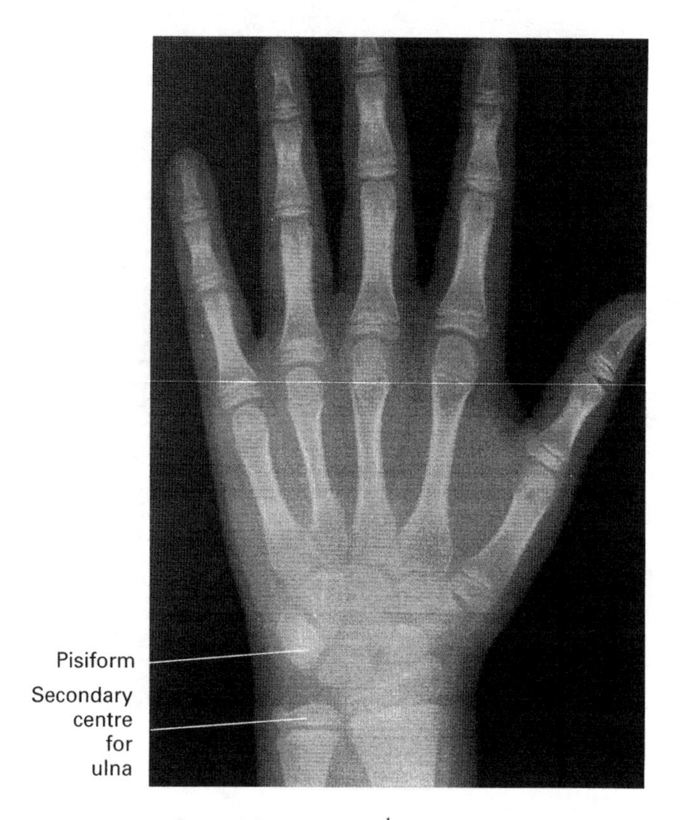

Fig. 6.36 Ossification of the hand: $11\frac{1}{2}$ years

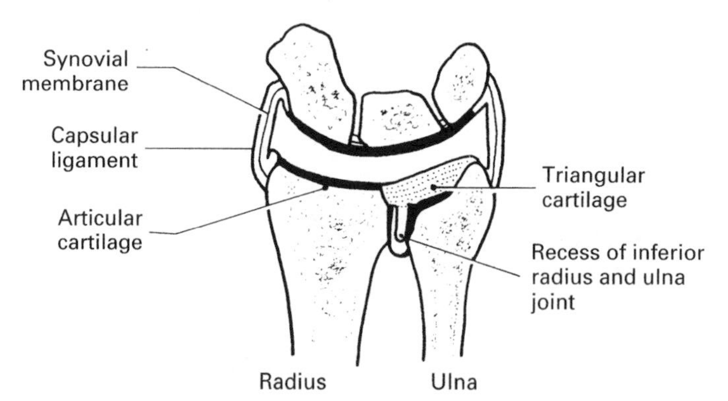

Fig. 6.37 Schematic diagram of the wrist joint

Collateral
- *radial*—attached to radial styloid, scaphoid and trapezium
- *ulnar*—attached to ulnar styloid and to triquetrum and pisiform.

Blood supply: Anterior and posterior carpal branches of radial and ulnar arteries.

Nerve supply: Branches of radial and ulnar nerves (interosseous nerves).

Movements and muscles:

Flexion—flexor carpi radialis, flexor carpi ulnaris, flexor digitorum superficialis and profundus, palmaris longus.
Extension—extensor carpi radialis longus and brevis, extensor carpi ulnaris, extensor digitorum communis.
Abduction—flexor and extensor carpi radialis.
Adduction—flexor and extensor carpi ulnaris.
Circumduction—combination of all these movements.

Flexor carpi radialis:
origin—medial epicondyle of humerus.
insertion—bases of 2nd and 3rd metacarpals anteriorly.
function—flexion of wrist (with flexor carpi ulnaris), abduction of wrist.
nerve supply—median nerve.

Flexor carpi ulnaris:
origin—medial epicondyle of humerus, medial side of proximal olecranon and posterior proximal ulna.
insertion—pisiform, hamate and 5th metacarpal anteriorly.
function—flexion of wrist, adduction of hand.
nerve supply—ulnar nerve.

Flexor digitorum superficialis:
origin—medial epicondyle of humerus, medial coronoid process of ulna and oblique line of radius.
insertion—medial and lateral aspects of middle phalanges of 4 fingers.
function—flexion of proximal and middle phalanges and of wrist.
nerve supply—median nerve.

Flexor digitorum profundus:
origin—anterior interosseous membrane, proximal anterior ulna.
insertion—bases of distal phalanges of 4 fingers (passes between diverging tendons of flexor digitorum superficialis).
function—flexion of wrist, flexion of distal phalanges.
nerve supply—median nerve and medial part of ulnar nerve.

Palmaris longus:
origin—medial epicondyle.
insertion—palmar aponeurosis and flexor retinaculum.
function—flexion of wrist.
nerve supply—median nerve.

Extensor carpi ulnaris:
origin—lateral epicondyle of humerus, posterior of ulna.
insertion—base of 5th metacarpal.
function—extension of wrist, adduction of hand.
nerve supply—posterior interosseous (deep branch of radial) nerve.

Extensor carpi radialis:
origin
- longus—lateral supracondylar ridge of humerus.
- brevis—lateral epicondyle of humerus.

insertion—bases of 2nd (longus) and 3rd (brevis) metacarpals.
function—extension and abduction of wrist.
nerve supply
- longus—radial nerve.
- brevis—posterior interosseous (deep branch of radial) nerve.

Extensor digitorum communis:
origin—lateral epicondyle of humerus.
insertion—bases of middle and distal phalanges of 4 fingers.
function—extension of fingers.
nerve supply—posterior interosseous (deep branch of radial) nerve.

Radiographic appearances of the wrist joint (Figs 6.38 to 6.45)

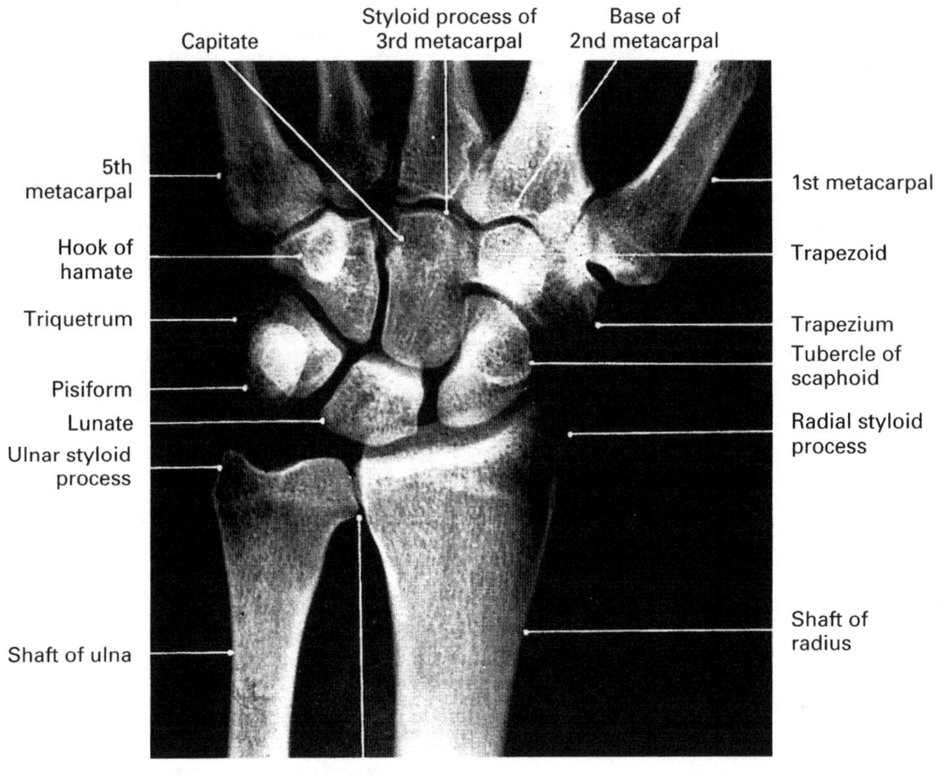

Fig. 6.38 Left wrist joint: posteroanterior view

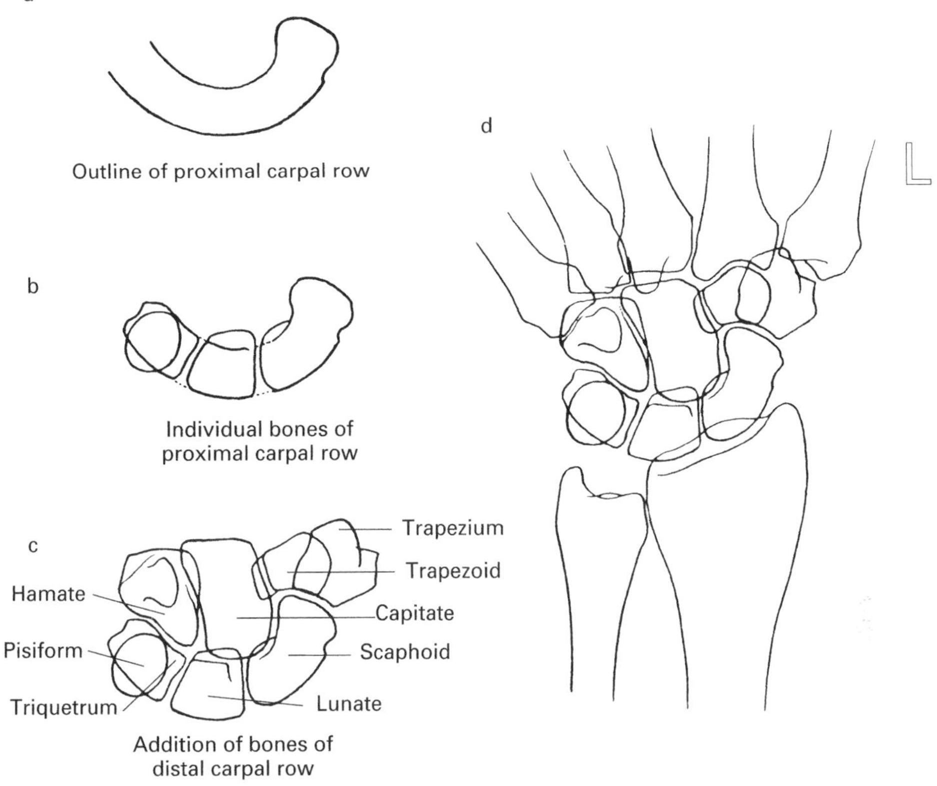

Fig. 6.39 Left wrist joint: posteroanterior view. Construction of diagram

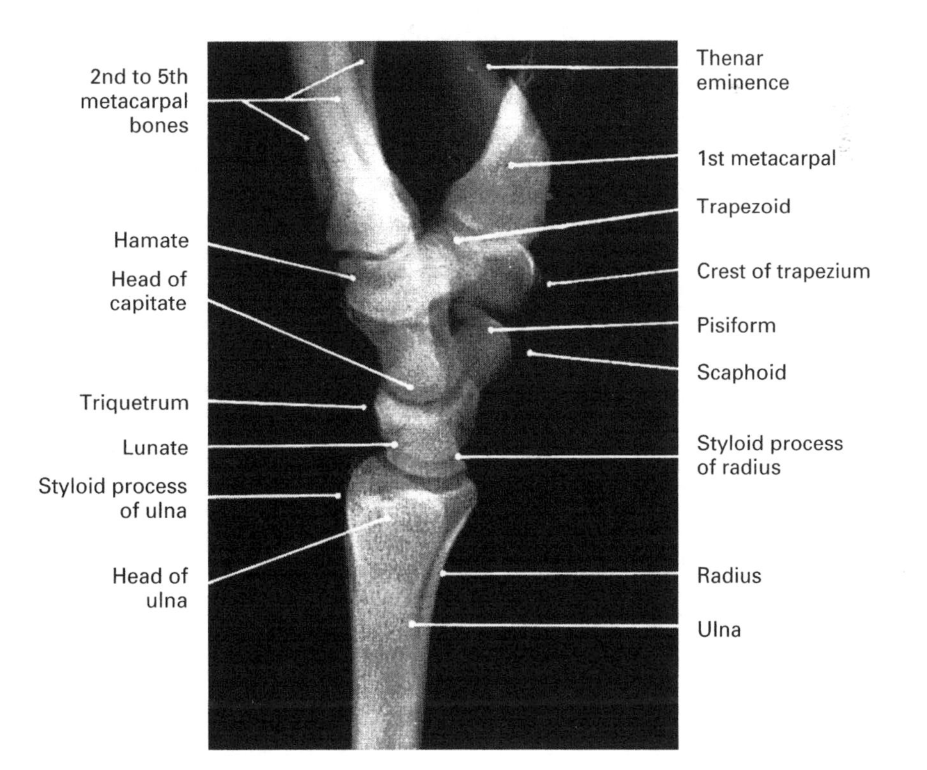

Fig. 6.40 Left wrist joint: lateral view

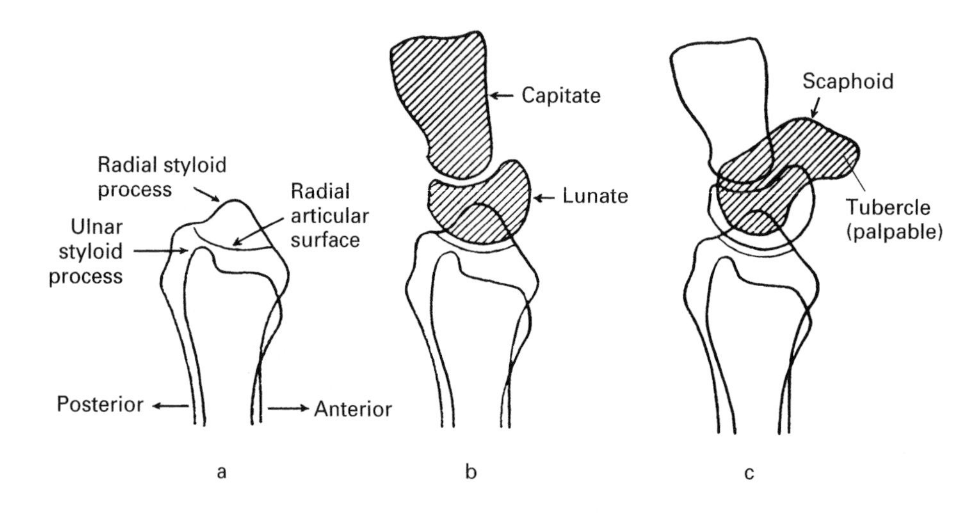

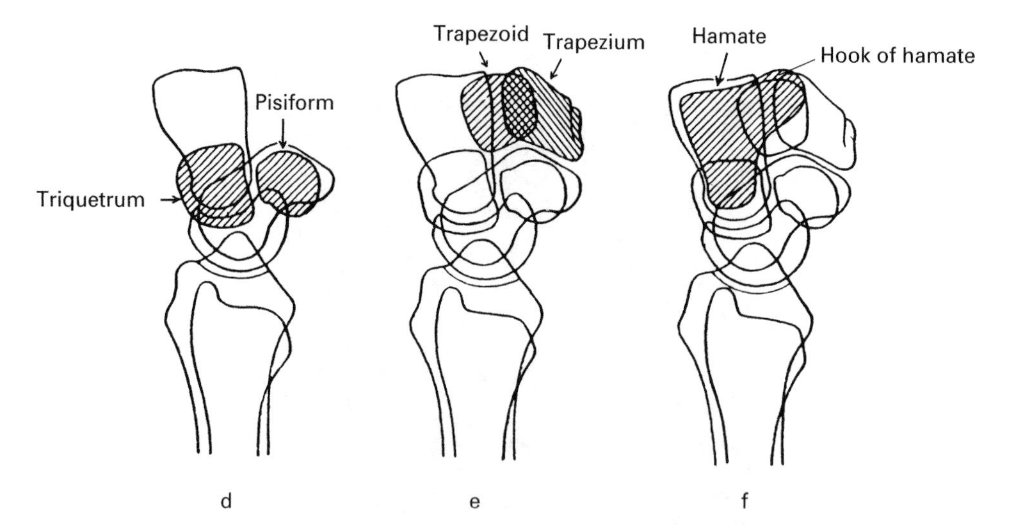

Fig. 6.41 Left wrist joint: lateral view. Construction of diagram. a: Relative levels of radial and ulnar styloid processes. Slightly concave radial articular surface. b: Semi-lunar shape of lunate; rounded head of capitate fits into concave upper surface of lunate and convex lower surface articulates in concave surface of radius. c: Shape and anterior inclination of scaphoid. d: Round pisiform rests on anterior surface of triquetrum. e: Trapezium with its prominent ridge forwards; trapezoid partly overlaps trapezium and capitate. f: Hamate is overshadowed by capitate; hook of hamate projects anteriorly

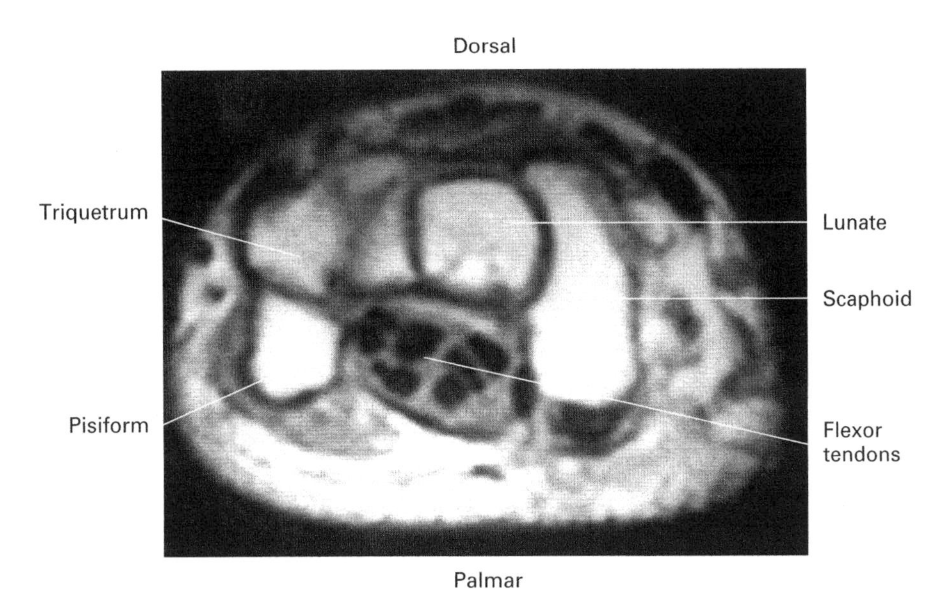

Fig. 6.42 Wrist: MR scan

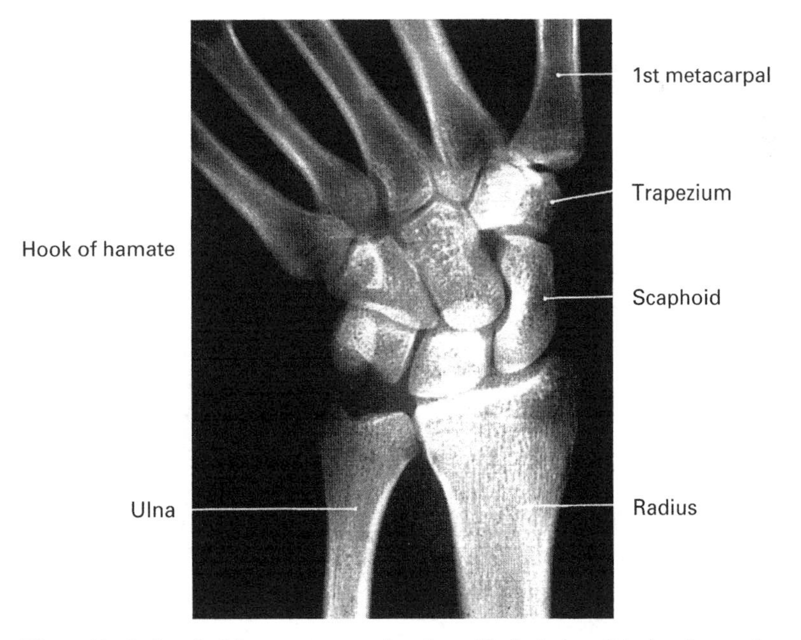

Fig. 6.43 Left wrist joint: posteroanterior view with deviation of the hand towards the ulna

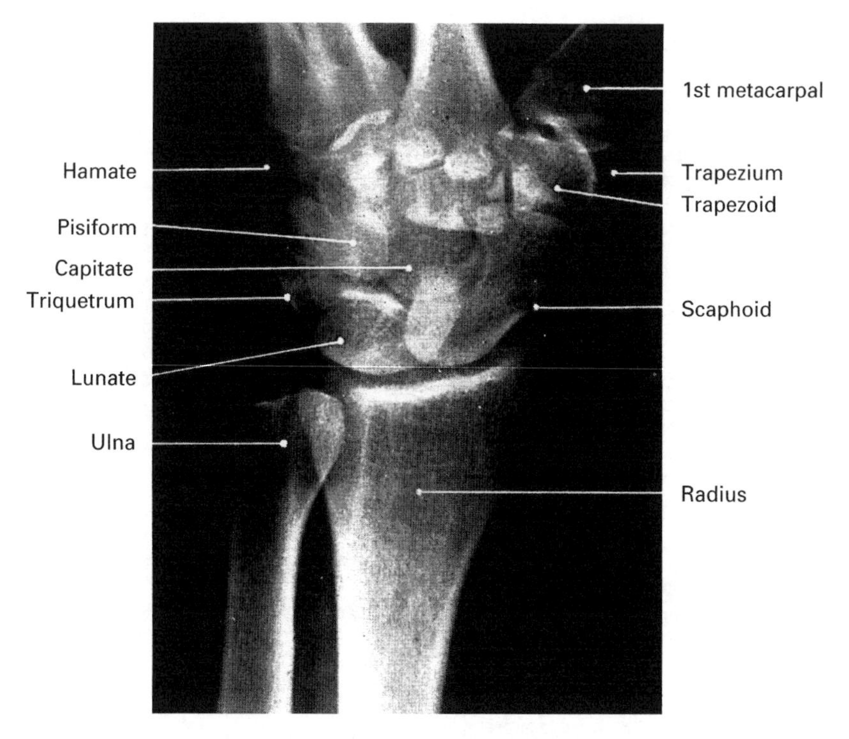

Fig. 6.44 Left wrist joint: posteroanterior oblique view

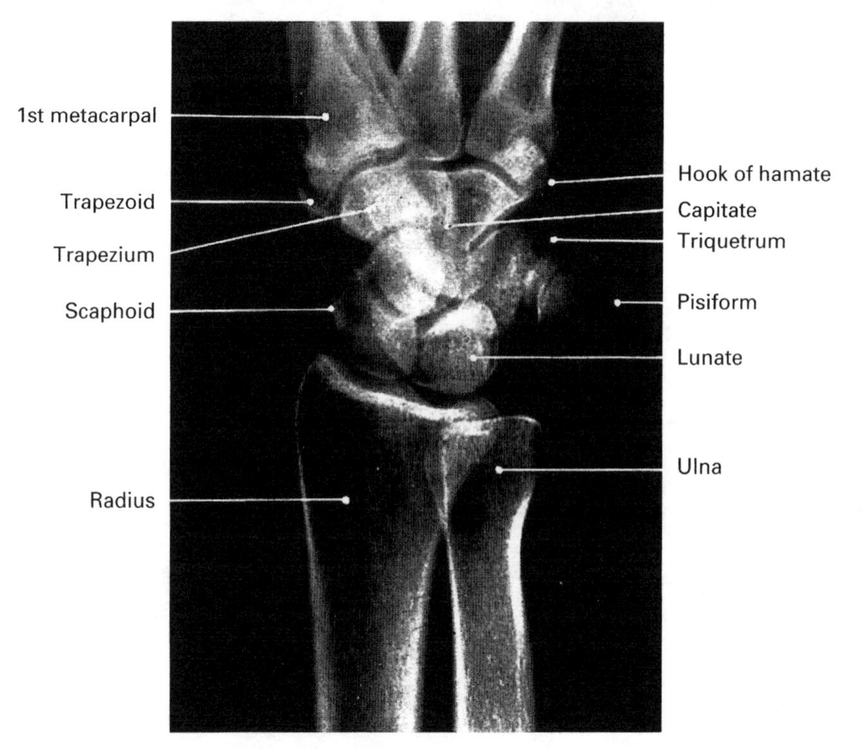

Fig. 6.45 Left wrist joint: anteroposterior oblique view

INTERCARPAL JOINTS

The intercarpal joints comprise joints:

- between the bones of the proximal row, i.e. the scaphoid, lunate and triquetrum
- between the bones of the distal row, i.e. the capitate, hamate, trapezium and trapezoid
- between the two rows of carpal bones—the midcarpal joint.

The bones of the proximal and distal row are joined by dorsal, palmar and interosseous ligaments.

The midcarpal joints comprise:

- an ellipsoid joint between the capitate and hamate distally, and scaphoid, lunate and triquetrum proximally
- a plane joint between the trapezium and trapezoid distally and the scaphoid proximally.

The midcarpal joints are strengthened by dorsal, palmar and medial and lateral collateral ligaments. The pisiform is joined to the hook of the hamate by the pisohamate ligament and to the 5th metacarpal bone by the pisometacarpal ligament. The capsular ligament joins it to the triquetrum.

The midcarpal joint often communicates with the 2nd to 5th carpometacarpal joints.

ACCESSORY LIGAMENTS

The flexor retinaculum is a thickened band of deep fascia forming a tunnel for the flexor tendons (carpal tunnel, p. 151 and Figs 6.28 and 6.42). It is attached to the pisiform, the hook of the hamate, the scaphoid tubercle and the anterior trapezium. It is continuous with the deep fascia and with the palmar aponeurosis into which the palmaris longus muscle inserts.

The flexor retinaculum also gives origin for most of the short muscles of the thumb anteriorly (at the thenar eminence) and for those of the little finger antero-medially (at the hypothenar eminence).

The extensor retinaculum is wider and longer than the flexor retinaculum but it is weaker. By confining the extensor tendons to the back of the wrist it performs a function similar to the flexor retinaculum.

The extensor retinaculum runs obliquely from its medial attachment at the triquetrum and the ulnar styloid process to the anterolateral surface of the distal radius. Fibrous septa pass from five points along the retinaculum, dividing it into six compartments for the extensor tendons.

SYNOVIAL TENDON SHEATHS

Flexor: Synovial tendon sheaths surround the individual tendons as they pass beneath the retinacula of the wrist. They act as long, lubricated channels. A common sheath envelops the tendons of flexores digitorum superficialis and profundus, commencing about 250 mm proximal to the flexor retinaculum. A *separate* sheath encloses the tendon of flexor pollicis longus. The sheaths of the digital flexors terminate at the mid-metacarpal region as blind-ending pouches, except for the little finger sheath which is continuous with its digital synovial sheath (i.e. that which surrounds the digital portion of the tendon). The other fingers also have digital synovial sheaths but they are separate from the flexor sheaths.

Extensor: The extensor retinaculum is divided into six compartments, each containing a synovial sheath. A common sheath is shared by extensor indicis and extensor digitorum longus but abductor pollicis longus, extensores pollicis longus and brevis, extensor digiti minimi and extensor carpi ulnaris have separate sheaths. These commence just proximal to the extensor retinaculum and terminate near the metacarpal bases.

CARPOMETACARPAL JOINTS

Movements of the digits (except the thumb) are all related to the middle finger, so fingers on either side moving away from it are said to abduct, while fingers moving towards it, adduct. Thus the axis of movement is in relation to the 3rd metacarpal and its phalanges (whereas the axis for movement in the foot is in relation to the 2nd toe and 2nd metatarsal bone).

FIRST CARPOMETACARPAL JOINT

Type: Synovial, saddle.

Articular surfaces: Base of 1st metacarpal and the trapezium.

Synovial membrane: Separate from that of the other carpometacarpal joints.

Ligaments:
Capsular—thick but loose, attached to articular margins.
Anterior, *Posterior*, *Lateral* — between corresponding surfaces of trapezium and 1st metacarpal bone.

Movements and muscles:
Flexion—flexores pollicis longus and brevis.
Extension—extensores pollicis longus and brevis.
Abduction—abductores pollicis longus and brevis.
Adduction—transverse and oblique fibres of adductor pollicis.
Circumduction—combination of all these muscles.
Opposition—opponens pollicis.

SECOND TO FIFTH CARPOMETACARPAL JOINTS

Type: Synovial, plane.

Articular surfaces: Base of metacarpal bones and distal row of carpus.

Synovial membrane: Usually continuous with intercarpal articulations.

Ligaments:
Capsular—surrounds articular surfaces.
Dorsal, *Palmar* — join carpals to metacarpals on the dorsal and palmar aspects. The dorsal ligament is the stronger.
Interosseous—short fibres, only at 3rd and 4th joints, i.e. between capitate and 3rd metacarpal and between hamate and 4th metacarpal.

INTERMETACARPAL JOINTS

The bases of the 2nd to 5th metacarpal bones form synovial joints with each other. They are joined together by dorsal, palmar and interosseous ligaments.

METACARPOPHALANGEAL JOINTS

Type: Synovial, ellipsoid.

Articular surfaces: Convex heads of metacarpal bones with concave bases of proximal phalanges and with palmar ligaments.

Ligaments:
Capsular—attached at articular margins, incomplete posteriorly where extensor tendon expansion (dorsal digital expansion) bridges the defect. A bursa lies between joint and dorsal digital expansion.
Palmar—thick fibrocartilaginous plate, strongly bound to phalangeal bases on palmar aspect and to collateral ligaments on each side. Loosely attached to metacarpal heads. Continuous with deep transverse metacarpal ligament. The fibrocartilaginous plate is grooved on anterior surface where flexor tendons cross the joints. Tendon sheaths are bound to margins of these grooves.
Collateral—strong, cylindrical bands attached between posterior tubercle of heads of metacarpals and anterolateral aspect of phalangeal bases.
Deep transverse—flat, broad fibrous bands of great importance joining metacarpals 2–5, i.e. 3 bands, easily palpable between 'knuckles'.

Movements and muscles:
Flexion (the movement most easily performed)—flexor digitorum profundus, flexor digitorum superficialis, lumbricals, interossei.
Thumb—flexor pollicis longus and brevis, 1st palmar interosseous.
Little finger—flexor digiti minimi.
Extension

- Thumb—extensores pollicis longus and brevis
- Index finger—extensor indicis
- Ring and middle fingers—extensor digitorum
- Little finger—extensor digiti minimi.

Abduction—dorsal interossei when fingers extended; abduction not possible in full flexion due to line of action of long flexor tendons.

Adduction—palmar interossei when fingers extended; flexor tendons maintain adduction in full flexion.

Flexion at metacarpophalangeal joints when the distal joints are extended is brought about by the lumbricals and interossei. Flexion of metacarpophalangeal joints when 'gripping' also requires extension of the wrist joint.

INTERPHALANGEAL JOINTS

Type: Synovial, hinge.

Articular surfaces: Concave notch on head of proximal phalanx articulates with base of middle phalanx which has concave facets on either side of a crest. Concave notch on head of middle phalanx articulates with base of distal phalanx which has concave facets on either side of a crest also.

Ligaments:

Capsular—attached at articular margins.

Palmar, *Collateral* — These are similar to those of the metacarpophalangeal joints (p. 166) except that there is no interconnection of the digits by a ligament equivalent to the deep transverse ligament.

Movements and muscles:

Flexion—
- distal joint—flexor digitorum profundus.
- middle joint—flexor digitorum superficialis and profundus.
- thumb—flexor pollicis longus.

Extension—extensor digitorum; thumb—extensor pollicis longus.

MUSCLES OF THE HAND

The palmar aponeurosis is a triangular, fibrous sheet with its apex at the wrist where the palmaris longus tendon inserts. The base is divided into four bands which insert into the deep fascia of the fingers. Fibrous septa pass between the aponeurosis and the thenar and hypothenar eminences, dividing the palm into three compartments: thenar, hypothenar and intermediate.

1. Thenar

Abductor pollicis brevis:
origin—flexor retinaculum, scaphoid tubercle and trapezium.
insertion—base of proximal phalanx, laterally.
function—abduction of thumb.
nerve supply—median nerve.

Flexor pollicis brevis:
origin—flexor retinaculum and trapezium.
insertion—lateral and medial sides of base of proximal phalanx.
function—flexion of thumb.
nerve supply—median nerve, deep branch of ulnar nerve.

Opponens pollicis:
origin—flexor retinaculum and trapezium beneath abductor pollicis brevis.
insertion—1st metacarpal, radial border.
function—opposition of thumb, i.e. abducts, flexes and rotates.
nerve supply—median nerve.

Adductor pollicis (lies deeply in palm). Two heads:
origin
- oblique—bases of 2nd, 3rd metacarpals and capitate
- transverse—shaft of 3rd metacarpal.

insertion—medial side of base of proximal phalanx.
function—adduction of thumb.
nerve supply—deep branch of ulnar nerve.

2. Hypothenar: superficial muscles

Abductor digiti minimi:
origin—pisiform.
insertion—base of proximal phalanx, medial side, of little finger.
function—abduction of little finger, flexion of proximal phalanx.
nerve supply—ulnar nerve.

Flexor digiti minimi:
origin—flexor retinaculum and hook of hamate.
insertion—base of proximal phalanx, medial side, of little finger.
function—flexion of proximal phalanx.
nerve supply—ulnar nerve.

Opponens digiti minimi (deep to abductor and flexor):
origin—flexor retinaculum and hook of hamate.
insertion—medial side of 5th metacarpal shaft.
function—opposition of little finger.
nerve supply—ulnar nerve.

3. Intermediate: deep muscles

Lumbricals (1st and 2nd):
origin—radial side of flexor digitorum profundus (1st) of index and (2nd) of middle finger.
insertion—lateral side of dorsal digital expansion (extensor expansion).
function—flexion of proximal phalanx, extension of middle and distal phalanges.
nerve supply—median nerve.

Lumbricals (3rd and 4th):
origin—by two heads, adjacent sides of tendons (3rd) of middle and ring fingers, and (4th) of ring and little fingers.
insertion—lateral side of dorsal digital expansion.
function—flexion of proximal phalanx, extension of middle and distal phalanges.
nerve supply—deep branch of ulnar nerve.

Interossei

There are eight interossei in two groups, palmar and dorsal. No palmar interossei act on middle finger.

Palmar interossei:
origin—1st and 2nd—medial side of 1st and 2nd metacarpals.
insertion—1st and 2nd—medial side of dorsal digital expansion of thumb and index finger.
origin—3rd and 4th—lateral side of 4th and 5th metacarpals.
insection—3rd and 4th—lateral side of dorsal digital expansion.
function—adduction of fingers.

Dorsal interossei (dorsal interossei 2 and 3 act on corresponding sides of middle finger):
origin—1st and 2nd—by two heads, adjacent sides of 1st and 2nd metacarpals (1st), 2nd and 3rd metacarpals (2nd).
insertion—lateral side of base of proximal phalanx of index and middle fingers.
origin—3rd and 4th—adjacent sides of 3rd and 4th metacarpals (3rd), 4th and 5th metacarpals (4th).
insection—medial side of base of proximal phalanx of middle and ring fingers.
function—abduction of fingers.

7. The pelvic girdle

The pelvic girdle is a strong, basin-shaped bony ring composed of two innominate bones (hip bones), the sacrum and the coccyx. The innominate bones articulate with each other anteriorly in the midline at the symphysis pubis (p. 178) and they form the anterior and lateral parts of the pelvis. Posteriorly the pelvic ring is completed by the sacrum and coccyx. The sacrum articulates with the innominate bones at the sacroiliac joints. The sacrum and coccyx are described on pages 78–81 and the sacroiliac joints on page 178.

The main function of the pelvic girdle is to transmit the body weight to the lower limbs. It is therefore very sturdily constructed and to it are attached the powerful muscles of the trunk and lower limbs. In the standing position, the weight of the body is transmitted through the upper part of the sacrum, the sacroiliac joints and the strong medial borders of the ilia to the acetabula and thence to the heads of the femora. In the sitting position, the weight is transmitted from the medial borders of the ilia to the ischial tuberosities. Another important function of the pelvis is to provide protection to the various organs of the lower abdominal and pelvic cavities.

INNOMINATE BONE (HIP BONE) (Figs 7.1 to 7.4)

The innominate bone is a large, irregularly-shaped bone with a narrow middle section in which is situated a cup-shaped articular area—the acetabulum. The head of the femur articulates with the acetabulum to form the hip joint (p. 186). Below and in front of the acetabulum there is a large gap, oval or triangular in shape, the obturator foramen, through which are transmitted the obturator vessels and nerves.

The innominate bones consists of three parts:

- the ilium
- the ischium
- the pubis.

In childhood the three parts are separated from each other by cartilage (the 'Y-shaped cartilage') but in the adult they are united by bone.

The ilium consists of two parts—an expanded and slightly curved upper part and a smaller lower part. The upper part forms a wide flattened plate with a long, curved, superior border—the iliac crest—to which are attached the muscles of the abdominal wall and back. The curve of the iliac crest ends anteriorly at the anterior superior iliac spine which is easily palpable.

Behind the anterior iliac spine is a bony prominence—the tubercle of the crest—which lies at the level of the 5th lumbar vertebra. Posterior to the tubercle, the highest part of the crest may reach to the level of the 4th lumbar vertebra. The anterior border of the ilium extends downwards from the anterior superior iliac spine to the acetabulum. Just above the acetabulum is a small bony prominence, the anterior inferior iliac spine.

The curve of the iliac crest ends posteriorly at the posterior superior iliac spine which is not easily palpable but whose position can be seen as a small depression in the skin of the buttock about 40 mm from the midline at the level of the second spinous tubercle of the sacrum.

The posterior border of the ilium commences at the posterior superior iliac spine and extends

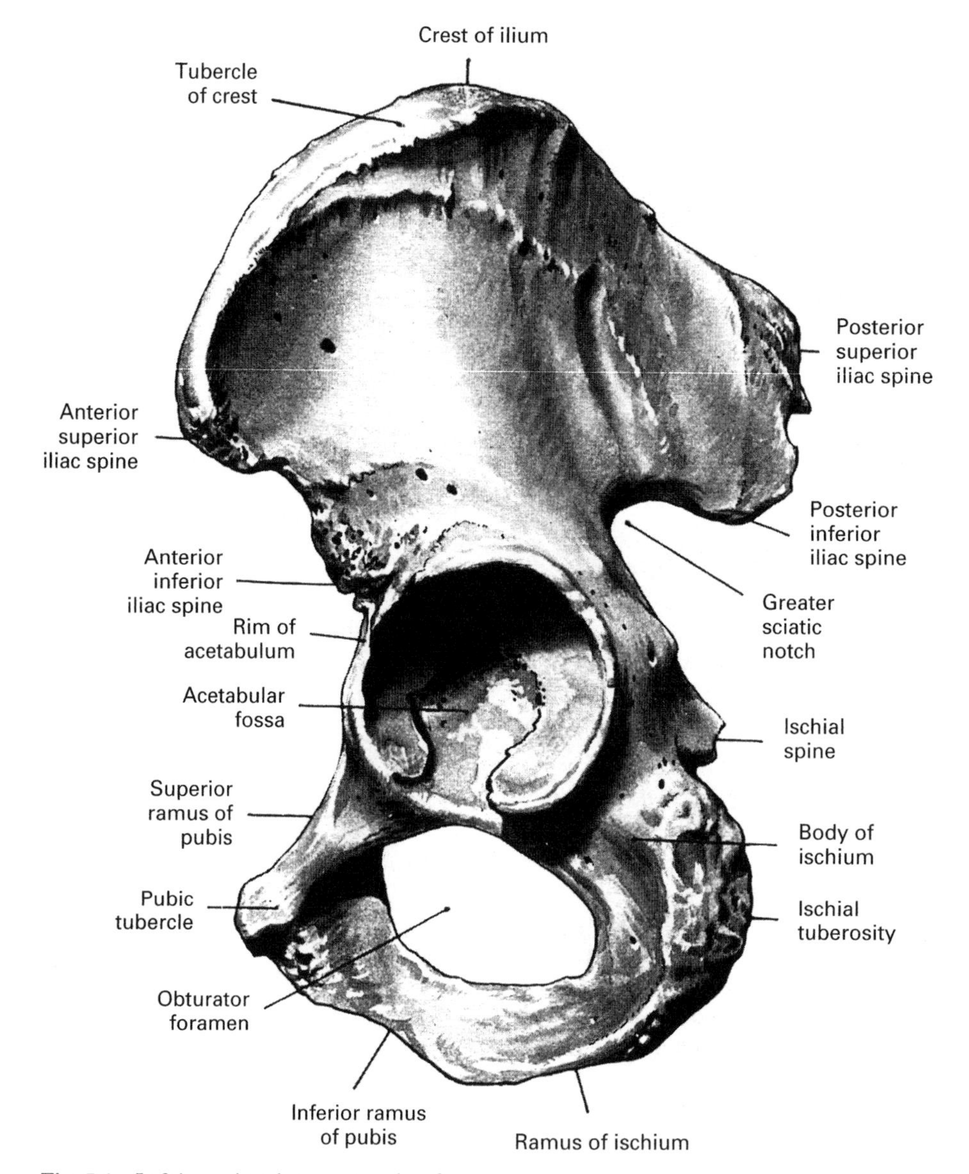

Fig. 7.1 Left innominae bone: external surface

downwards and forwards to become continuous with the posterior border of the ischium. The posterior inferior spine lies 25 mm below the posterior superior spine and from this point the posterior border turns sharply forwards and downwards to form the upper part of the greater sciatic notch through which the sciatic nerve passes out of the pelvis into the thigh and leg. The sciatic notch is converted into a foramen by ligaments.

The medial (internal) surface is divided into two parts, the iliac fossa and the pelvic surface, by the medial border. This runs obliquely forwards and downwards across the bone from the posterior part of the iliac crest to the ileopubic (ileopectineal) eminence which marks the union of the ilium and the pubis. The lower portion of the medial border forms the arcuate line.

The iliac fossa lies between the medial and anterior borders and is a large, shallow, concave surface. It forms the posterolateral wall of the greater (or true) pelvis.

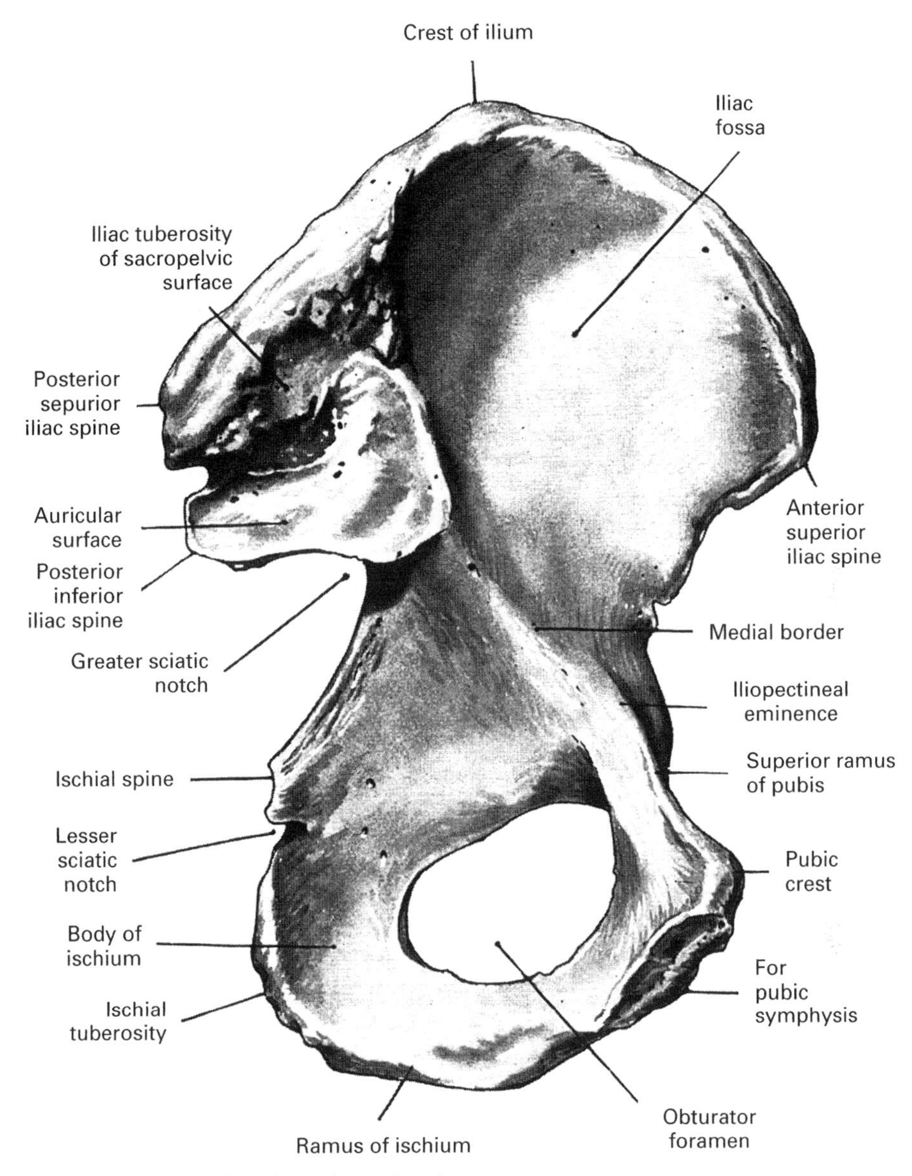

Fig. 7.2 Left innominate bone: internal surface

The sacropelvic surface lies between the medial and posterior borders. The upper part, the iliac tuberosity, is roughened and pitted for ligamentous attachment. The middle part bears a large, curved articular surface for articulation with the sacrum. The lower part is smooth and is bounded posteriorly by the greater sciatic notch; it forms the lateral wall of the true pelvis.

The external surface of the ilium is divided into a large gluteal surface and a smaller acetabular portion. The gluteal surface is bounded above by the iliac crest and below by the acetabulum. The surface is roughened and is marked by three uneven ridges—the anterior, posterior and inferior gluteal lines which divide the surface into areas of attachment for the gluteal muscles (p. 188).

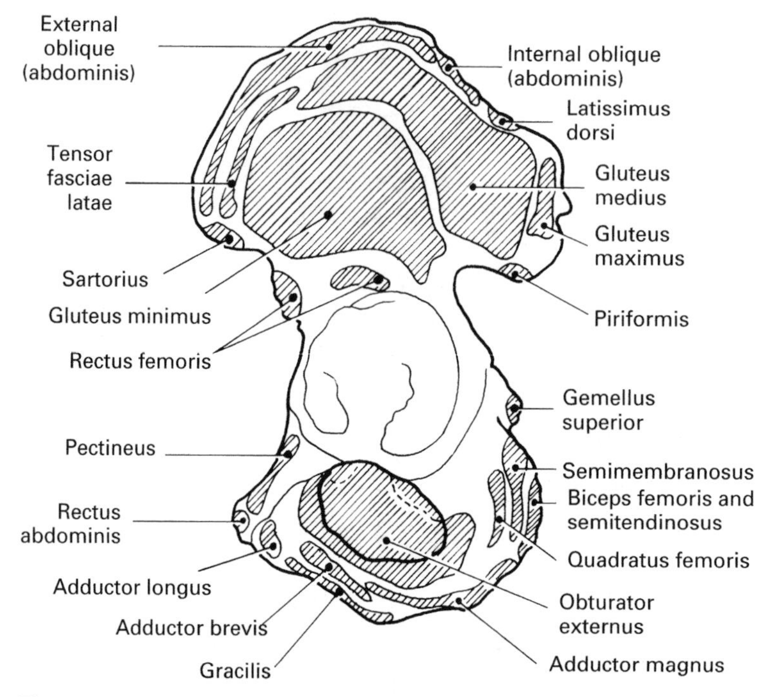

Fig. 7.3 Left innominate bone: external surface showing muscle attachments

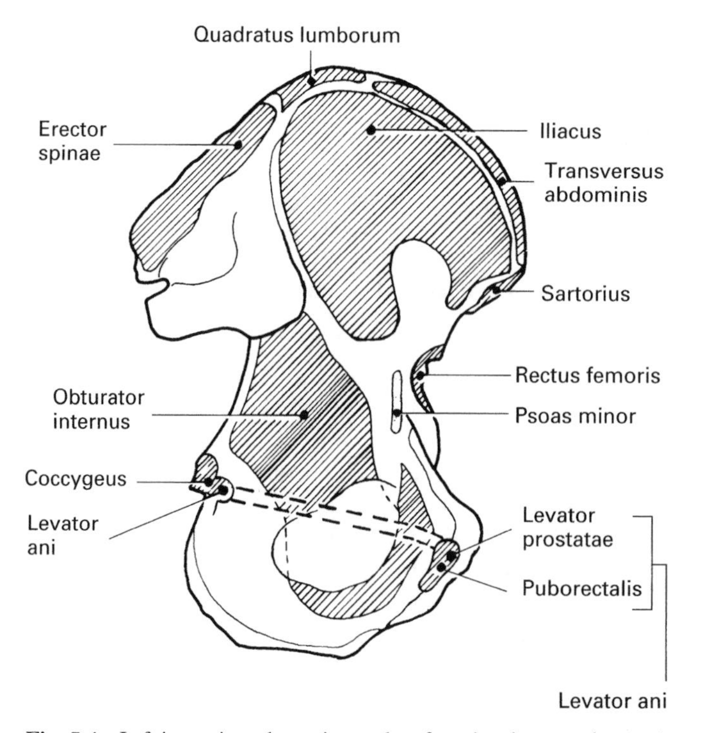

Fig. 7.4 Left innominate bone: internal surface showing muscle attachments

The ischium forms the inferior and posterior parts of the innominate bone. The ischium consists of:

- the body
- the ramus.

The upper part of the **body** forms two-fifths of the acetabulum. The upper part of the posterior border of the body completes the greater sciatic notch; this ends in a well-marked protuberance, the ischial spine, below which is the lesser sciatic notch. In the living subject, the greater and lesser sciatic notches are converted into foramina by the sacrotuberous and sacrospinous ligaments.

The **ramus** of the ischium is thin and flat. It extends slightly upwards to join the inferior ramus of the pubis.

The obturator foramen is situated inferior and anterior to the acetabulum and is enclosed by the body of the ischium, the ischial ramus and the superior and inferior rami of the pubis. A fibrous membrane enclose the foramen, except at its upper edge where an opening transmits the obturator vessels and nerve from the pelvis to the thigh. The foramen is large and oval in the male but is smaller and nearly triangular in the female.

The pubis forms the anterior part of the innominate bone and it joins the pubis of the opposite side in the midline, the symphysis pubis. Together they form the anterior wall of the pelvis. The pubis consists of:

- the body
- the superior ramus
- the inferior ramus.

The **body** is flat and in the standing position its anterior surface faces downwards and forwards. The posterior wall faces backwards and upwards and forms part of the anterior wall of the true pelvis where it is closely related to the urinary bladder. The elongated medial surfaces of the body face one another and are joined by a cartilaginous disc. The rounded upper border of the body, the pubic crest, projects forwards and overhangs the upper part of the anterior surface. Its lateral extremity forms a small prominence, the pubic tubercle, which gives attachment to the inguinal ligament whose lateral end is inserted into the anterior superior iliac spine.

The **superior ramus** extends laterally and slightly upwards to join the ilium and the ischium at the acetabulum. The iliopectineal (iliopubic) eminence is situated at its junction with the ilium.

The **inferior ramus** passes backwards, downwards and laterally to join the ramus of the ischium on the medial side of the obturator foramen.

The acetabulum is an approximately hemispherical cavity on the external surface of the innominate bone. All three parts of the innominate bone contribute to the formation of the acetabulum. It is directed laterally, downwards and forwards. It is surrounded by an irregularly prominent rim which is deficient inferiorly. This gap is called the acetabular notch. The cup-shaped cavity is further deepened by a cartilaginous rim, the acetabular labrum, which is attached to the bony rim. The ligament of the head of the femur, ligamentum teres, which arises from the side of the acetabular notch, is inserted into the fovea of the femoral head. The transverse ligament of the acetabulum bridges the acetabular notch, forming the acetabular foramen through which nerves and blood vessels enter the joint.

The hip joints transmit the weight of the trunk and upper limbs to the lower limbs. For this reason they are strong, stable joints. At the same time, considerable freedom of movement is needed to ensure mobility of the lower limbs.

Radiographic appearances of the pelvis (Figs 7.5 to 7.10)

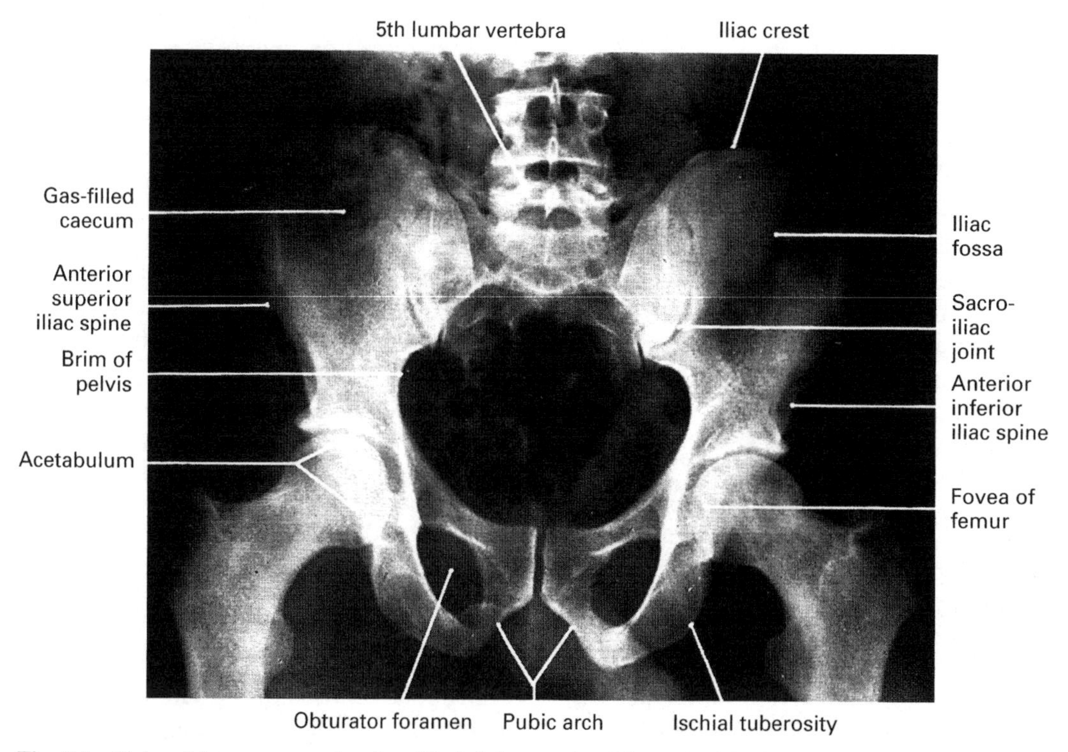

Fig. 7.5 Male pelvis: anteroposterior view. The inferior margin of the neck of the femur and the superior medial margin of the obturator foramen form a continuous arc—Shenton's line

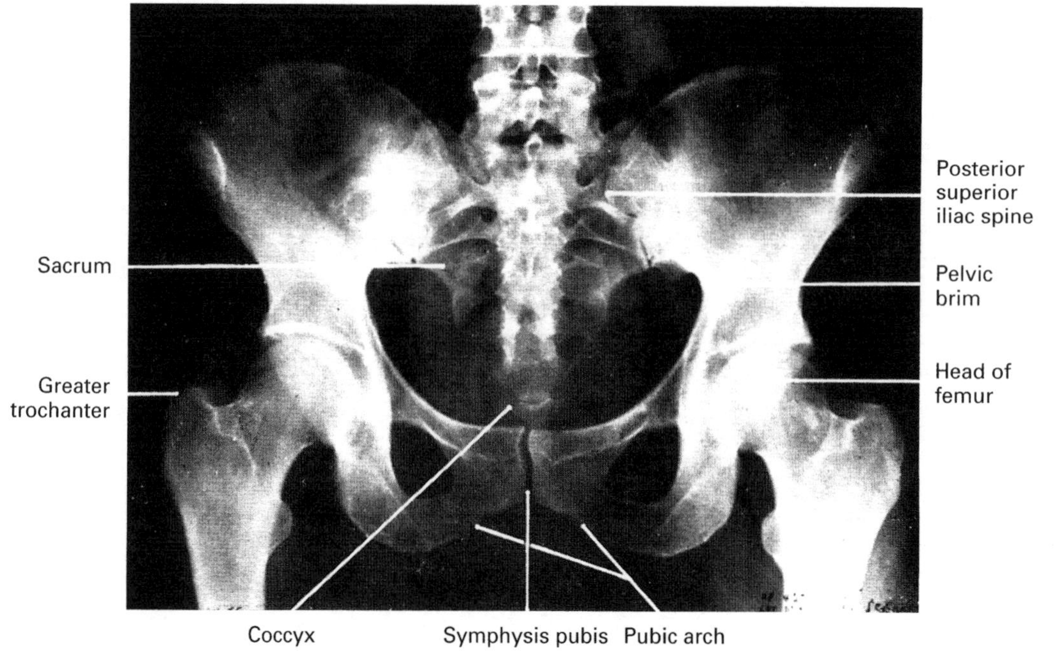

Fig. 7.6 Female pelvis: anteroposterior view

DIFFERENCES BETWEEN MALE AND FEMALE PELVIS (compare Fig. 7.5 and Fig. 7.6)

In the male:

- the bones are larger and stronger
- the false pelvis is narrower
- the inlet of the true pelvis is narrower and triangular
- the ischial spines are closer together, so the outlet is narrower
- the pubic arch is narrower
- the obturator foramen is large and oval whereas in the female it is small and triangular.

MUSCLES OF THE PELVIS (Fig. 7.10)

The floor of the pelvis is a muscular diaphragm. It is formed by two main muscles on each side—the levator ani and the coccygeus.

The levator ani arises from the anterior and lateral walls of the true pelvis between the pubis, obturator fascia, ischial spine, sacrum and

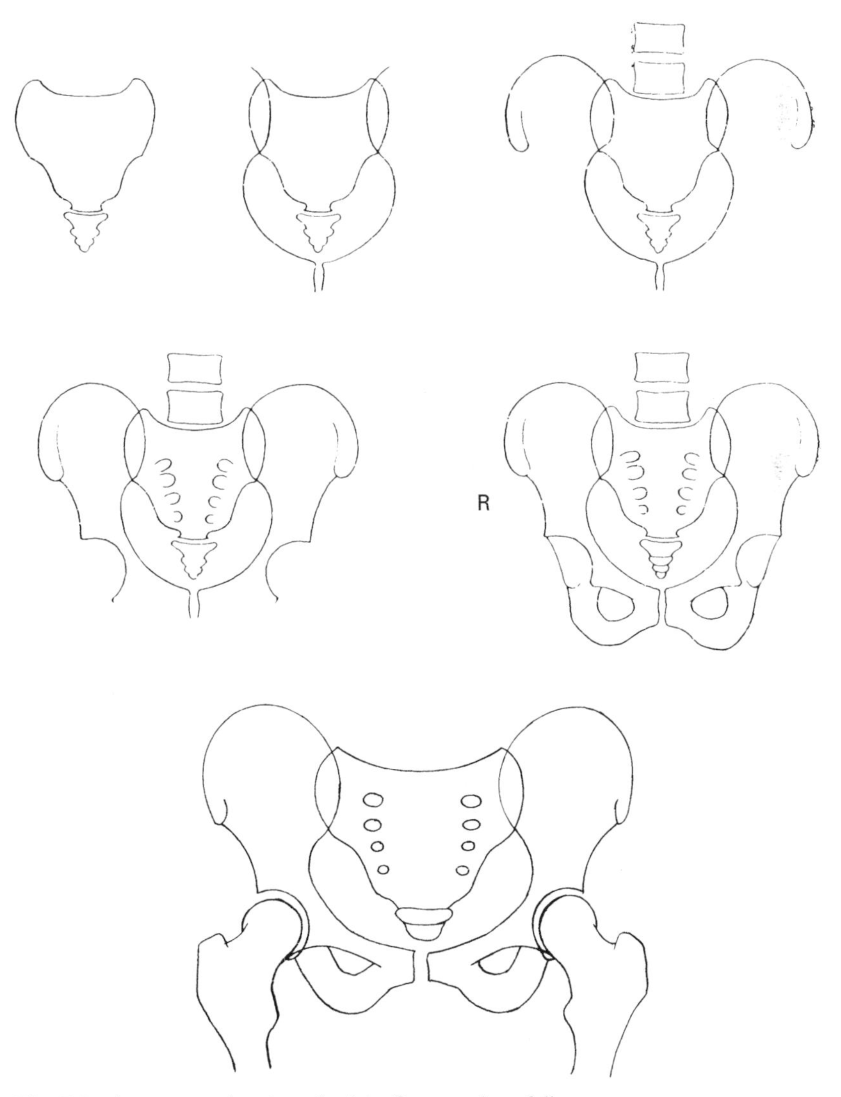

Fig. 7.7 Anteroposterior view of pelvis. Construction of diagram

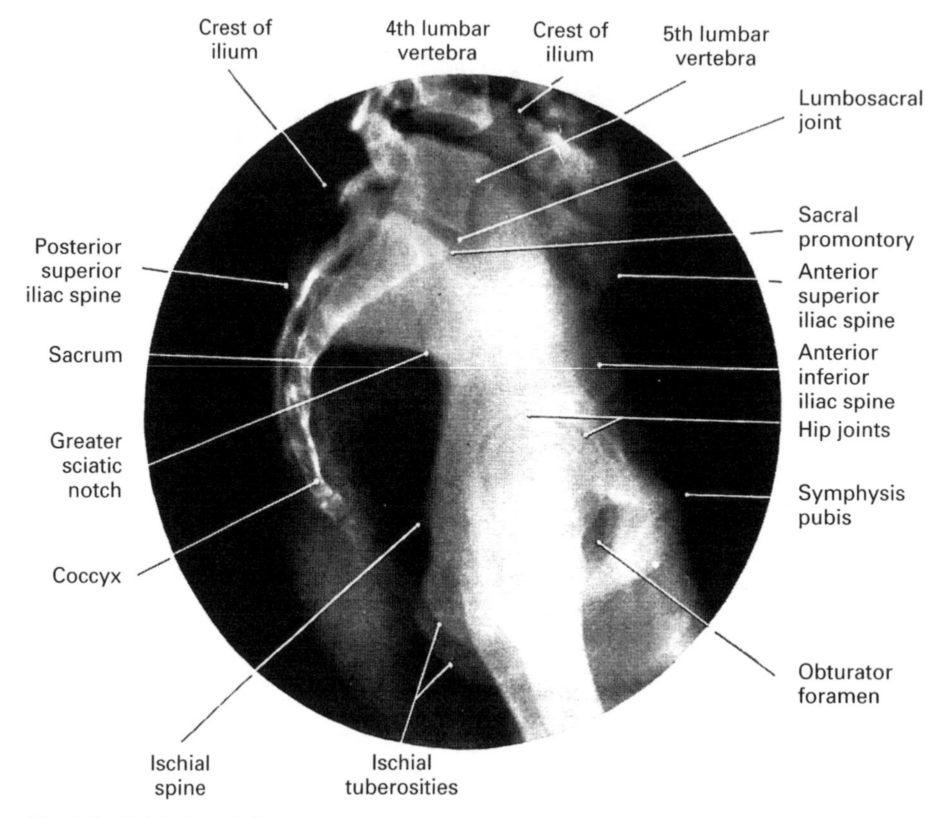

Fig. 7.8 Pelvis: lateral view

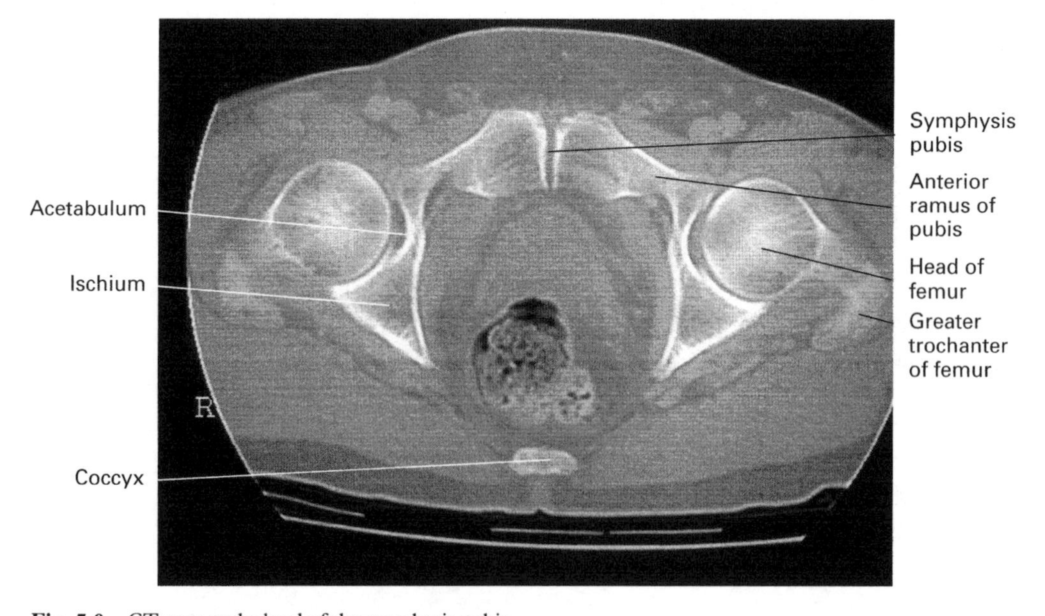

Fig. 7.9 CT scan at the level of the symphysis pubis

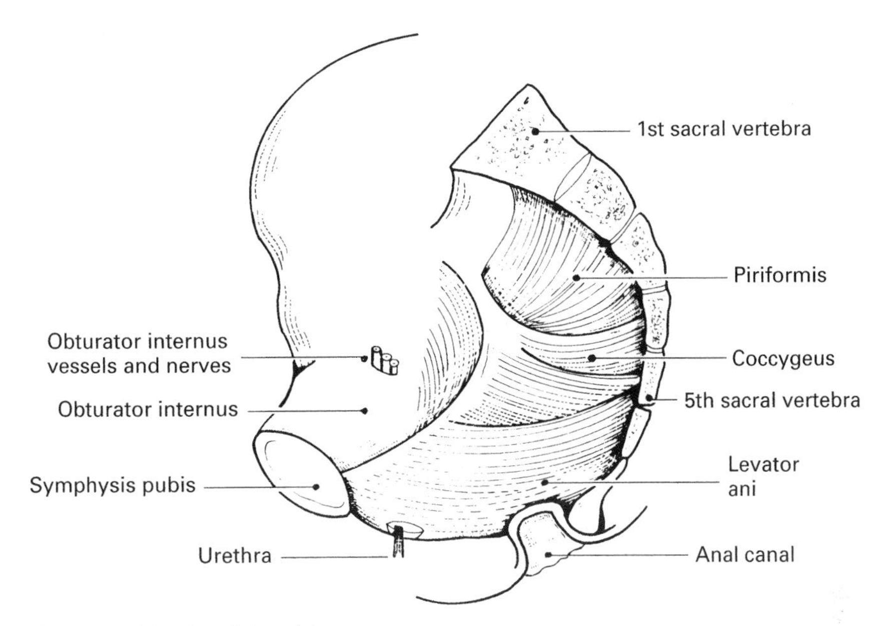

Fig. 7.10 Muscles of the pelvis

coccyx. It unites with the muscle from the other side, in the midline, by a raphé.

The coccygeus is a small triangular muscle which lies posterior to the levator ani. It arises from the ischial spine and is inserted into the base of the coccyx.

The muscles are covered by a sheet of fibrous tissue—the pelvic fascia. They are perforated in the midline by the urethra, vagina and anus in the female and by the urethra and anus in the male.

The obturator internus and the piriformis muscles arise from the lateral and posterior walls of the true pelvis. They pass through the lesser and the greater sciatic foramina.

LIGAMENTS OF THE PELVIS

The sacrospinous ligament is triangular. It is attached by its apex to the spine of the ischium and by its base to the lateral aspect of the lower sacral and coccygeal vertebrae.

The sacrotuberous ligament is attached to the posterior iliac spines, lateral sacral tubercles, lateral coccyx and the medial aspect of the ischial tuberosities.

The sacrospinous and sacrotuberous ligaments convert the greater and lesser sciatic notches into foramina.

Accessory ligaments

The inguinal ligament is attached between the anterior superior iliac spine and the pubic tubercle. It is formed by the inferior border of the external oblique muscle of the abdomen.

The pectineal ligament passes along the upper border of the superior pubic ramus (the pecten pubis). It is firmly attached to the periosteum. It meets the inguinal ligament medially at the pubic tubercle and gives origin to the pectineus muscle.

The lacunar ligament is a short triangular ligament which joins the medial part of the inguinal ligament and the pecten pubis.

The obturator membrane is a fibrous membrane which closes the obturator foramen except for a small opening superiorly. Through this opening the obturator vessels and nerve pass from the pelvis to the thigh. It gives origin to the obturator internus muscle within the pelvis and to the obturator externus on its outer surface.

JOINTS OF THE PELVIS

In childhood and early adolescence the three bones making up the innominate bone (ilium, ischium and pubis) are joined by primary cartilaginous joints (synchondroses). These three bones meet in the centre of the acetabulum. The union of the bones, to form synostoses, occurs at about 16 years of age.

The joints between the two halves of the pelvis are the symphysis pubis and the sacroiliac joints. The hip joint is described on page 186.

SYMPHYSIS PUBIS (Fig. 7.9)

Type: Secondary cartilaginous (symphysis or amphiarthrosis) with an intervening disc of fibrocartilage.

Articular surfaces: Pubic bodies and articular disc. Hyaline cartilage covers the pubic articular surfaces.

Ligaments:

Anterior, *Arcuate* — strong, passing between pubic bones and attached to disc. The anterior ligament passes between the pubic tubercles. The arcuate joins inferior pubic rami.

Superior, *Posterior* — weaker than the above.

Movements: The joint takes up the stresses applied to the pelvis, notably from the hip joints. Slight gliding movement and widening of the joint occur during pregnancy.

SACROILIAC JOINTS (Figs 3.35, 7.11, 7.12)

Type: Synovial, paired, plane. More irregular in adulthood. 'Locks' like a jigsaw.

Articular surfaces: Auricular (earshaped) surfaces of ilium and sacrum. Sacral surface is covered by fibrocartilage. Iliac surface is covered by hyaline cartilage.

Ligaments:

Capsular—attached to margins of auricular surfaces.

Interosseous—Deep and superficial parts forming the major strength of the joint. Unites bone directly between the roughened area posterior and superior to auricular surfaces.

Anterior—connects periphery of auricular surfaces. It is a thickening of the capsule.

Posterior—short and long portions. Superficial to interosseous ligament. Short fibres unite ilium to upper transverse tubercles of sacrum. Long fibres pass more vertically between posterior superior iliac spine and lower transverse tubercles of sacrum.

Movements: A small rotational movement occurs through the transverse axis of the joint in response to movements of the vertebral column. Pregnancy causes increased movements.

Radiographic appearances of the sacroiliac joint (Figs 7.11, 7.12; see also Fig. 3.35)

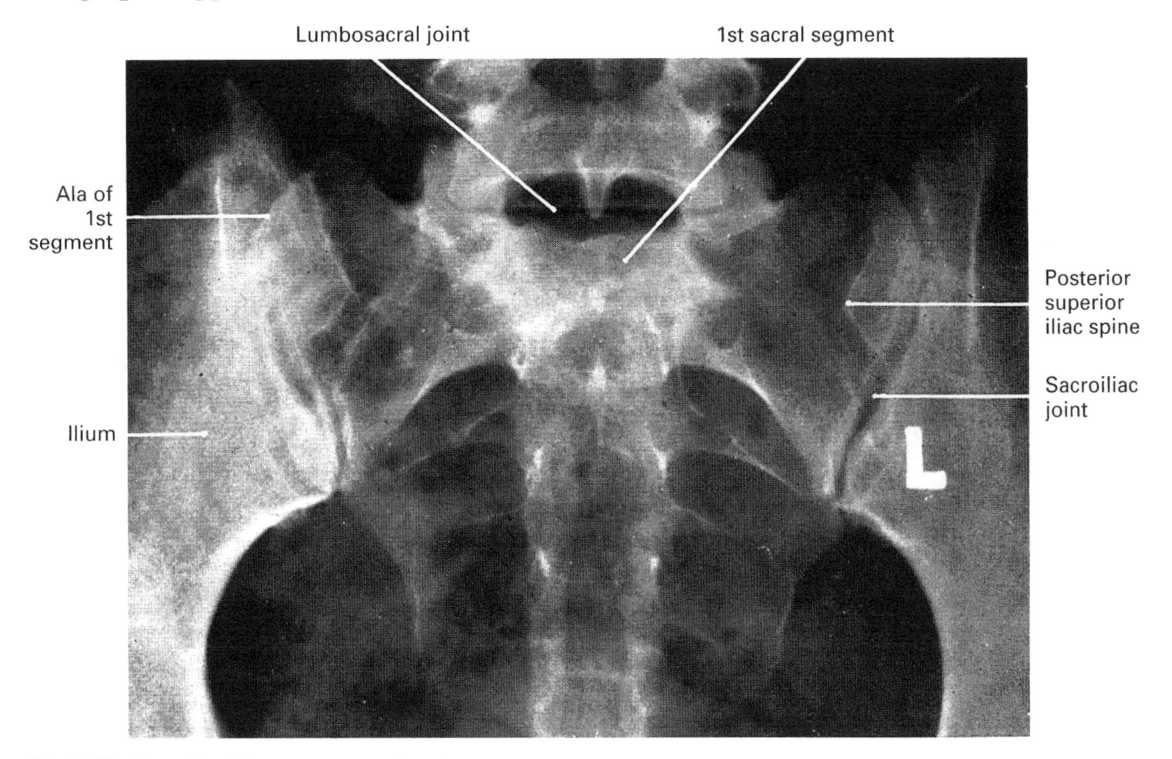

Fig. 7.11 Sacroiliac joints: anteroposterior view

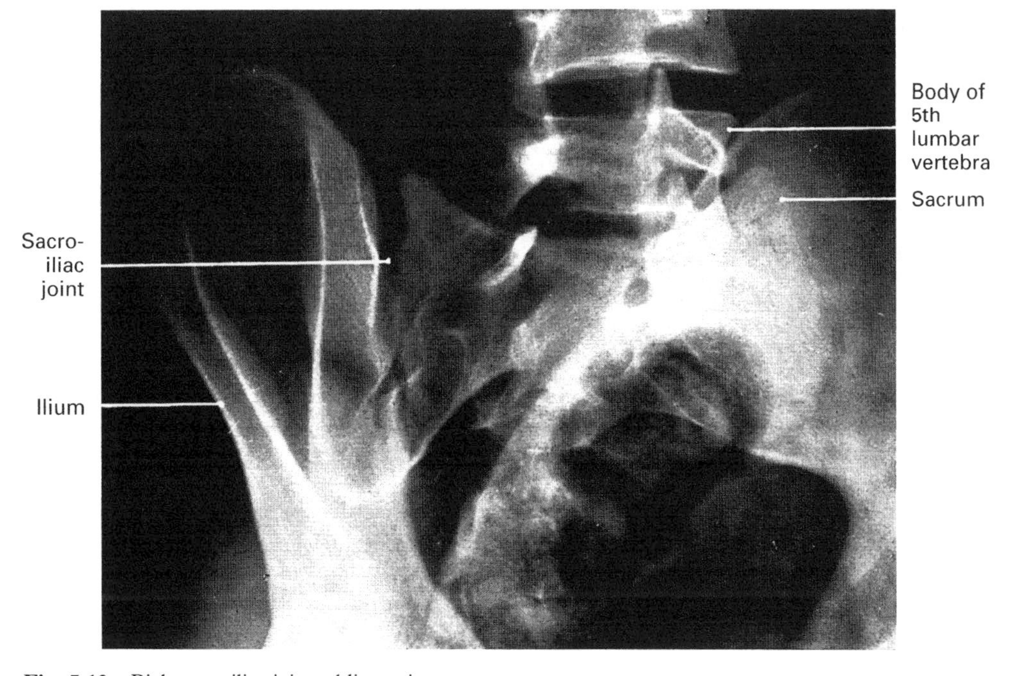

Fig. 7.12 Right sacroiliac joint: oblique view

OSSIFICATION OF THE INNOMINATE BONE (Figs 7.13–7.16)

The three components of the innominate bone—ilium, ischium and pubis—ossify from separate primary centres in early intrauterine life. At first they are separated in the acetabulum by the 'Y-shaped cartilage'.

Primary centres

- for the ilium, appears at about week 8 of intrauterine life
- for the ischium, at about week 12 of intrauterine life
- for the pubis, at about week 18 of intrauterine life.

Secondary centres

- for iliac crest.
- for anterior inferior iliac spine.
- for ischial tuberosity.
- for symphysis pubis.
- in the 'Y-shaped cartilage' (one or more centres).

These epiphyses appear about puberty and unite with the innominate bone between the 20th and 23rd years, with the exception of those in the Y-shaped cartilage which unite soon after puberty.

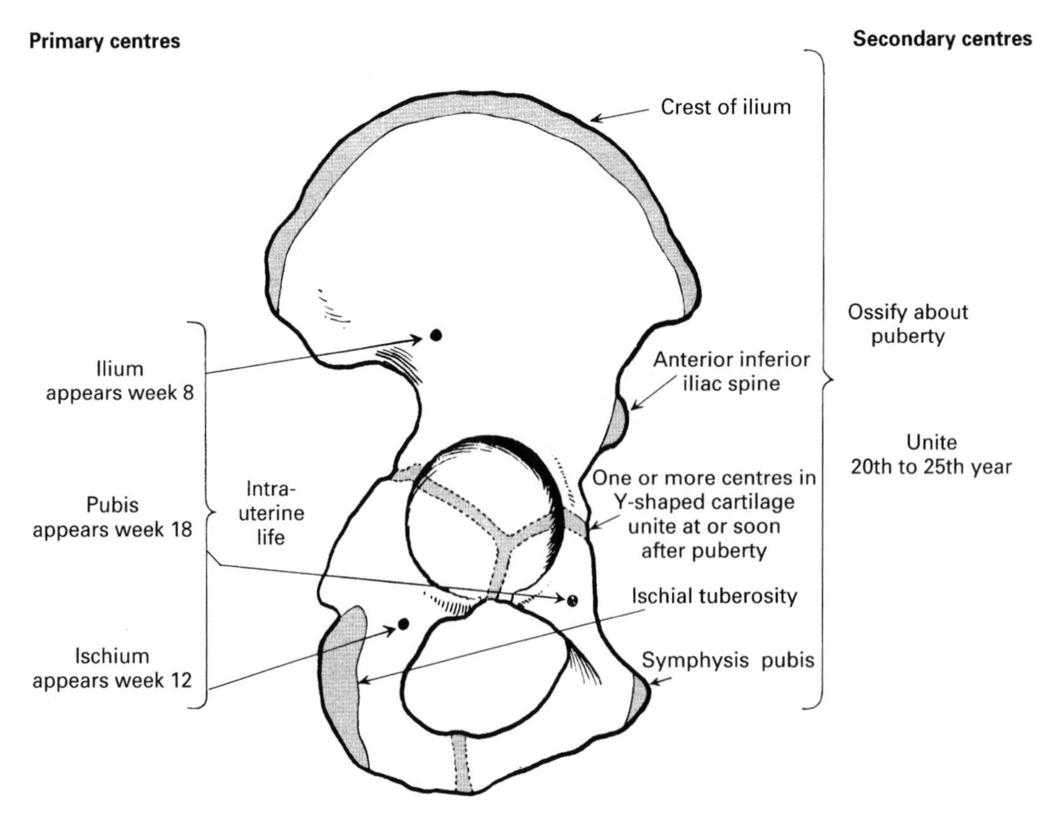

Fig. 7.13 Ossification of the innominate bone

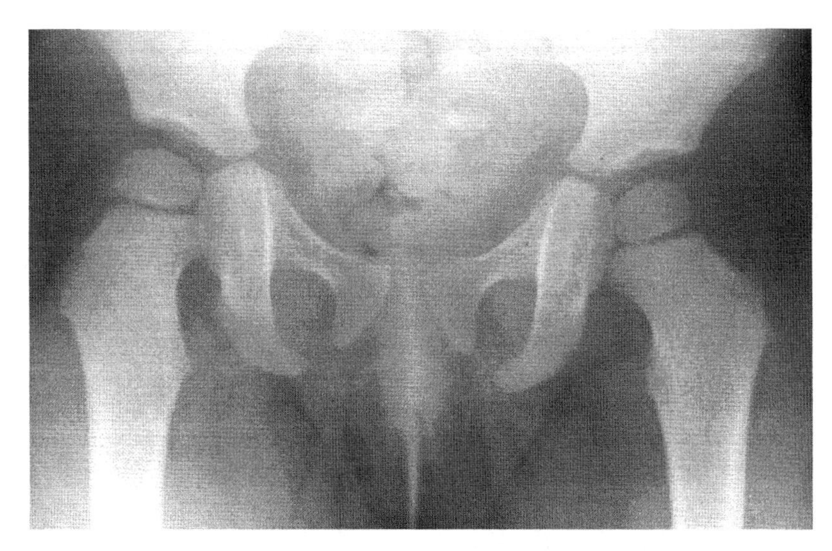

Fig. 7.14 Ossification of the innominate bone; 18 months female. There is complete bony separation of the three parts of the hip bone. The acetabulum and femoral head are still cartilaginous

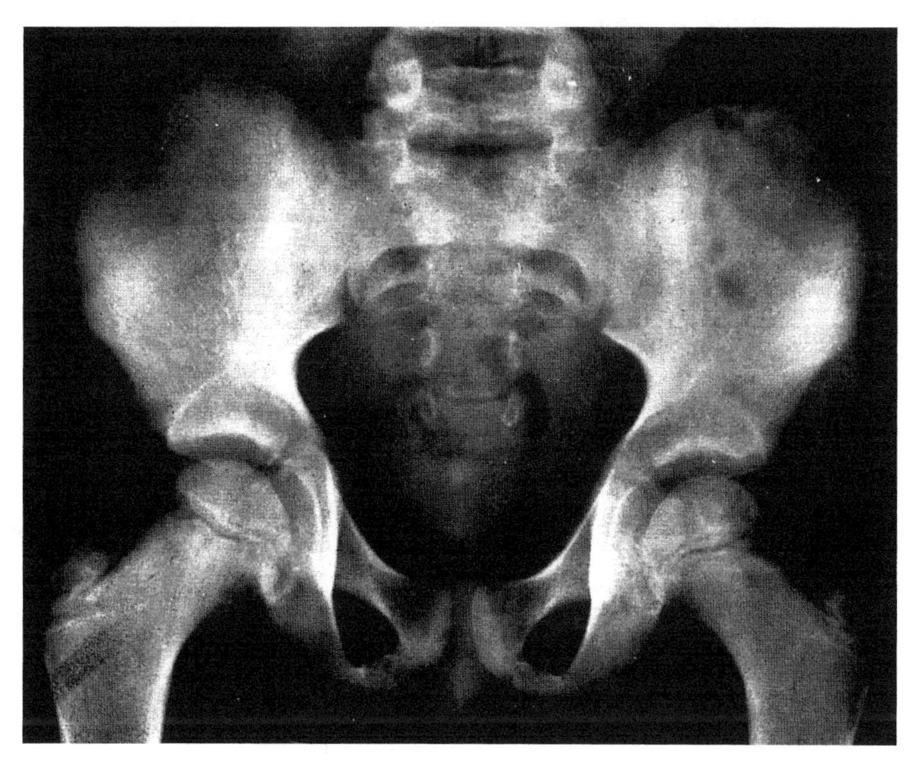

Fig. 7.15 Ossification of the innominate bone; 7 years. The 'Y-shaped cartilage' is still present and the femoral head and the acetabulum are well formed. Secondary centres have not yet appeared

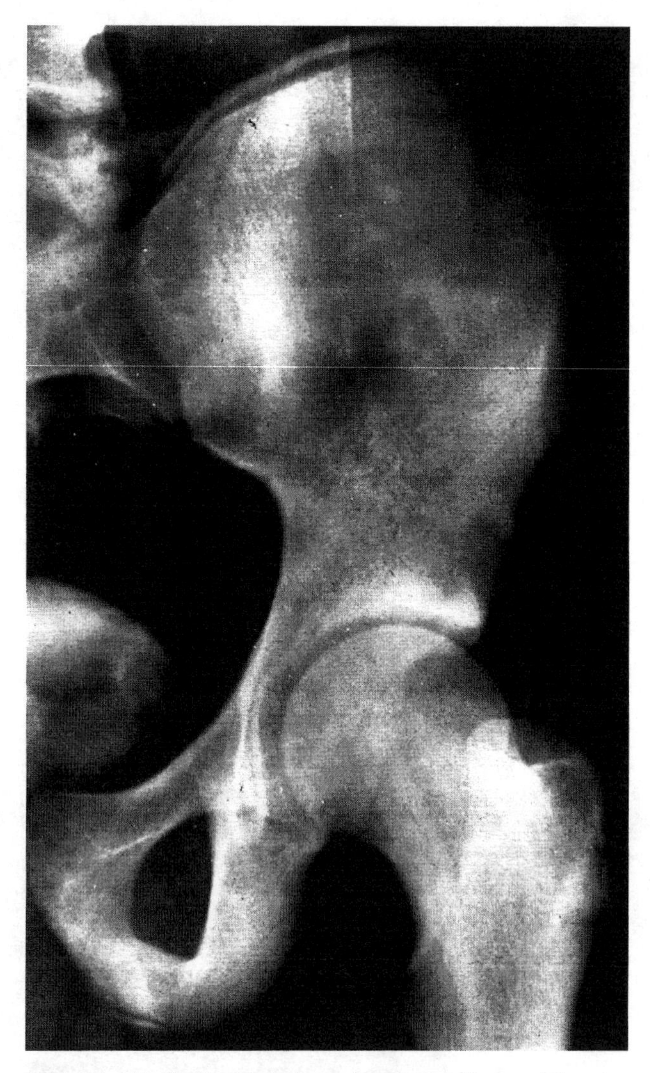

Fig. 7.16 Ossification of the innominate bone; 18 years. The epiphysis for the iliac crest is still present

THE SKELETAL PELVIS (Figs 7.17 to 7.22)

The pelvis is usually considered as divisible into two parts:

- the greater (or false) pelvis.
- the lesser (or true) pelvis.

The two parts are divided by an oblique plane, the pelvic brim. The false pelvis lies above the pelvic brim, the true pelvis below it. The two parts are in fact continuous, as are the pelvic viscera they contain.

The false pelvis

The cavity of the false pelvis is part of the abdomen. Because of the angle of the pelvis, the false pelvis has very little bony wall anteriorly.

The true pelvis

It is smaller than the false pelvis and it contains the internal organs of reproduction, the urinary bladder, the pelvic colon and the rectum. The size and shape of the true pelvis in the female

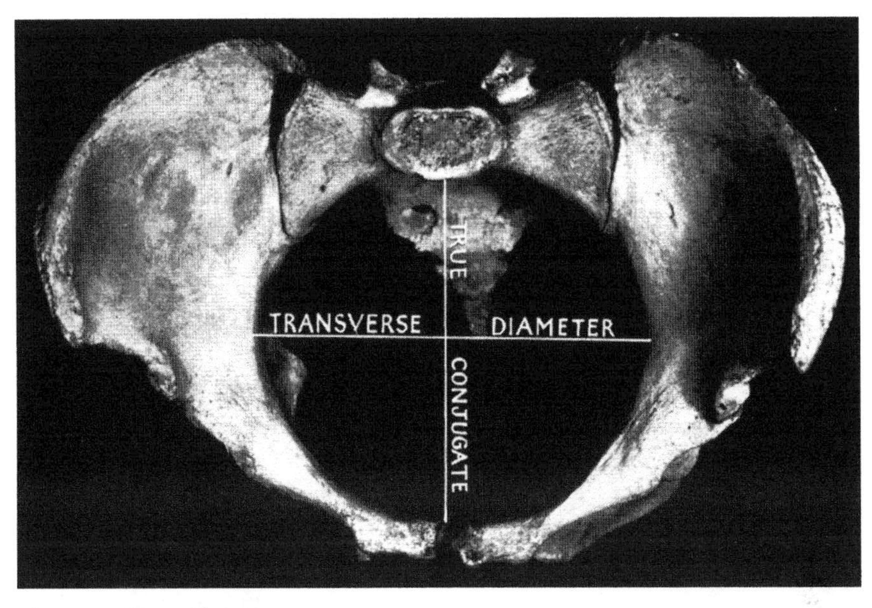

Fig. 7.17 The pelvic inlet

exert a great influence on the mechanism of labour.

The true pelvis can be divided into:

- the pelvic inlet
- the pelvic cavity
- the pelvic outlet.

The pelvic inlet (superior pelvic aperture) (Figs 7.17 and 7.20) is rounded or oval in shape. Its boundaries are the sacral promontory, the anterior border of the alae of the sacrum, the iliopectineal line, the pubic crest and the superior border of the posterior surface of the symphysis pubis.

The pelvic cavity (Figs 7.18 and 7.21) is a curved canal extending from the pelvic inlet above to the pelvic outlet below. It is deeper posteriorly than anteriorly and it contains the pelvic organs. It is bounded anteriorly by the pubic bones, posteriorly by the sacrum and coccyx and laterally by the ilia and ischia.

The pelvic outlet (inferior pelvic apertur) (Figs 7.19 and 7.22) is less smooth than the inlet because it is indented by the ischial tuberosities and by the sacrum and coccyx. It is roughly diamond-shaped. It is bounded by the apex of the sacrum, the sacrotuberous ligament, the ischial tuberosities, the ischial rami, the inferior pubic rami and the posterior surface of the inferior border of the symphysis pubis.

There are three corresponding planes (Fig. 7.21):

Plane of the inlet (also known as the inclination of the pelvis). In the standing subject this plane is inclined at an angle of 55° to the horizontal.

Midplane of the pelvic cavity. This lies on a line from the centre of the symphysis pubis to the centre of the third sacral vertebra.

Plane of the outlet. This extends from the inferior border of the symphysis pubis to the apex of the sacrum.

MEASUREMENTS IN OBSTETRIC RADIOGRAPHY

In obstetric radiology, measurements of the pelvic inlet, cavity and outlet are made to assess the dimensions and shape of the pelvic canal. These measurements are taken in the three planes described above. The most important are:

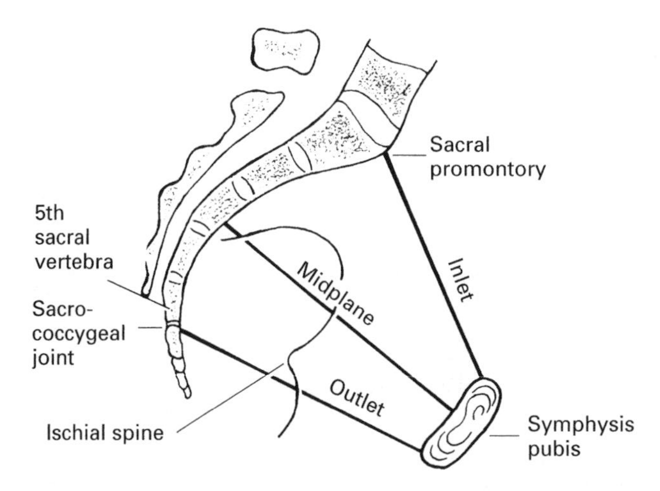

Fig. 7.18 Diagram of the pelvic cavity

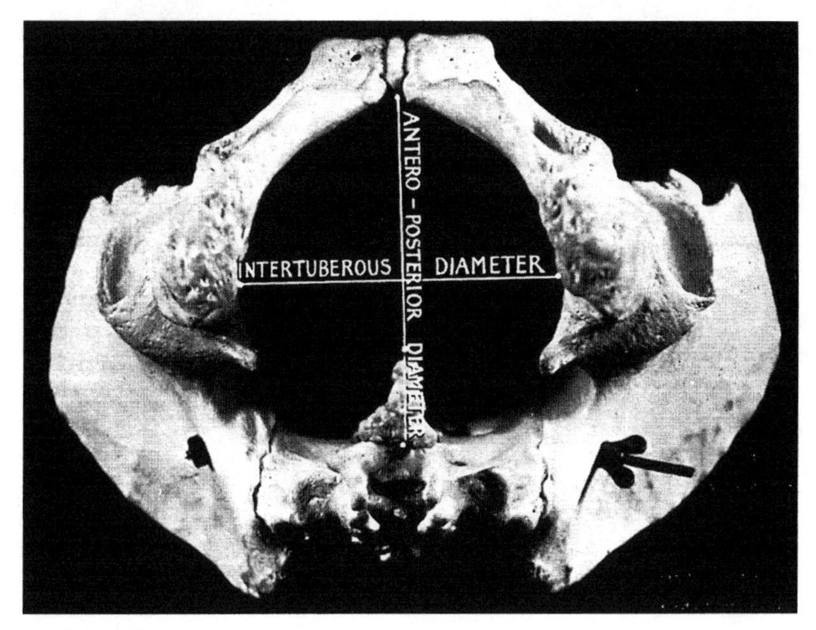

Fig. 7.19 The pelvic outlet

Plane of the inlet

Anteroposterior diameter (true conjugate)—measured from the sacral promontory to the posterior edge of the upper border of the symphysis pubis (112 mm).

Transver diameter—the widest transverse measurement of the pelvic brim (125 mm).

Plane of the outlet

Anteroposterior diameter—measured from the mid-point of the fifth sacral segment to the posterior surface of the lower border of the symphysis pubis (130 mm).

Transverse diameter (intertuberous diameter)—the distance between the medial surfaces of the ischial tuberosities (118 mm).

These measurements represent the average, but there are many variations in size and shape of the female pelvis. Four main types are usually recognized, as shown in Table 7.1 (p. 186).

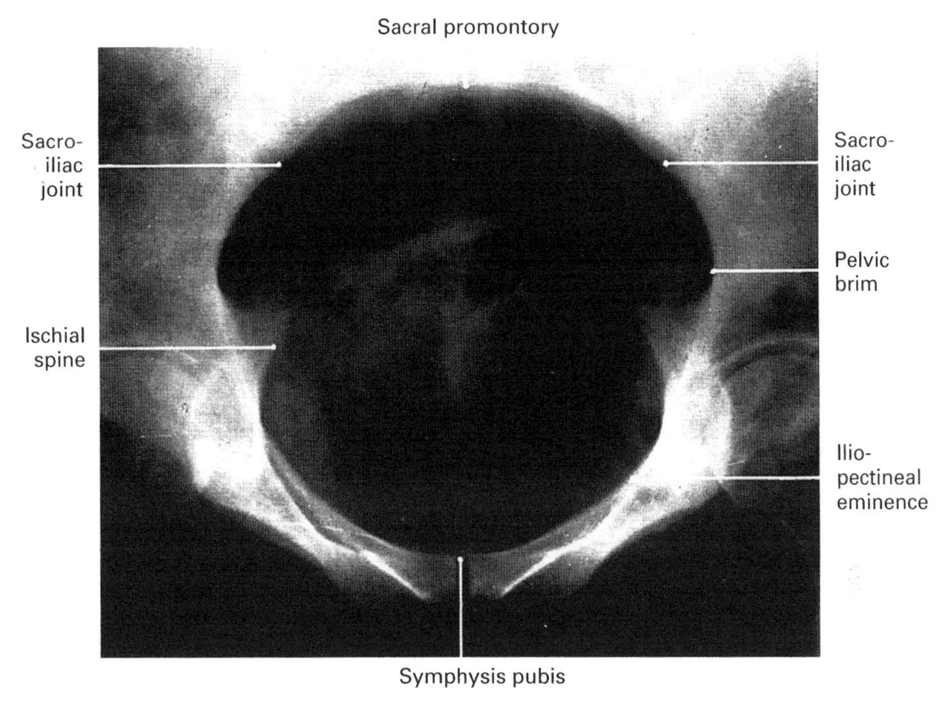

Fig. 7.20 Pelvic inlet: superoinferior view

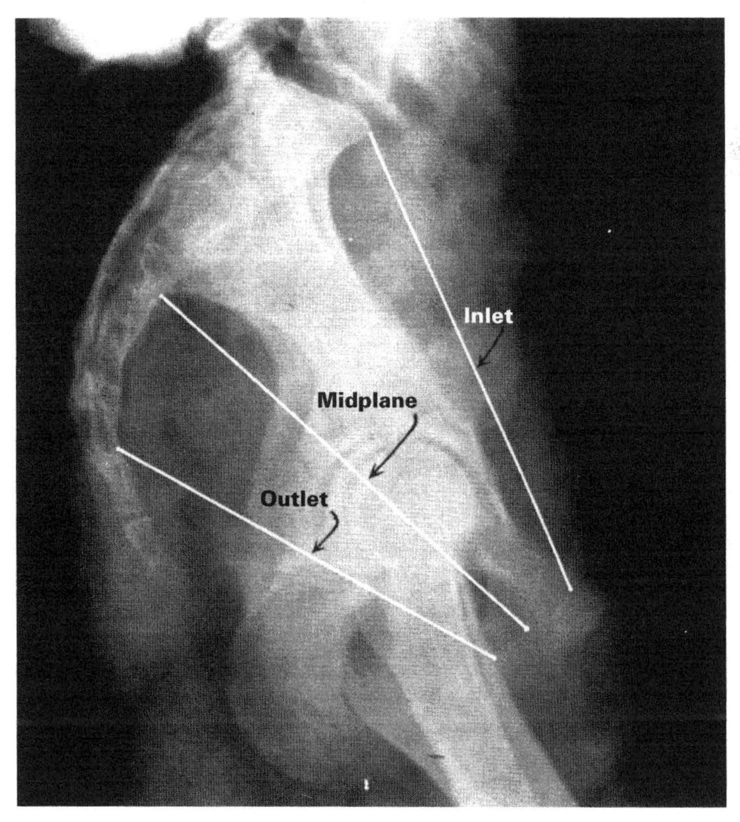

Fig. 7.21 Pelvic canal: lateral view

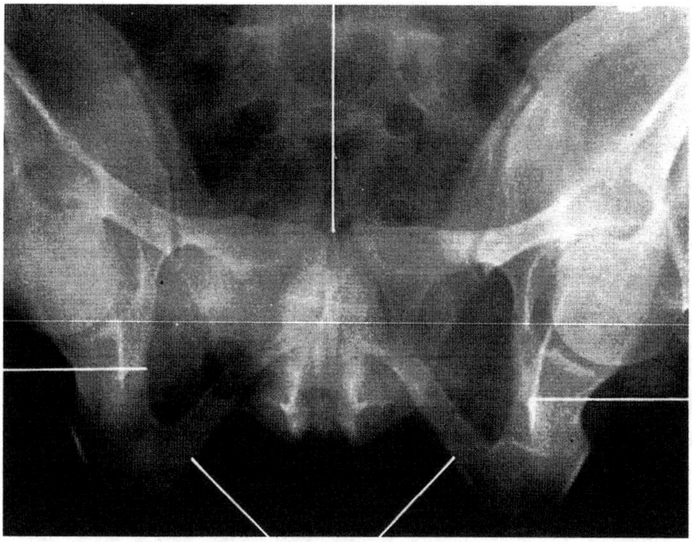

Symphysis pubis

Lateral border of obturator foramen

Pelvic surface of ischium

Pubic arch

Fig. 7.22 Pelvic outlet: superoinferior view

Table 7.1

	Shape of inlet	*Midplane of cavity*	*Outlet*	
			Subpubic angle	*Greater sciatic notch*
Gynaecoid (typical female type)	Rounded. Transverse diameter slightly greater than antero-posterior diameter	Side walls straight	Wide	Average to wide
Platypelloid (flat pelvis)	Anteroposterior diameter much shorter than transverse diameter	Side walls usually straight	Average to wide	Narrow
Anthropoid	Transversely narrow	Side walls straight or divergent	Average to wide	Wide
Android (typical male type)	Wedge-shaped or triangular	Side walls convergent	Narrow	Narrow

These four main types form a basis for classification only and there are many intermediate forms, e.g. a pelvis may have a large, rounded inlet (gynaecoid) with a cavity funnelling to a narrow outlet (android).

HIP JOINT (Figs 7.23 to 7.30)

Type: Synovial, ball and socket.

Articular surfaces: The head of femur and acetabulum of innominate bone. The depth of the acetabulum is deepened by a fibrocartilaginous acetabular labrum.

Synovial membrane: Lines capsular ligament, medial part of femoral neck up to acetabular margin of femoral head and acetabular labrum.

Ligaments:

Capsular—thick and strong, particularly anteriorly where it is attached to trochanteric line. Posteriorly, attached to junction of middle and distal thirds of

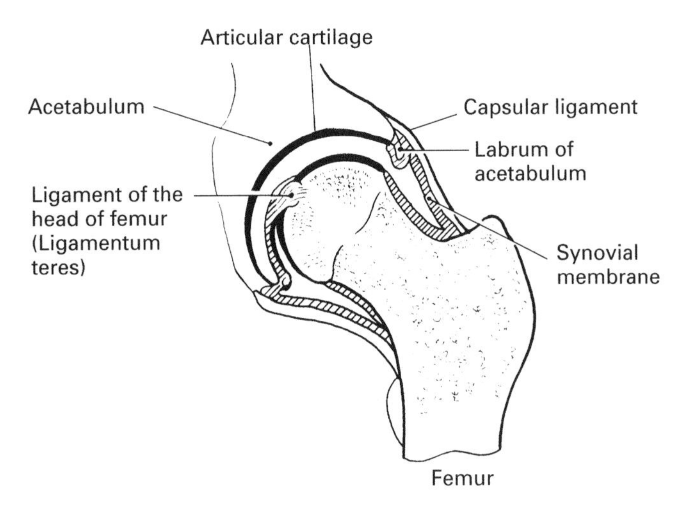

Fig. 7.23 Schematic diagram of the hip joint

femoral neck medial to trochanteric crest and to superior and inferior surfaces of neck, close to trochanters. Medially attached to acetabular rim, acetabular labrum and transverse acetabular ligament. Covers that part of femoral neck within joint.

Iliofemoral—Y-shaped and very strong. Attached above to anterior inferior iliac spine and passes to trochanteric line of femur.

Ischiofemoral—posterior. Spiral course. Attached to ischium behind acetabulum. Passes to neck of femur leaving an area medial to trochanteric crest. Synovial membrane may protrude laterally.

Pubofemoral—inferior, attached to iliopectineal eminence of innominate bone, to superior ramus of pubis and obturator crest. Joins capsular ligament inferiorly.

Ligamentum teres—attached to fovea of head of femur and to acetabular rim inferiorly. Is intracapsular and ensheathed by synovial membrane.

Transverse acetabular—bridges acetabular notch, forming foramen for vessels and nerves.

Zona orbicularis—irregular fibres which encircle femoral neck superficial to synovial membrane.

Retinacular fibres—longitudinal fibres running parallel with femoral neck and firmly bound to it. Essential vessels to the femoral head accompany them.

(It should be noted that the ilio-, ischio- and pubofemoral ligaments form a spiral around the femoral neck so that they tighten when the femur is extended, thus limiting this movement.)

Blood supply: Superior and inferior gluteal, obturator and medial femoral circumflex arteries.

Nerve supply: Branches from femoral, obturator and superior gluteal nerves.

Movements and muscles:

Flexion—psoas major, rectus femoris, (pectineus, sartorius).

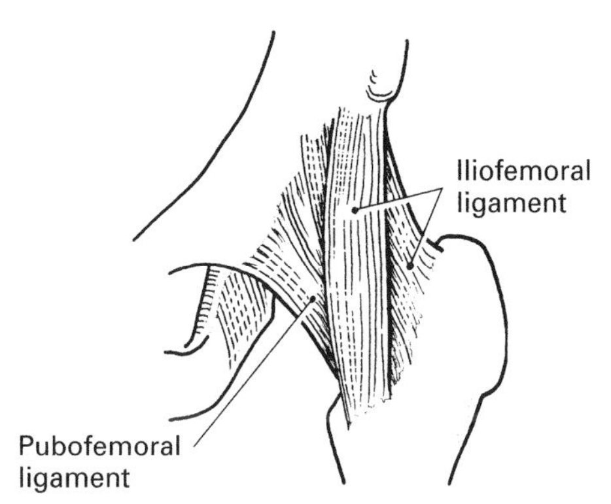

Fig. 7.24 Ligaments of the hip joint: anterior view

Extension—gluteus maximus, hamstrings.
Abduction—gluteus medius, gluteus minimus (tensor fasciae latae).
Adduction—adductus longus, adductor magnus, adductor brevis (pectineus, gracilis).
Medial rotation—gluteus medius, gluteus minimus.
Lateral rotation—obturator externus and internus, quadratus femoris, (gluteus maximus, piriformis, sartorius).
Circumduction—combination of all these movements.

Flexors

Psoas major:
origin—transverse processes of L1–L5, lateral side of vertebral bodies and intervertebral discs of T12–L4.
insertion—lesser trochanter of femur, with tendons of iliacus.
function—flexion and lateral rotation of hip, lateral flexion of spine.
nerve supply—from lumbar nerves.

Iliacus:
origin—iliac crest and iliac fossa.
insertion—joins tendon of psoas major and also inserts directly into lesser trochanter.
function—flexion and lateral rotation of hip.
nerve supply—femoral nerve.

Rectus femoris (part of Quadriceps group):
origin—anterior inferior iliac spine.
insertion—superior border of patella.
function—flexion of hip and extension of knee.
nerve supply—femoral nerve.

Extensors

Hamstrings: These comprise the biceps femoris, semitendinosus and semimembranosus.

Biceps femoris:
origin
- long head—ischial tuberosity with semimembranosus and sacrotuberous ligament.
- short head—linea aspera, lateral aspect.

insertion—head of fibula, lateral condyle of tibia.

Semitendinosus:
origin—ischial tuberosity, medial aspect.
insertion—medial aspect of tibia, below condyle.

Semimembranosus:
origin—ischial tuberosity, superior aspect.
insertion—medial tibial condyle posteriorly.
nerve supply (all hamstrings)—tibial branch of sciatic nerve.

Gluteus maximus (forms the prominence of the buttock. It is the most superficial of the gluteal muscles):
origin—outer surface of iliac bone.
insertion—quadrate tubercle of upper part of femur posteriorly.
function—extension and lateral rotation of thigh, particularly in standing erect and climbing.
nerve supply—inferior gluteal nerve.

Abductors

Gluteus minimus:
origin—outer surface of anterior ilium between anterior and inferior gluteal lines.
insertion—greater trochanter of femur anteriorly, separated from it by a bursa.
function—abduction and medial rotation of hip.
nerve supply—superior gluteal nerve.

Gluteus medius:
origin—outer surface of ilium, between anterior and posterior gluteal lines and over lying fascia.
insertion—lateral surface of greater trochanter (separated by a bursa).
function—abduction and medial rotation of hip.
nerve supply—superior gluteal nerve.

Tensor fasciae latae:
origin—anterior iliac crest and area between superior and inferior iliac spines.
insertion—via iliotibial tract to lateral tibial condyle.
function—abduction, flexion, medial rotation of hip. Aids maintenance of extension of knee.
nerve supply—superior gluteal branch of sciatic plexus.

Adductors

Adductor longus:
- origin—anterior pubis, below pubic tubercle.
- insertion—medial aspect of linea aspera on posterior femur.
- function—adduction, flexion and lateral rotation.
- nerve supply—anterior division of obturator nerve.

Adductor magnus:
- origin—inferior ischiopubic ramus.
- insertion—medial aspect of linea aspera down to adductor tubercle of medial femur.
- function—adduction and some flexion.
- nerve supply—anterior division of lumbosacral plexus.

Adductor brevis:
- origin—inferior pubic ramus.
- insertion—proximal linea aspera and pectineal line of posterior femur.
- function—adduction, lateral rotation.
- nerve supply—obturator nerve.

Pectineus:
- origin—upper border of pubis (pecten pubis).
- insertion—pectineal line of proximal posterior femur.
- function—adduction, lateral rotation and flexion.
- nerve supply—femoral nerve.

Gracilis:
- origin—inferior pubic rami.
- insertion—proximal medial tibia below condyle.
- function—adduction of thigh, flexion and medial rotation of tibia.
- nerve supply—obturator nerve.

Lateral rotation

Obturator internus:
- origin—inner aspect of ischiopubic ramus and obturator membrane.
- insertion—tendon emerges via lesser sciatic foramen. Separated from ischium by a bursa and inserts medially on greater trochanter.
- nerve supply—from sciatic plexus.

Obturator externus:
- origin—outer aspect of ischiopubic ramus and obturator membrane.
- insertion—trochanteric fossa.
- nerve supply—obturator nerve.

Quadratus femoris:
- origin—ischial tuberosity.
- insertion—quadrate tubercle of trochanteric crest.
- nerve supply—from sciatic plexus.

Piriformis:
- origin—anterior sacrum between anterior foramina.
- insertion—emerges through greater sciatic foramen; tendon inserts into superior aspect of greater trochanter.
- nerve supply—from sciatic plexus.

Radiographic appearances of the hip joint (Figs 7.25 to 7.30)

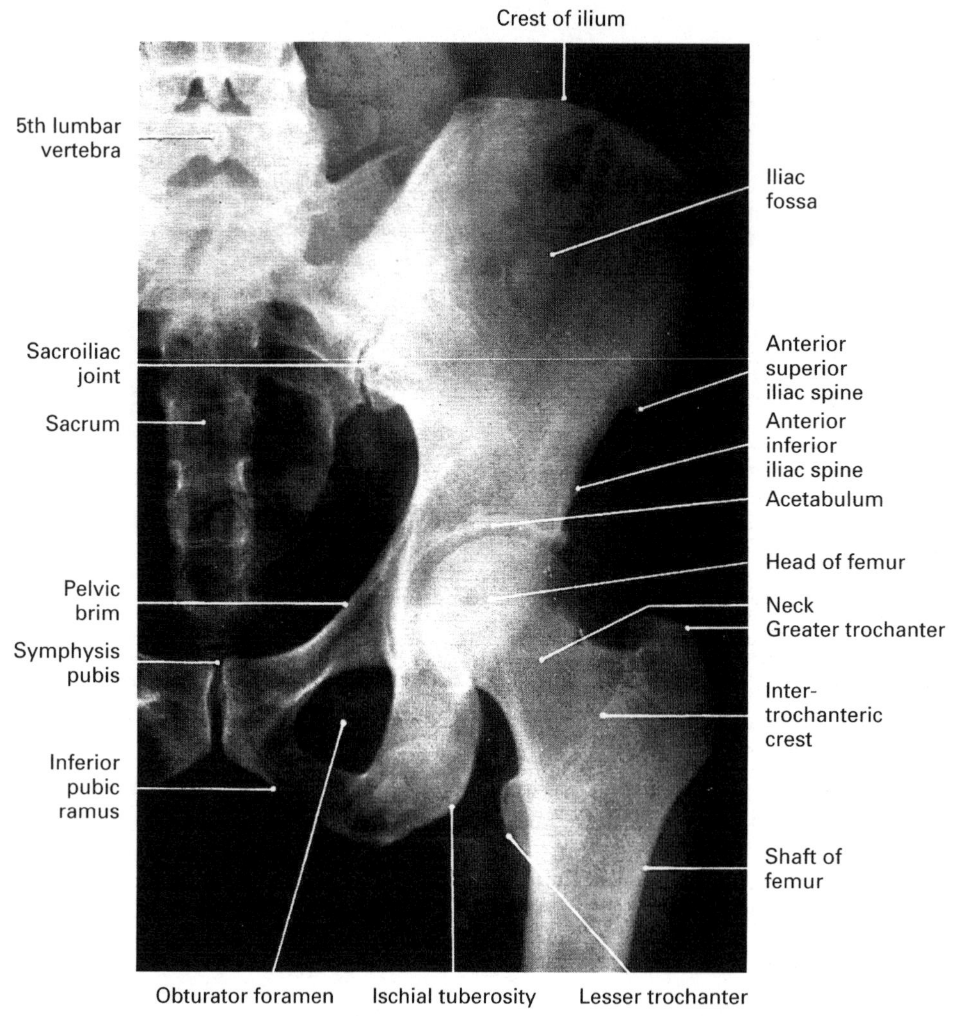

Fig. 7.25 Left hip joint: anteroposterior view

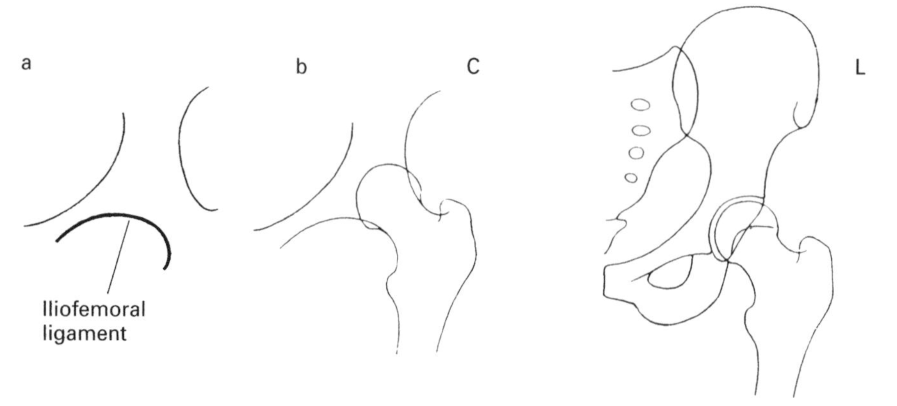

Fig. 7.26 Anteroposterior view of hip. Construction of diagram. (a) Three guide lines are used—i. the curve of the anterior border of the ilium and the upper border of the neck of the femur, ii. the upper border of the obturator foramen and the lower border of the neck of the femur (Shenton's line) and iii. the pelvic brim. (b) The head, neck and trochanters are added. (c) The rest of the innominate bone and sacrum are drawn

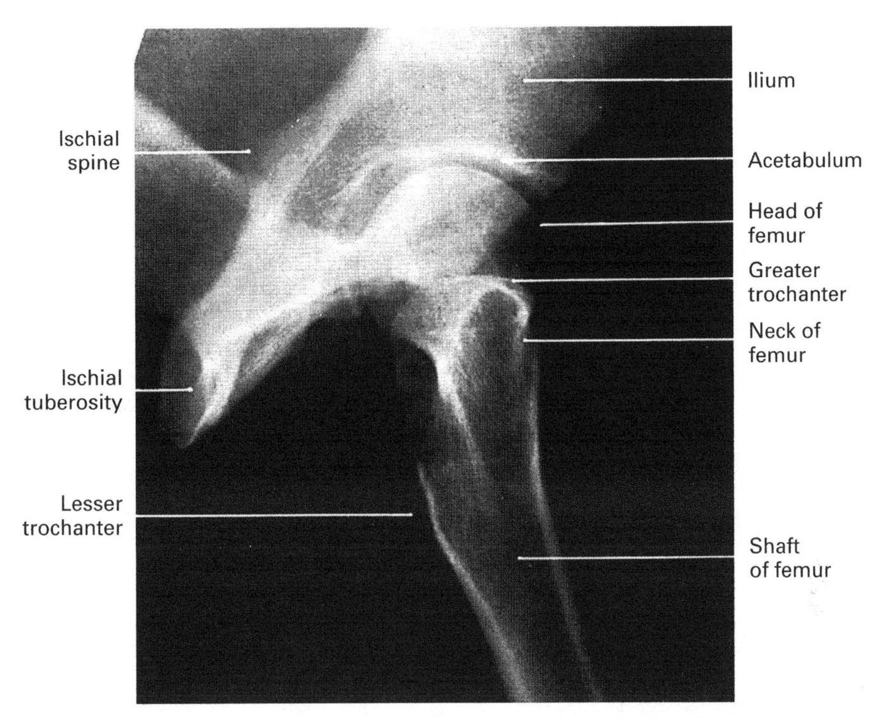

Fig. 7.27 Left hip joint: lateral view

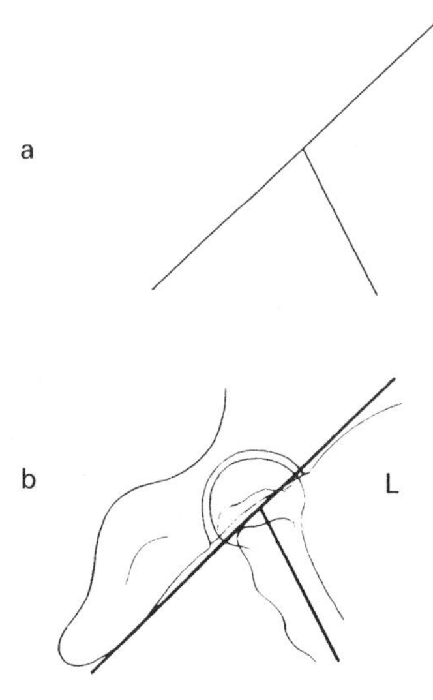

Fig. 7.28 Lateral view of hip. Construction of diagram. (a) Two guide lines are used. The first represents the general line of the hip bone. The second forms the head, neck and shaft of the femur. (b) The acetabulum and femoral head are drawn at the junction of the lines. The neck, greater trochanter and shaft are drawn round the second guide line

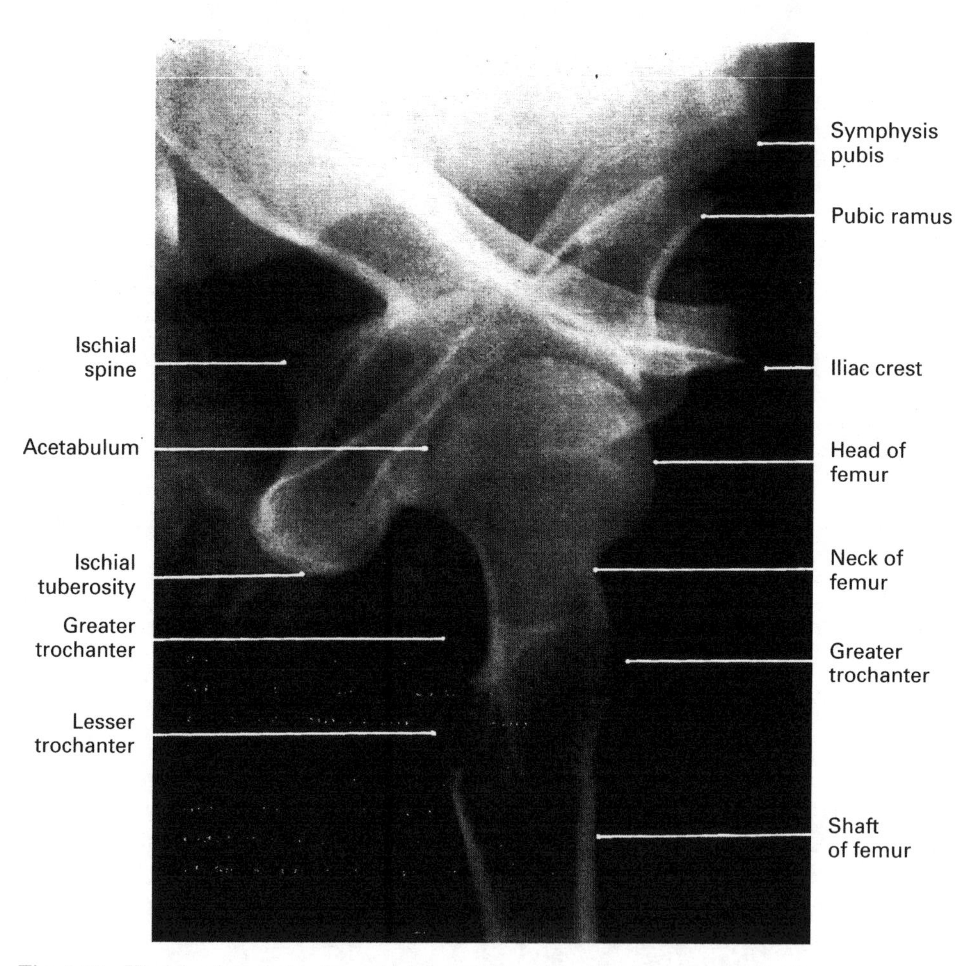

Fig. 7.29 Hip joint: lateral view for neck of femur. The neck of the femur is now clearly demonstrated as it is not superimposed by the greater trochanter as in Fig. 7.27

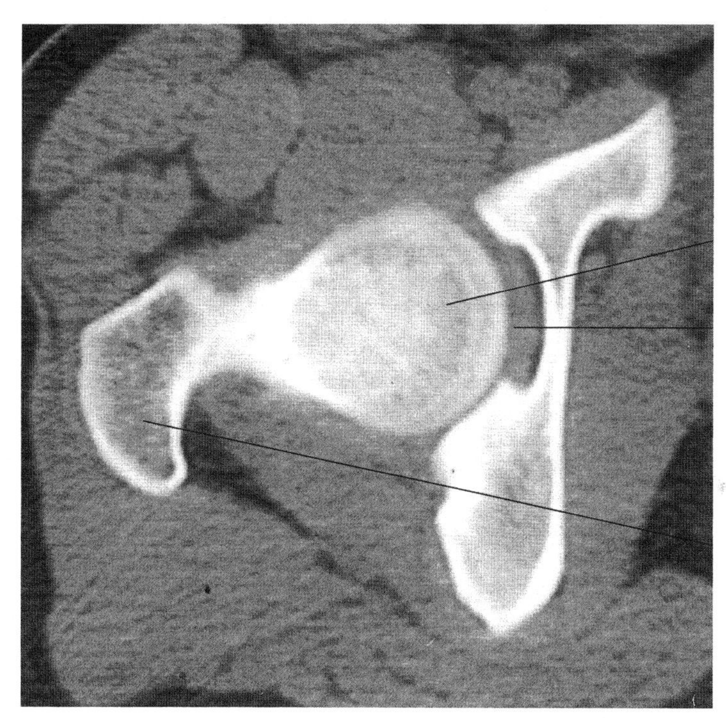

Fig. 7.30 Right hip joint: CT scan

8. The lower limb

The bones of the lower limb are those of the thigh, leg and foot, with the intervening joints of the hip, knee, ankle and foot.

THIGH

FEMUR (Figs 8.1 to 8.7)

The femur, or thigh bone, is the longest and strongest bone in the body. It consists of:

- the upper end
- the shaft
- the lower end.

The upper end has a rounded head for articulation with the acetabulum of the pelvis at the hip joint (p. 186). The lower end articulates with the tibia at the knee joint (p. 205). In the standing position, the femora are oblique inferiorly and medially. The heads are separated by the width of the lesser pelvis and the shafts slope medially so that the medial sides of the knees are almost in contact. The degree of medial obliquity varies in different individuals; it is greater in women because of the greater width of the female pelvis.

The upper end consists of:

- the head
- the neck
- the greater trochanter
- the lesser trochanter.

The head forms about two-thirds of a sphere and is directed upwards and medially to articulate with the acetabulum of the hip bone. Its surface is smooth and is covered with articular cartilage, except for a small roughened depression, the fovea, below the centre. The ligament of the head of the femur, ligamentum teres, extends from the fovea to the sides of the acetabular notch.

The neck is about 50 mm long and it joins the shaft at an angle of approximately 125°—the angle of inclination—which ensures that the lower limb swings free of the pelvis. The neck is slightly flattened giving an anterior and a posterior surface and upper and lower borders. The upper border of the neck is short, almost horizontal and slightly concave. The lower border is longer and runs obliquely downwards.

On the anterior surface of the junction of the neck and the shaft is a bony ridge—the intertrochanteric line. On the posterior surface is the prominent intertrochanteric crest. The trochanters are situated one at each end of this crest. In the middle of the intertrochanteric crest there is a small bony elevation, the quadrate tubercle.

The strong capsular ligament of the hip joint, which arises from the acetabulum, is attached below to the base of the neck, with a more extensive attachment to the anterior surface than to the posterior.

The greater trochanter is a prominence projecting upwards and laterally from the junction of the neck and the shaft. It is quadrilateral in shape and has a roughened surface for insertion of most of the muscles of the buttock. The upper posterior part of the greater trochanter extends medially and overhangs a roughened, depressed area—the trochanteric fossa—into which is inserted the obturator externus muscle of the pelvis.

The lesser trochanter is a small conical prominence projecting medially from the lower end of the intertrochanteric crest at the junction of the

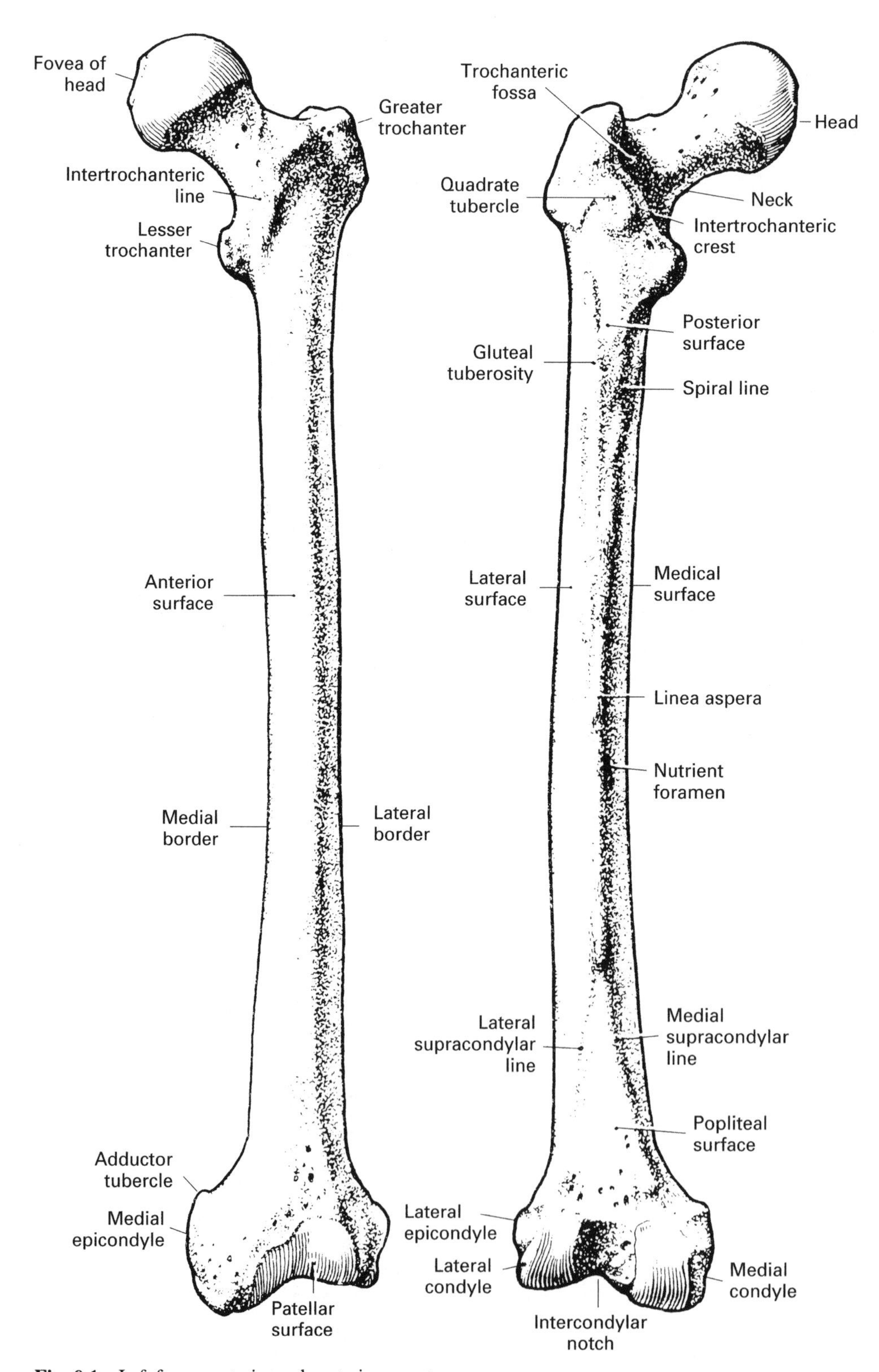

Fig. 8.1 Left femur: anterior and posterior aspects

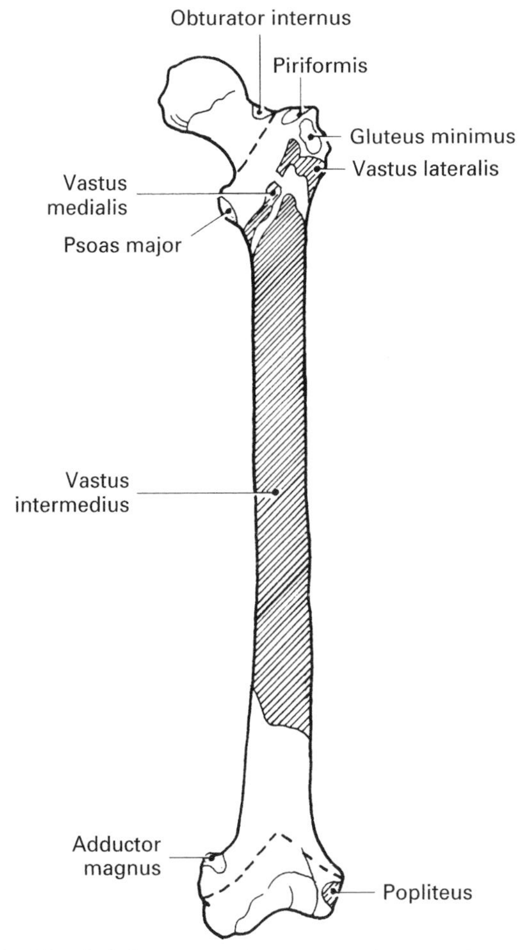

Fig. 8.2 Left femur: posterior aspect, to show muscle attachments

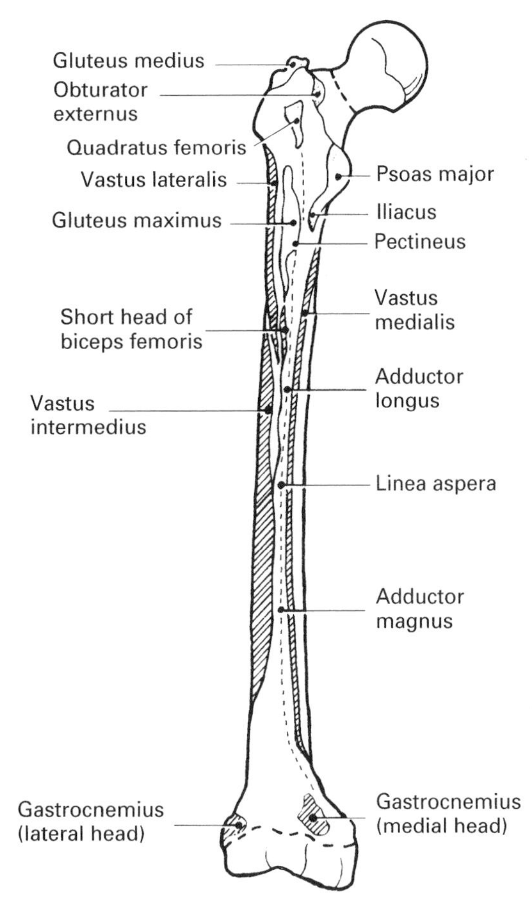

Fig. 8.3 Left femur: anterior aspect, to show muscle attachments

neck and shaft. The posterior surface is smooth but the upper and anterior surfaces are roughened for the attachment of the psoas major and iliacus muscles.

The shaft is long and cylindrical, narrowest at the middle and widest at the lower end. Its long axis is at an angle of about 10° from that of the tibia. The shaft has a slight forward convexity and there is a well-marked bony ridge on the posterior aspect—the linea aspera—to which are attached the adductors, vasti and the short head of the biceps femoris muscles. Foramina for nutrient arteries are close to the linea aspera.

The middle part of the shaft has three surfaces and three borders, anterior, medial and lateral. The anterior and medial borders are ill-defined. The anterior surface is smooth and convex. A nutrient artery is present in the middle third of the medial surface. The posterior border is formed by the linea aspera.

The upper posterior end of the shaft widens to form a V-shaped posterior surface, bounded by two bony ridges which converge inferiorly to become continuous with the linea aspera. The ridge on the lateral side is the gluteal tuberosity, to which is attached the gluteus maximus muscle. The medial ridge is called the spiral line and is continuous above with the lower end of the intertrochanteric line.

In the lower third of the posterior surface of the shaft, the linea aspera divides into medial and lateral supracondylar lines which run obliquely to the femoral condyles. The triangular area—the popliteal surface—between the supracondylar lines

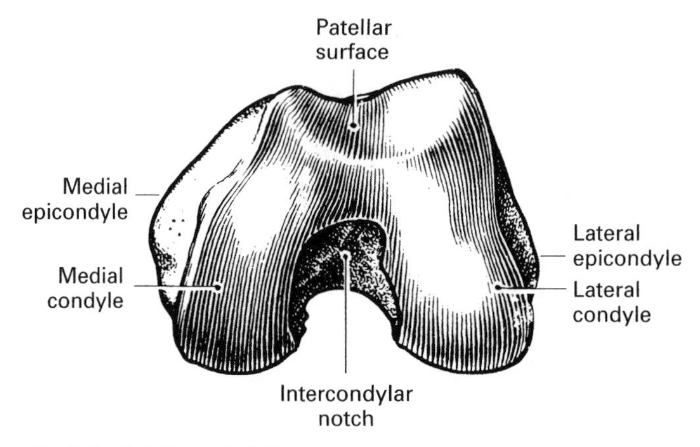

Fig. 8.4 Left femur: inferior aspect

forms the floor of the upper part of the popliteal fossa. The popliteal artery passes over the popliteal surface and is separated from the bone by a variable amount of fat.

The lower end is expanded from side to side and forms a weight-bearing surface for transference of body weight to the tibia. It is formed by two bony prominences:

- the medial condyle
- the lateral condyle.

Anteriorly, the condyles present a united surface, in line with the anterior surface of the shaft of the femur. Posteriorly they project beyond the line of the shaft and are separated by a deep intercondylar notch (Fig. 8.4). The condyles are markedly convex from back to front and each condyle is slightly convex from side to side. The articular surface is continuous across the anterior surface of the condyles, with a shallow median groove with which the patella articulates. The articular surface extends inferiorly and posteriorly on each condyle and is divided by the intercondylar notch.

The femoral condyles do not conform in shape exactly to the shallow concave facets of the tibial condyles but this is partly compensated for by the interposed semilunar cartilages (p. 205).

The medial condyle has a roughened prominence—the medial epicondyle—on its medial aspect. Above this is the adductor tubercle into which is inserted the adductor magnus muscle of the inner side of the thigh. The roughened lateral wall of the condyle forms one wall of the intercondylar notch and gives insertion to the posterior cruciate ligament.

The lateral condyle is slightly larger and stronger than the medial condyle since it supports more of the body weight. A small bony prominence—the lateral epicondyle—is present on the lateral surface and below it is a groove to which is attached the popliteus muscle.

The cruciate ligaments are two strong, rounded ligaments which extend from the non-articular intercondylar notch of the femur to the non-articular area of the tibia. They form, with the tibial and fibular collateral ligaments, the main bonds between the femur and the tibia. See p. 206 and Fig. 8.16.

The semilunar cartilages (menisci) are two crescent-shaped wedges of fibrocartilage which lie between the femur and the tibia peripherally. See p. 205 and Fig. 8.16.

Radiographic appearances of the femur (Figs 8.5 to 8.7)

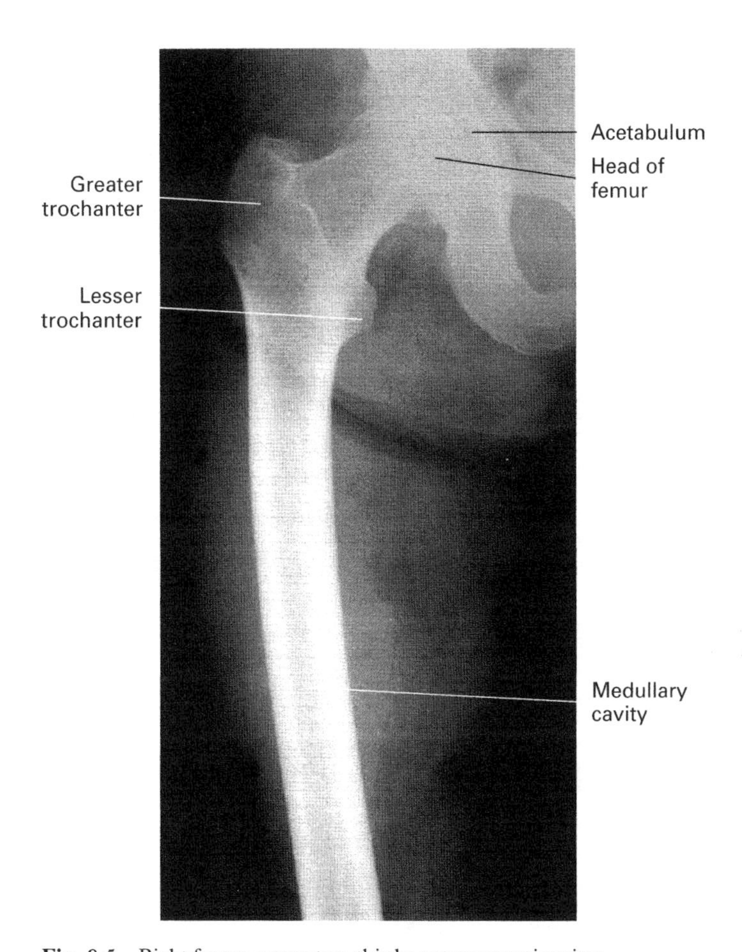

Fig. 8.5 Right femur: upper two-thirds; anteroposterior view

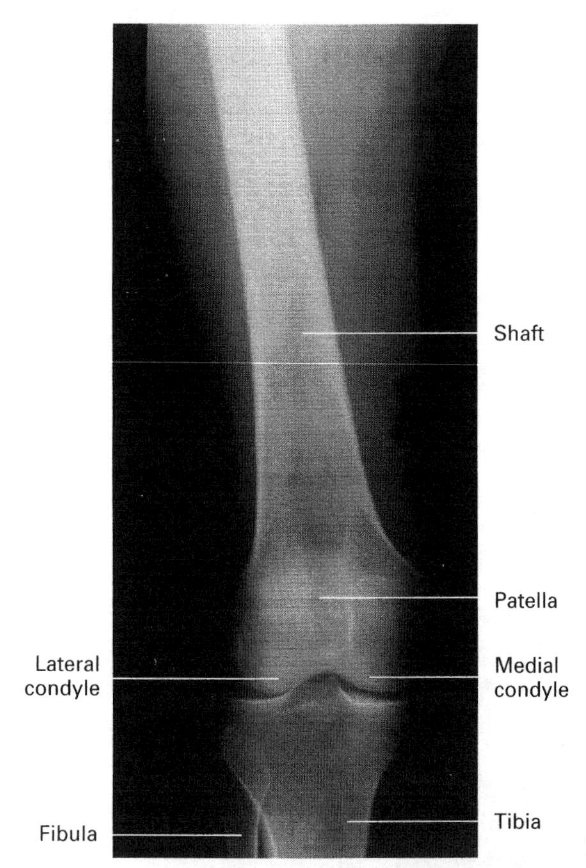

Fig. 8.6 Femur: lower third; anteroposterior view

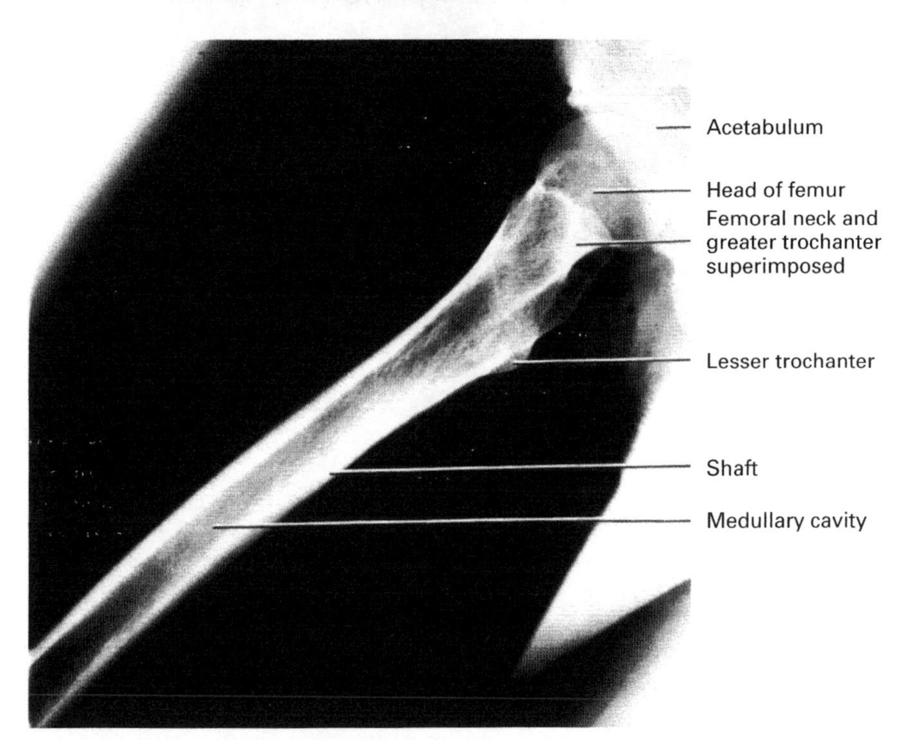

Fig. 8.7 Femur: upper two-thirds; lateral view (for lateral view of lower third see Fig. 8.18)

PATELLA (Figs 8.8 to 8.11)

The patella is the largest sesamoid bone in the body. It lies in the tendon of the quadriceps femoris muscle, on the anterior aspect of the knee joint and it articulates with the femoral condyles. It is roughly triangular in shape, with a thick superior border and thinner medial and lateral borders which converge to an apex. The apex is situated about 25 mm superior to the knee joint when the leg is extended. The patella has an anterior and a posterior surface.

The anterior surface (Fig. 8.8) is easily palpable. It has vertical grooves and ridges for attachment to the muscle tendon in which it is situated. It is separated from the skin by the prepatellar bursa.

The posterior surface (Fig. 8.9) is mainly covered with hyaline cartilage. Its surface is divided by a vertical ridge into two parts, the lateral part being the larger. Two horizontal lines divide each side into three—thus there are six areas of articulation plus a small area at the medial edge which is in contact with the medial femoral condyle in full flexion only. The insertion of the vastus medialis muscle prevents the patella's inclination to dislocate laterally.

Below the articular facets the apex of the patella points downwards and has a roughened surface for the attachment of the ligamentum patellae.

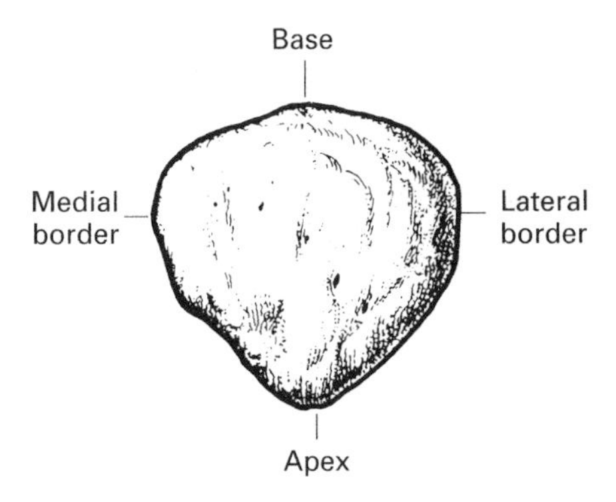

Fig. 8.8 Left patella: anterior aspect

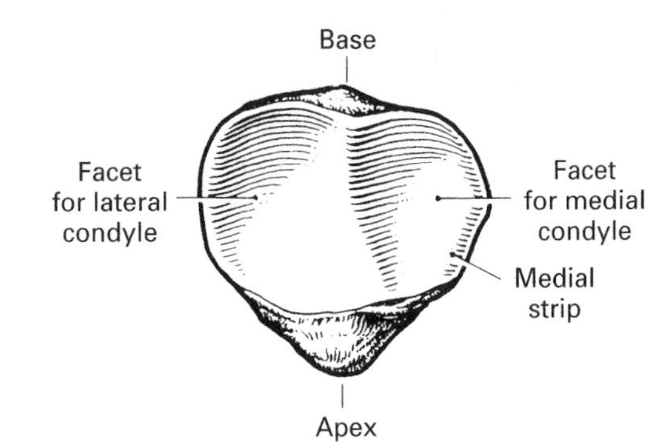

Fig. 8.9 Left patella: posterior aspect

Radiographic appearances of the patella (Figs 8.10 and 8.11)

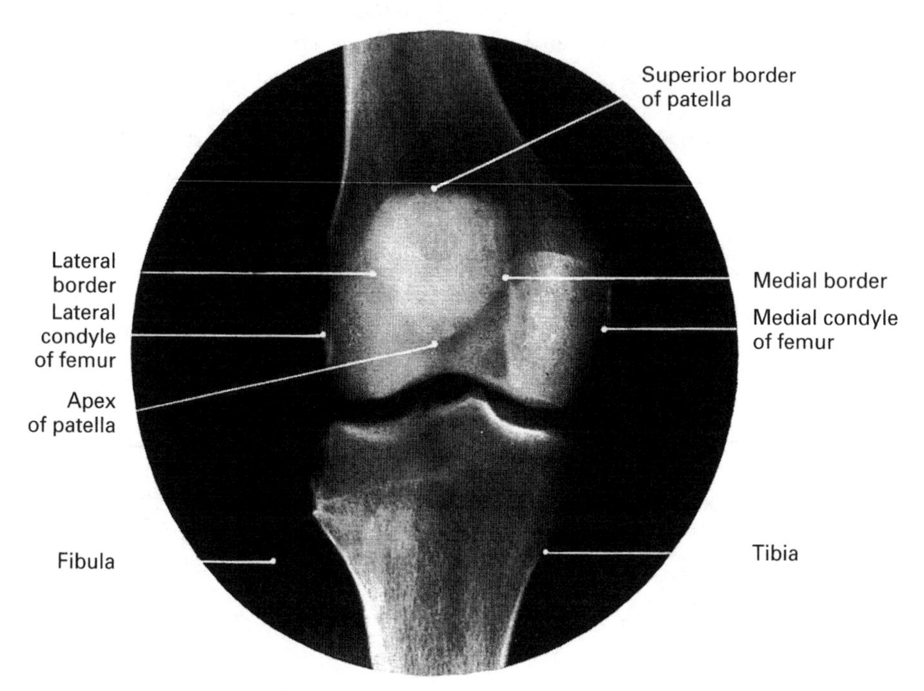

Fig. 8.10 Patella: postero-anterior view

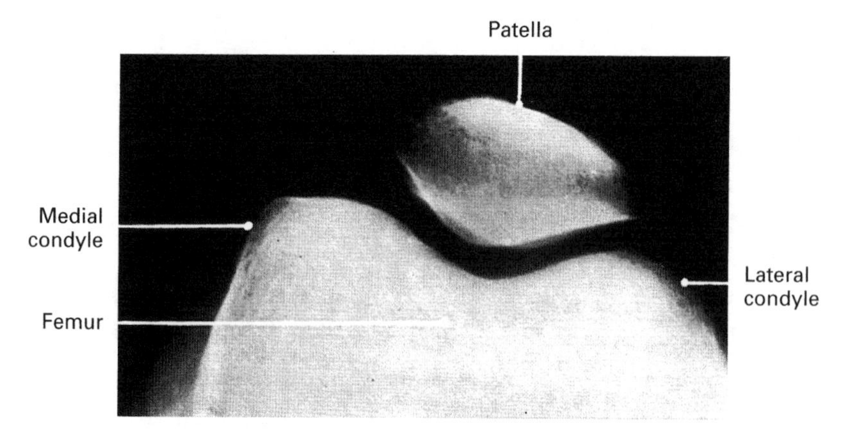

Fig. 8.11 Patella: infero-superior ('skyline') view

FABELLA

The fabella is an inconstant sesamoid bone in the lateral head of the gastrocnemius muscle. The fabella is found where the lateral head of the muscle passes over the lateral condyle of the femur.

OSSIFICATION OF THE FEMUR (Figs 8.12 to 8.15) (For ossification of the proximal end of the femur see Figs 7.14 and 7.15)

Primary centre

- for the shaft, appears about week 7 of intra-uterine life.

Secondary centres

- for the head appears at about 6th month
- for the greater trochanter, at about 4th year
- for lesser trochanter, at about 12–14th years
- for lower end, at about week 36 of intrauterine life.

  (Its presence on a radiograph of a fetus in utero indicates that the fetus is at, or near, term).

The epiphyses for the proximal end unite with the shaft between the 18th and 20th years. The epiphysis for the distal end units with the shaft in about the 20th year.

OSSIFICATION ON THE PATELLA (Fig. 8.12)

Primary centre appears between the 2nd and 6th year. Ossification is complete at puberty.

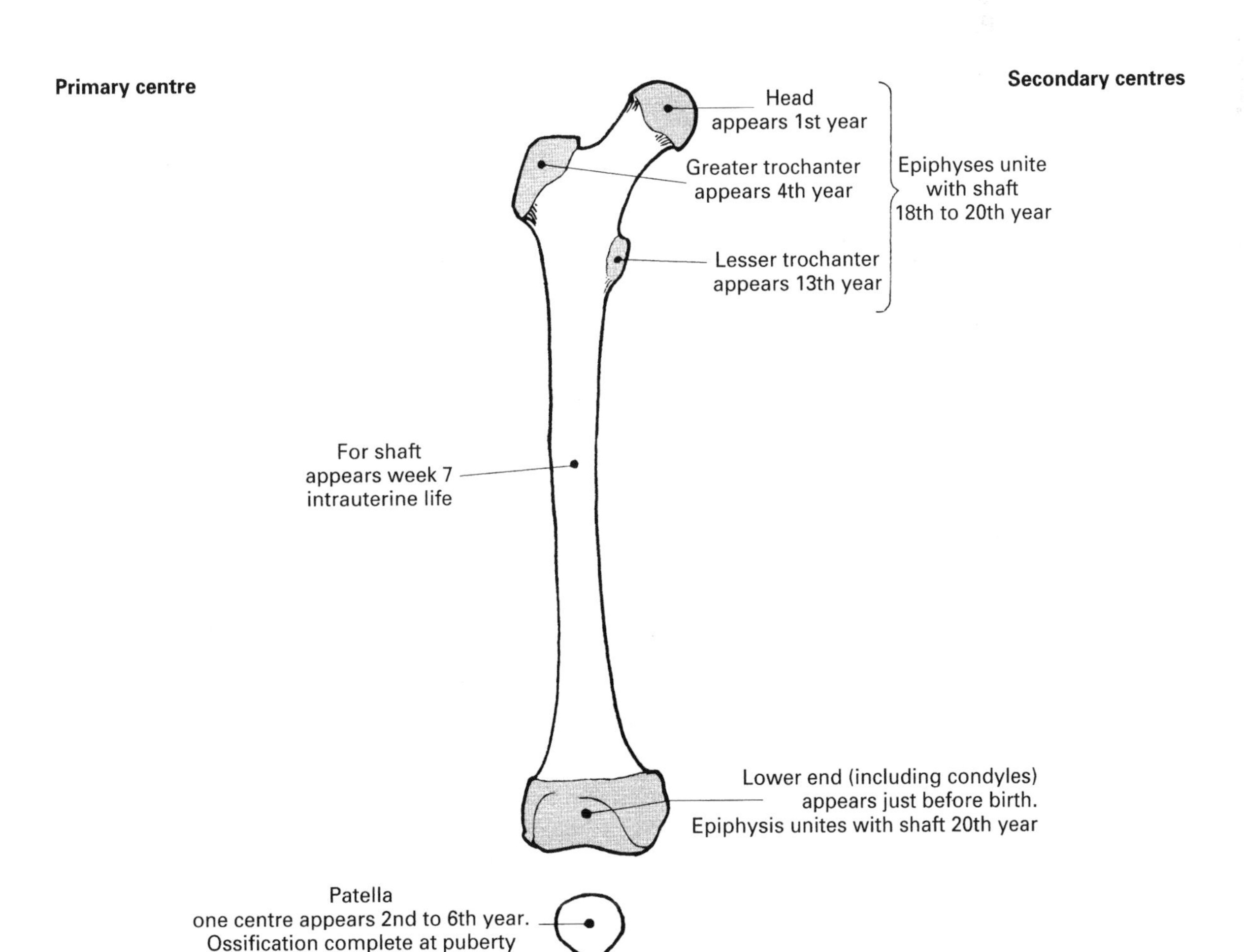

Fig. 8.12 Ossification of femur and patella

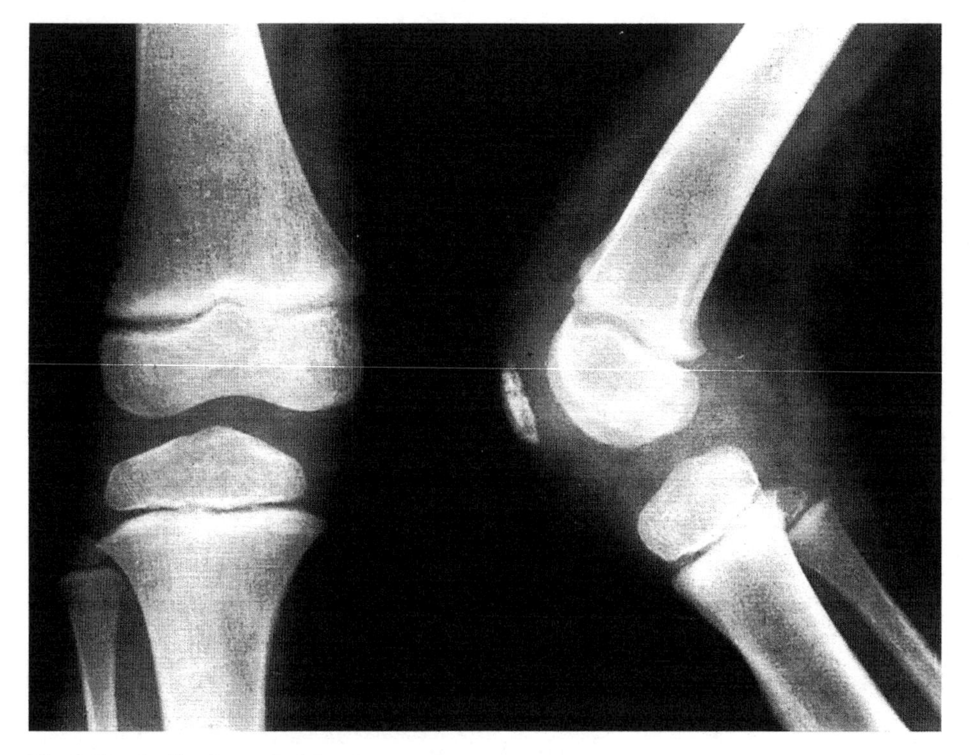

Fig. 8.13 Ossification at the knee: 4 years

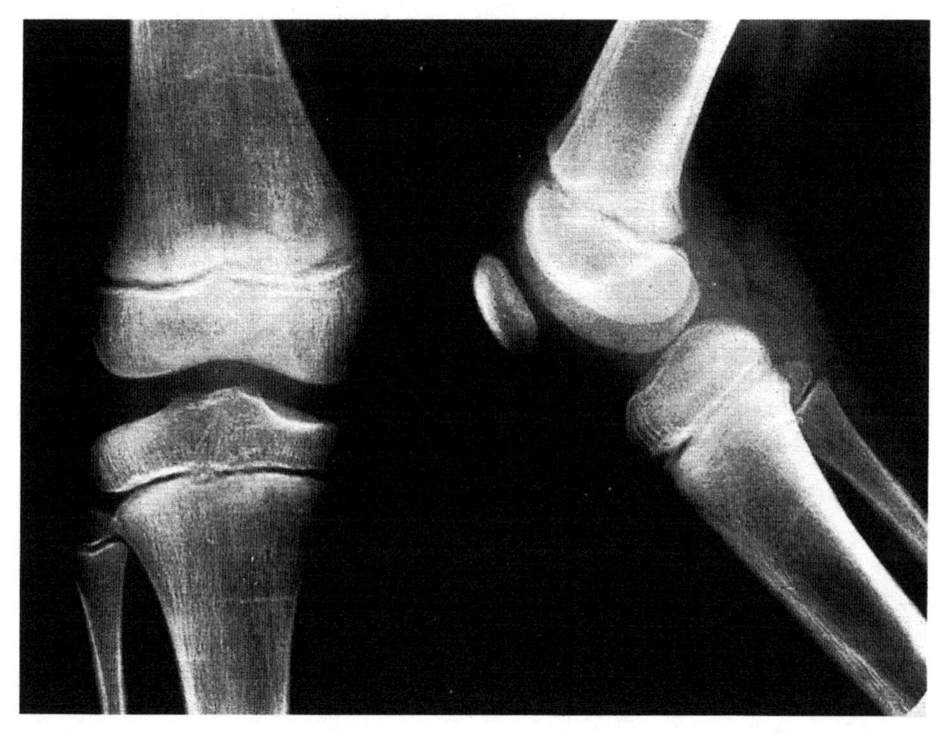

Fig. 8.14 Ossification at the knee: 7 years

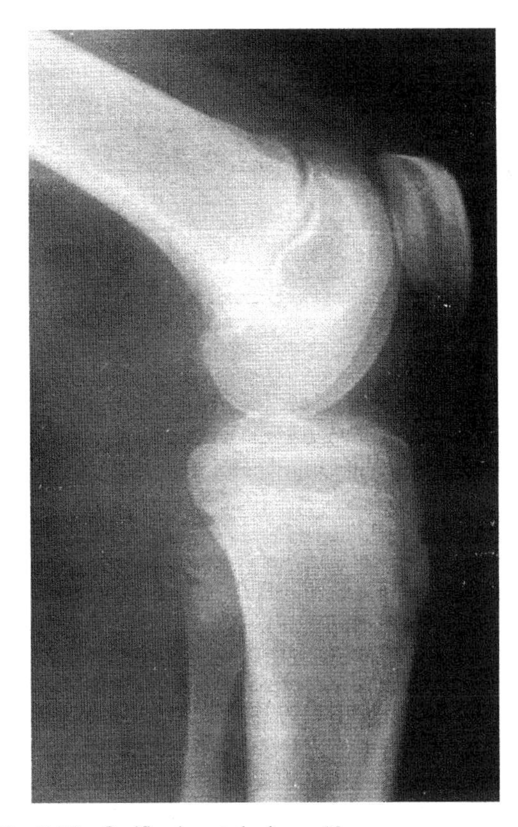

Fig. 8.15 Ossification at the knee: 13 years

KNEE JOINT (Figs 8.16 to 8.20)

Type: Synovial, condylar, compound.

Articular surfaces: Medial and lateral condyles of femur with medial and lateral condyles of tibia. Posterior surface of the patella articulates with the patellar articular area between femoral condyles. Articular surfaces are covered with hyaline cartilage. Semilunar fibrocartilages separate tibial and femoral condyles peripherally.

Synovial membrane: Most extensive of any joint. Lines the capsule. Anteriorly extends from the upper border of patella to form suprapatellar bursa. Covers anterior and lateral surfaces of cruciate ligaments, leaving posterior aspect covered only by the capsular ligament. Laterally it lines capsule, inferiorly is separated from the ligamentum patellae by infrapatellar fat pad. Behind lateral meniscus, between the meniscus and tendon of popliteus muscle, is a pouch of synovial membrane—subpopliteal recess.

Capsular ligament: Attached superiorly to femoral condyles, inferiorly to tibial condyles, laterally to lateral tibial condyle and to head of fibula. Anteriorly it is deficient above the patella where it is replaced by the quadriceps muscles.

Intracapsular structures:

Semilunar cartilages (menisci)—two crescent-shaped wedges of fibrocartilage lying between femur and tibia peripherally, deepening the articular surface of tibia. The medial meniscus is semicircular, wider posteriorly, firmly attached to tibia. The firm attachment makes it more liable to tearing because it moves little when the femur rotates on the tibia. The lateral meniscus is nearly circular and is capable of a little anteroposterior movement during 'locking' and 'unlocking' of the knee in extension.

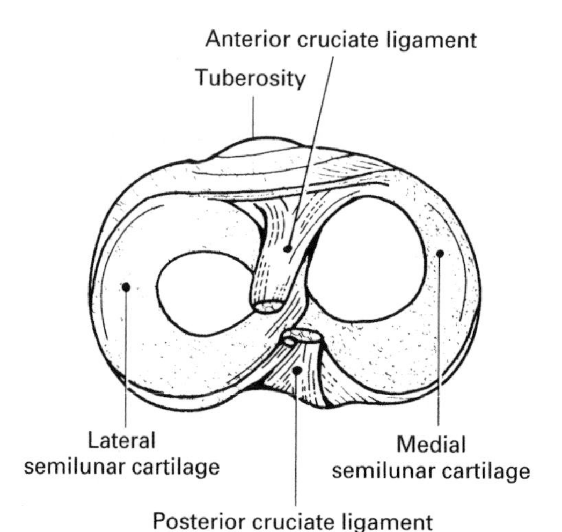

Fig. 8.16 Diagram of upper surface of tibia

Cruciate ligaments—two strong ligaments between the femur and tibia forming a cross. Anterior cruciate ligament is attached between anterior intercondylar region of tibia and medial surface of lateral femoral condyle. Posterior ligament is attached to posterior intercondylar region of tibia and lateral surface of medial femoral condyle. It lies posteromedial to the anterior cruciate ligament.

Bursae: 12 or 13 are recognized: 4 anterior, 4 lateral, 4 or 5 medially. They prevent friction between bones, ligaments, tendons or muscles. The most important are:

Suprapatellar—between lower end of femur and quadriceps femoris muscle. Communicates with the joint.

Prepatellar—between anterior surface of patella inferiorly and the skin.

Infrapatellar—subcutaneous part is between lower part of tibial tuberosity and the skin—deep part is between tibia and ligamentum patellae.

Gastrocnemius—between medial head of gastrocnemius and capsule, with an extension between the femur and the tendon of semimembranosus. This bursa frequently communicates with the joint.

Semimembranosus—between tendon and medial tibial condyle. Often communicates with gastrocnemius bursa.

Ligaments:

Tibial collateral—between lateral epicondyle of femur and medial condyle of tibia. Fuses with capsular ligament.

Fibular collateral—between lateral epicondyle of femur and head of fibula, by a strong cord.

Ligamentum patellae—between inferior margin of patella and tibial tuberosity. Part of tendon of quadriceps femoris muscle.

Oblique popliteal—between medial condyle of tibia, as prolongation of semimembranosus insertion. Supero-laterally attached to lateral condyle of femur.

Arcuate popliteal—forms an arc between head of fibula and posterior intercondylar region of tibia. Anterior band sometimes passes to lateral epicondyle of femur.

Transverse—between anterior margins of menisci.

Coronary—attaches periphery of each meniscus to tibia.

Cruciate—See above. The femur rotates on the tibia about an axis which passes through the lateral condyle, close to anterior cruciate ligament. This is an essential part of the movement of extension and locking of the knee.

Blood supply: Genicular branches of popliteal, femoral, anterior tibial and lateral circumflex femoral arteries.

Nerve supply: From branches of obturator, femoral, tibial and common peroneal nerves.

Movements and muscles:

Flexion—biceps femoris, gastrocnemius, semitendinosus, semimembranosus, popliteus, sartorius, gracilis.

Extension—quadriceps femoris, tensor fasciae latae.

Lateral rotation (femur fixed)—biceps femoris.

Medial rotation (femur fixed)—popliteus, semitendinosus, semimembranosus (sartorius and gracilis).

Quadriceps femoris: This is composed of 4 parts: rectus femoris, vastus lateralis, vastus medialis and vastus intermedius.

origin

- rectus femoris—anterior inferior iliac spine
- vastus lateralis } medial and lateral borders of linea
- vastus medialis } aspera on posterior femur
- vastus intermedius—from broad area on anterior femoral shaft. Lies deep to lateralis and medialis.

insertion—tubercle of tibia via patellar tendon.
function—main extensor of knee.
nerve supply—femoral nerve.

Gastrocnemius:
origin—by two heads—condyles of femur, posteriorly.
insertion—calcaneum via tendo calcaneus.
function—flexion of knee (and plantar flexion of foot) particularly in running, walking, etc.
nerve supply—tibial nerve.

Popliteus:
origin—by tendon from lateral femoral condyle and popliteal ligament.
insertion—proximal posterior tibia above soleal line.
function—flexion and medial rotation of knee. (Note that rotation of the knee may occur either with the femur fixed or with the tibia fixed, e.g. standing—tibia fixed; sitting (knee flexed)—femur fixed.)
nerve supply—tibial nerve.

Tensor fasciae latae: See under 'Hip Joint', p. 188.

Hamstrings: These consist of biceps femories, semitendinosus and semimembranosus. See under 'Hip Joint', p. 188.

Gracilis: this is the most superficial of the adductor group of muscles of the thigh. See under 'Hip Joint', p. 189.

Sartorius ('tailor's muscle)
origin—anterior superior iliac spine.
insertion—proximal medial tibia.
function—flexion of hip and knee, rotation of hip laterally and of knee medially.

Radiographic appearances of the knee joint (Figs 8.17 to 8.20)

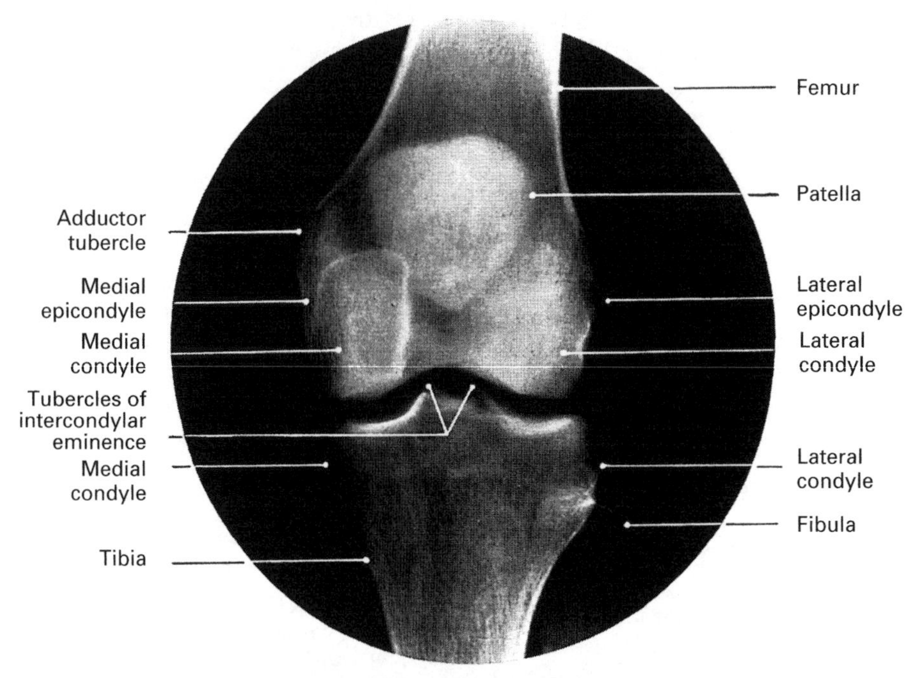

Fig. 8.17 Left knee: anteroposterior view

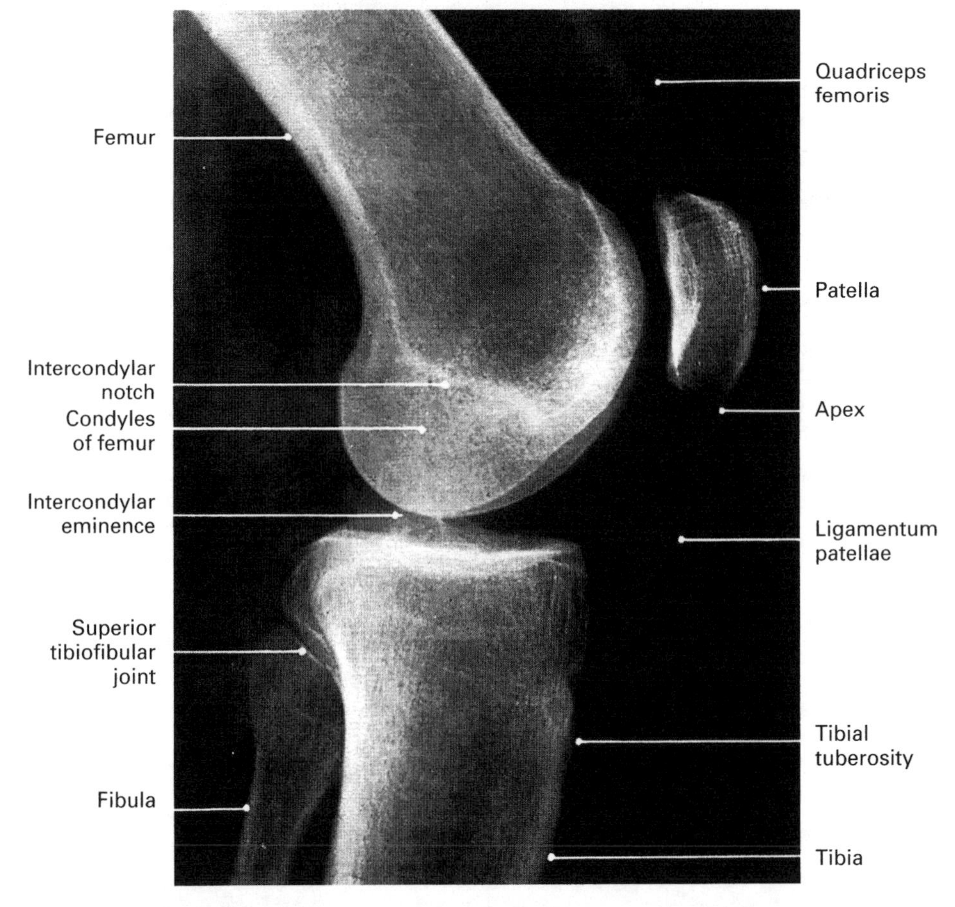

Fig. 8.18 Left knee: lateral view

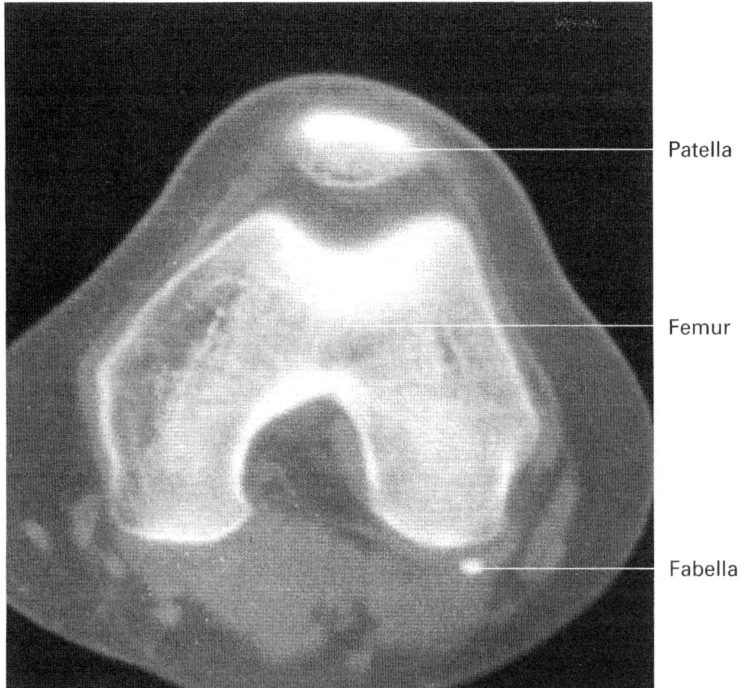

Fig. 8.19 Left knee: CT scan

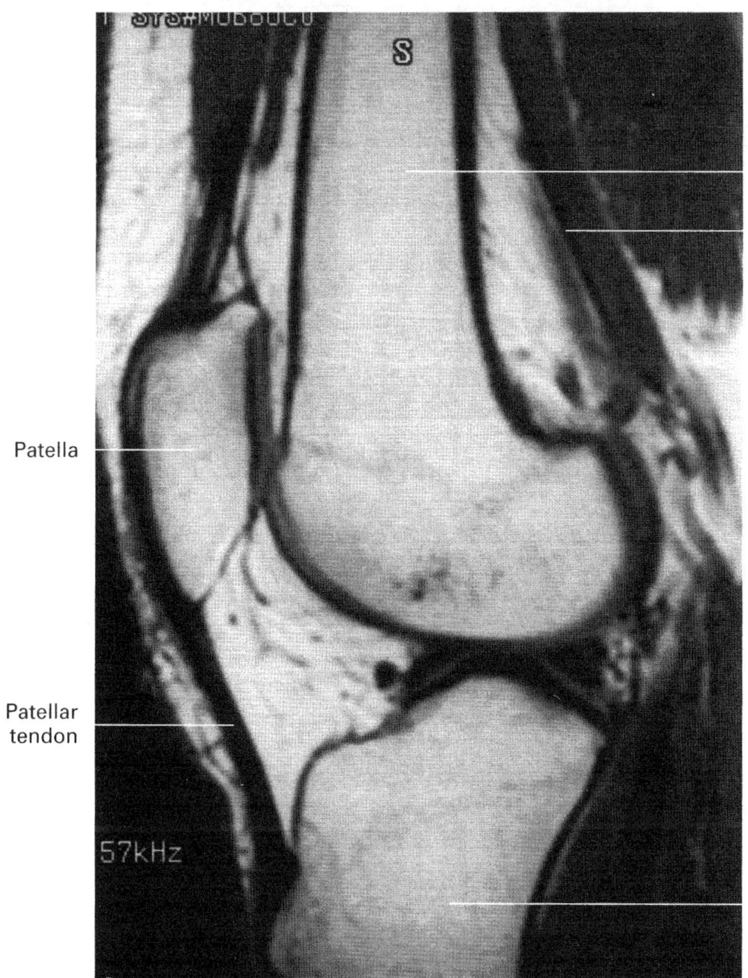

Fig. 8.20 Knee: MR scan

TIBIOFIBULAR JOINTS

Superior (Fig. 8.21)

Type: Synovial, plane

Articular surfaces: Posterior of lateral tibial condyle and head of fibula.

Synovial membrane: Attached to articular margins. Communicates with knee joint in 10 per cent of individuals by means of a pouch extension between lateral meniscus and popliteus tendon (subpopliteal recess).

Capsular ligament: Attached to articular margins.

Ligaments:
Anterior, *Superior* } pass from fibular head to anterior and posterior aspects of lateral femoral condyle.

Blood supply: Anterior tibial artery.

Nerve supply: Common peroneal branch of sciatic nerve.

Inferior (Fig. 8.31)

Type: Syndesmosis. Very strong.

Articular surfaces: Fibular notch of tibia and medial aspect of lower fibula, both of which

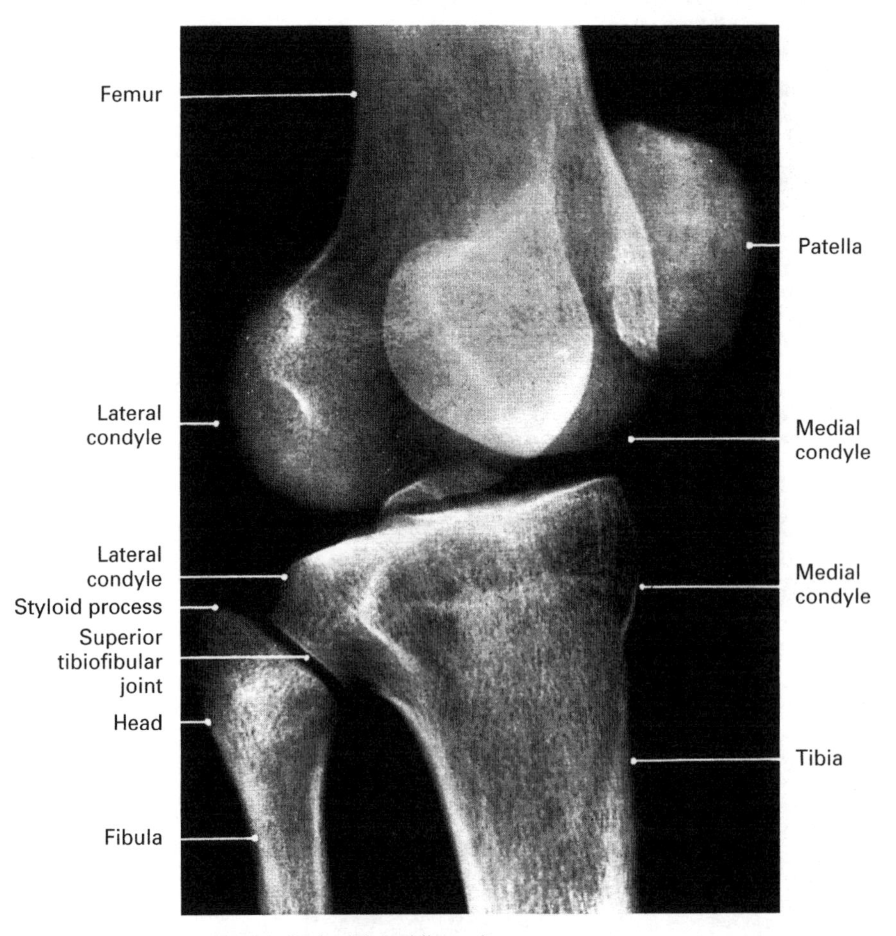

Fig. 8.21 Left superior tibiofibular joint: oblique view

are roughened by fibrous ligamentous attachments.

Ligaments:

Anterior tibiofibular and *Posterior tibiofibular* — run obliquely from fibula to tibia. The posterior is the stronger.

Inferior transverse—passes between malleolar fossa of fibula to just lateral to medial malleolus of tibia in a transverse direction.

Interosseous—an extension of interosseous membrane.

Blood supply: perforating branch of peroneal artery, malleolar branches of anterior and posterior tibial arteries.

Nerve supply: tibial, deep peroneal and saphenous nerves.

Interosseous

This is a fibrous membrane with the majority of fibres running inferiorly from medial to lateral. It has defects above and below for vessels and nerves to pass from anterior to posterior. It acts as a joint, being continuous with the fibrous syndesmosis of the inferior tibiofibular joint and it allows increased surface area for muscular origins.

LEG (Figs 8.22 to 8.26)

There are two bones in the leg—the tibia and the fibula (Fig. 8.22). They lie side by side and articulate with each other at the superior and inferior tibiofibular joints (see above). The tibia is much thicker and stronger than the fibula and it lies on the medial side of the leg with the fibula on the lateral side.

TIBIA

The tibia consists of:

- the upper end
- the shaft
- the lower end.

The upper end is large and articulates with the femur at the knee joint (p. 205). It is the widest part of the bone and consists of medial and lateral condyles which supply the weight-bearing surfaces of the knee joint.

The condyles overhang the shaft posteriorly. On the superior surface of each condyle is an articular surface for articulation with the corresponding condyle of the femur. Between the articular surfaces is a narrow, slightly irregular, non-articular area—the intercondylar area—upon which is the intercondylar eminence.

The medial condyle is larger than the lateral. Its articular area is oval and slightly concave. This concavity is accentuated laterally by the medial intercondylar tubercle, one of two tubercles of the intercondylar eminence. The anterior and medial surfaces of the condyle are roughened for muscle attachment. A horizontal groove is present on the posterior surface just below the articular margin.

The lateral condyle overhangs the shaft, particularly posterolaterally where a small articular facet is present for articulation with the fibula at the superior tibiofibular joint (p. 210). The superior articular surface of the lateral condyle is smaller than that of the medial condyle. It is nearly circular, slightly concave and extends on the medial side to the lateral intercondylar tubercle. The anterior, posterior and lateral surfaces of the condyle are roughened.

The semilunar cartilages (p. 205) deepen the concavity of the tibial condyles and adapt their superior surfaces to the shape of the rounded femoral condyles.

The tibial tubercle is a small bony projection on the anterior surface of the tibia, just distal to the condyles. Into the upper part of this tubercle is inserted the ligamentum patellae.

The shaft of the tibia is triangular in cross-section. It has three borders: anterior, medial and interosseous; and three surfaces: medial, lateral and posterior.

The curved anterior border begins below the tibial tubercle and extends to the anterior border of the medial malleolus. It is subcutaneous throughout and forms the crest of the shin.

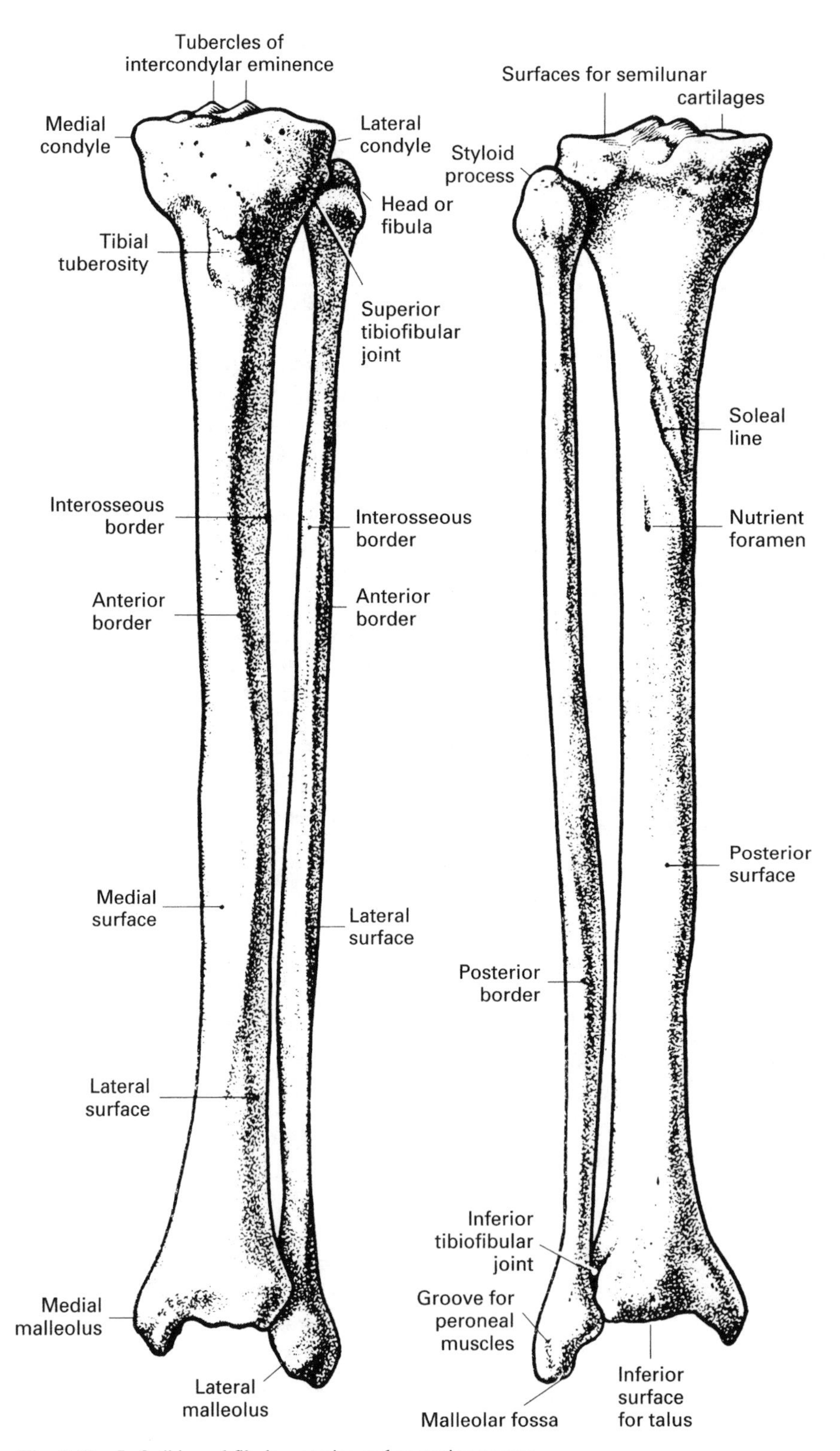

Fig. 8.22 Left tibia and fibula: anterior and posterior aspects

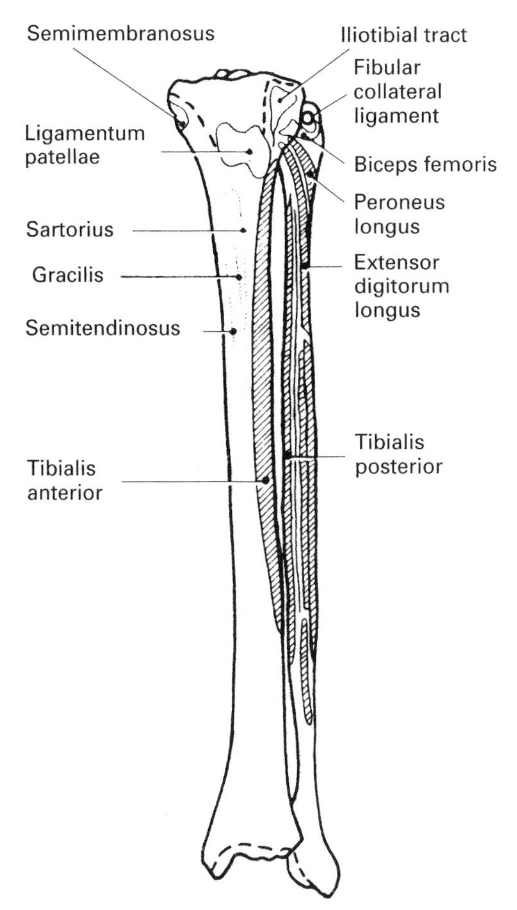

Fig. 8.23 Left tibia and fibula: anterior aspect to show muscle attachments

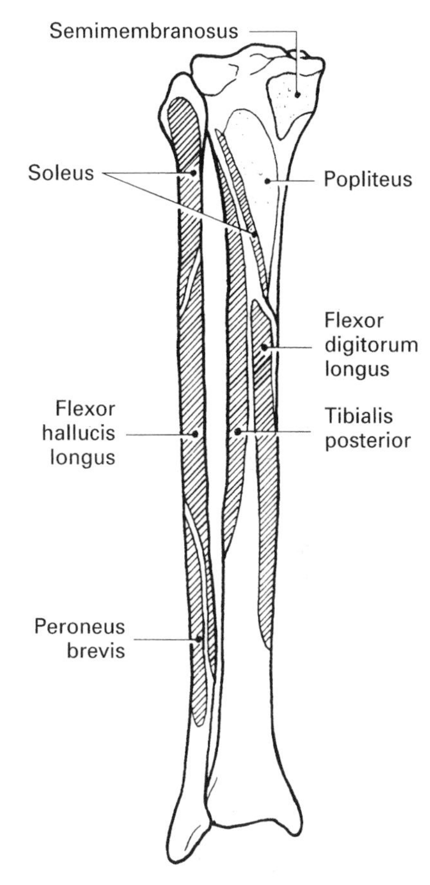

Fig. 8.24 Left tibia and fibula: posterior aspect to show muscle attachments

The medial border begins below the groove on the medial condyle and ends on the posterior surface of the medial malleolus.

The interosseous border gives attachment to the interosseous membrane, connecting the tibia to the fibula: it extends from in front of the fibular facet downward to the fibular notch on the lower end of the tibia.

The medial surface is smooth and subcutaneous for most of its length.

The upper end of the posterior surface is crossed obliquely by a ridge of bone—the soleal line—which gives attachment to some of the fibres of the soleus muscle of the calf.

The lower end is smaller than the upper end and articulates with the talus at the ankle joint (p. 217). It has five surfaces. The anterior and posterior surfaces are convex. A groove for the tendon of the tibialis posterior muscle is present on the posterior surface medially.

The medial surface is prolonged downward as the medial malleolus, which articulates with the medial side of the body of the talus.

The fibular notch forms the lateral surface. The lower end of the fibula articulates with this notch and is attached to the tibia by a strong interosseous ligament, forming the inferior tibiofibular joint (p. 210).

The inferior surface consists of an almost square articular facet, slightly broader anteriorly than posteriorly. It is concave from back to front and a little convex from side to side to conform to the trochlear articular surface of the talus with which it articulates at the ankle joint.

FIBULA

The fibula lies on the lateral side of the tibia. It is a long, thin bone and is not constructed to take any share in supporting the weight of the body. It does not take part in the knee joint, but its lower extremity—the lateral malleolus—forms an important part of the ankle joint.

The fibula consists of:

- the head
- the shaft
- the lower end.

The head is small and bears on its upper surface a facet for articulation with the lateral condyle of the tibia at the superior tibiofibular joint. This joint is situated on the postero-lateral surface of the lateral condyle and the head of the fibula can be felt about 25 mm inferior to the level of the knee joint.

The styloid process projects upwards from the posterior surface of the head. The peroneus longus muscle on the lateral aspect of the leg is attached to the head and upper shaft of the fibula. Immediately below the head is the neck of the fibula.

The shaft is triangular in cross-section and has anterior, posterior and lateral surfaces, separated by well-marked borders. The interosseous border separates the anterior and posterior surfaces and gives attachment to the interosseous membrane which connects the tibia and fibula throughout most of their length.

The lower part of the shaft fits into a fibular notch on the lateral surface of the tibia.

The lower end of the fibula projects downwards and is slightly expanded to form the lateral malleolus. On the medial surface of the lateral malleolus is a triangular facet for articulation with the lateral surface of the talus. Behind the facet there is a small depression—the malleolar fossa—to which is attached part of the lateral ligament of the ankle joint. A well-marked groove for the tendons of the peroneal muscles is present on the posterior aspect of the lateral malleolus.

Radiographic appearances of the tibia and fibula (Figs 8.25 and 8.26)

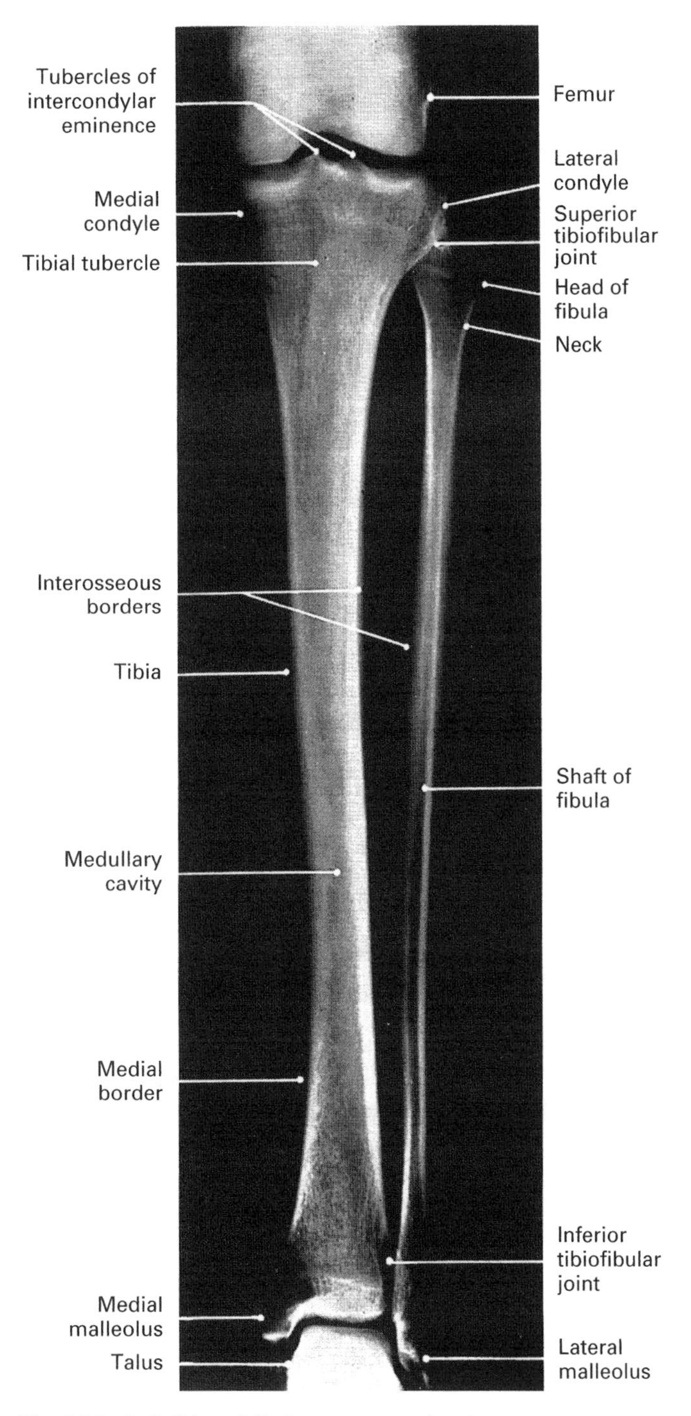

Fig. 8.25 Left tibia and fibula: anteroposterior view

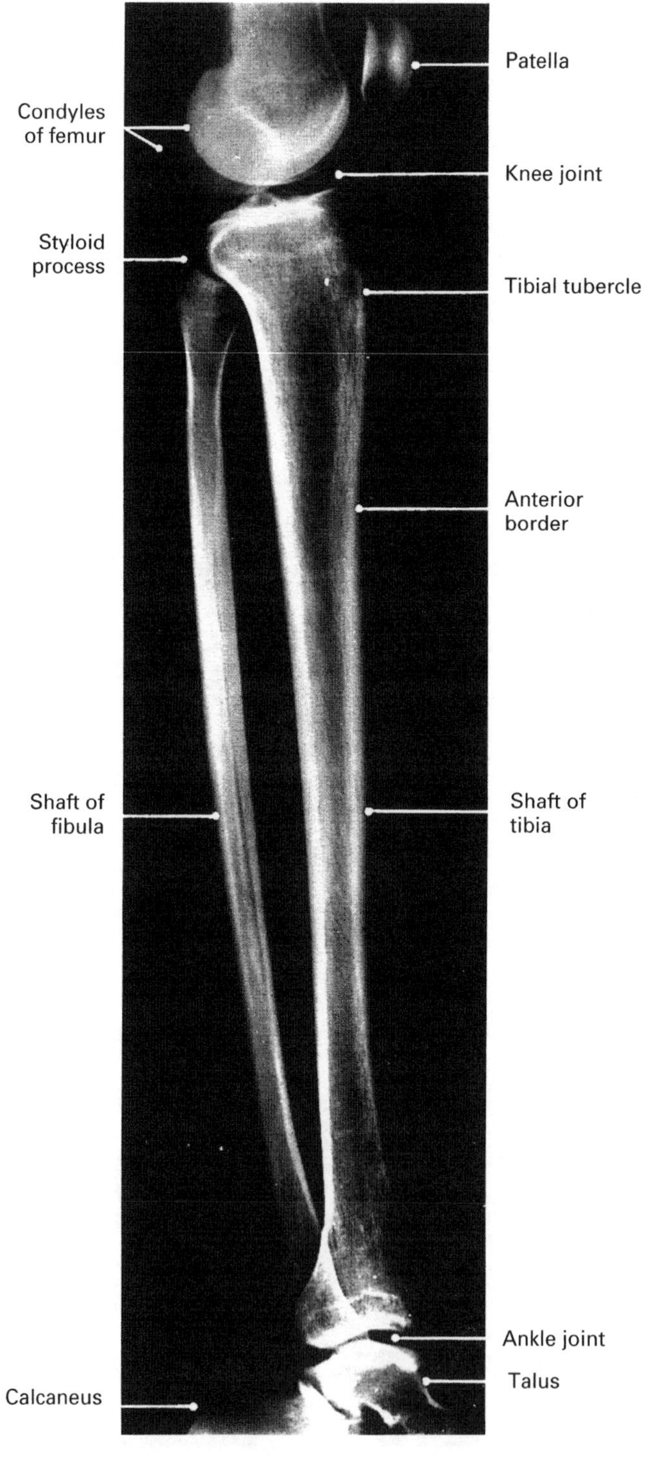

Fig. 8.26 Left tibia and fibula: lateral view

OSSIFICATION OF THE TIBIA AND FIBULA (Figs 8.27 & 8.28)

Primary centres

Tibia

- for the shaft, in about week 7 of intrauterine life.

Fibula

- for the shaft, in about week 8 of intrauterine life.

Secondary centres

Tibia

- for the proximal end, just before birth, or just after
- for the distal end, at about 2nd year.

Fibula

- for the proximal end, about 5th year
- for the distal end, about 2nd year.

The epiphyses for the distal ends of the bones unite with the shafts about the 18th year. The epiphyses for the proximal ends unite with the shafts later, about 20th year, because the knee joint area is the site of maximum growth in length of the lower limb.

The tibial tubercle may develop from a tonguelike extension of the proximal secondary centre or it may develop from an additional secondary centre appearing about the 12th year.

ANKLE JOINT (Figs 8.29 to 8.33)

Type: Synovial, hinge.

Articular surface: Articular surface of medial malleolus of tibia, inferior surface of tibia and articular surface of lateral malleolus of fibula together form a mortice for the trochlear surface of the body of the talus.

Ligaments:

Capsular—fibrous, surrounds joint. Attached above to margins of articular surface of tibia

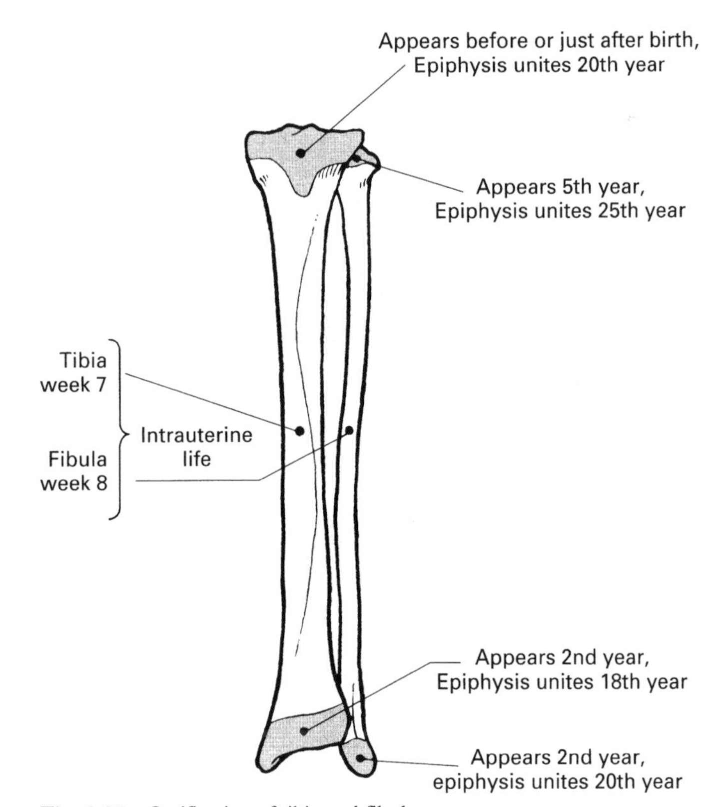

Fig. 8.27 Ossification of tibia and fibula

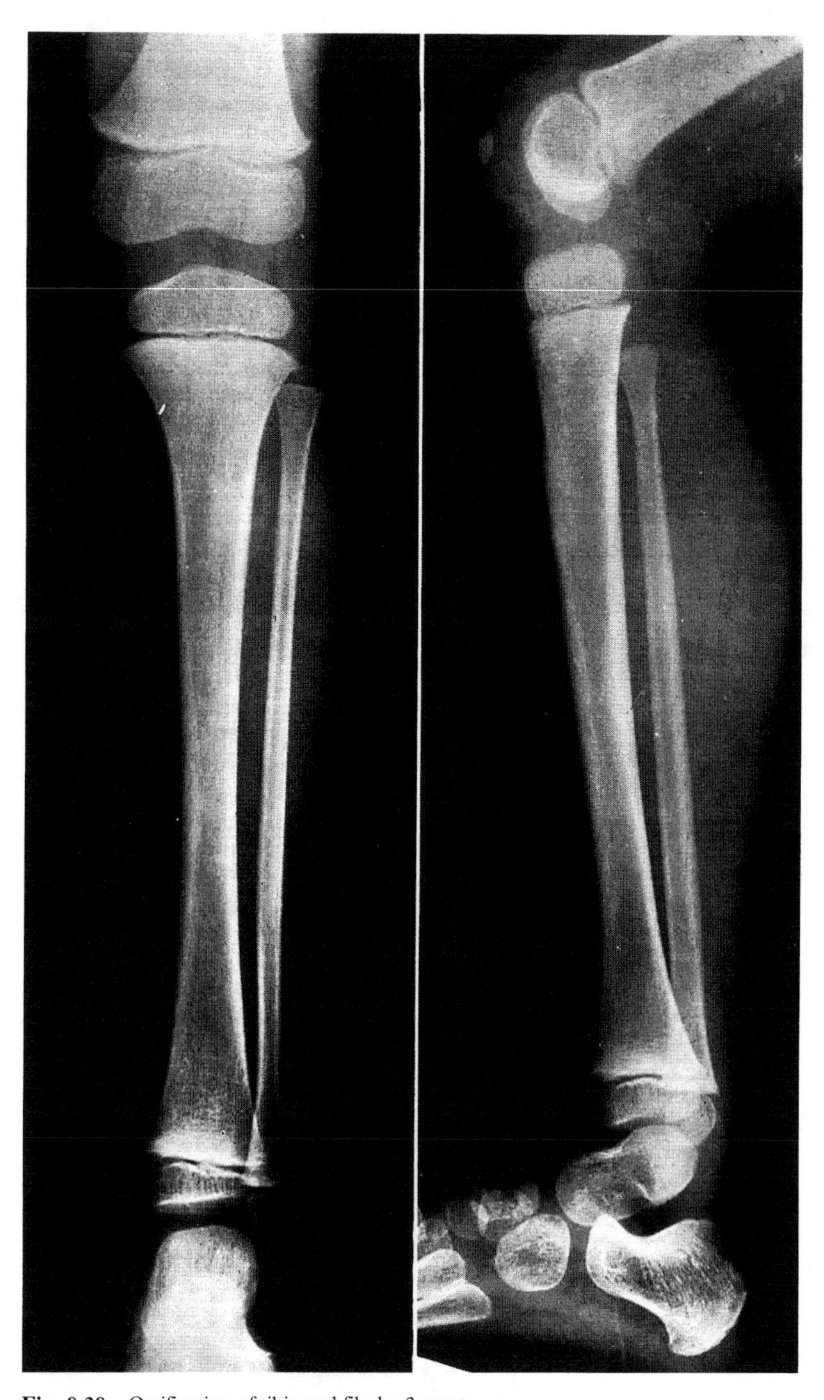

Fig. 8.28 Ossification of tibia and fibula: 3 years

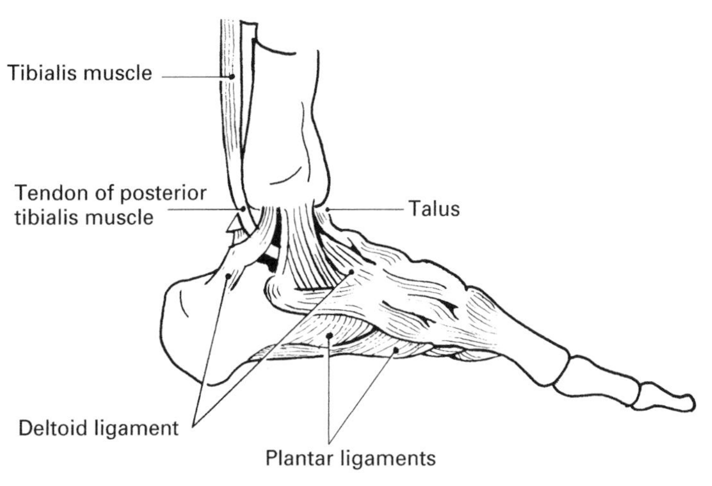

Fig. 8.29 Left ankle and foot: medial aspect, to show ligaments

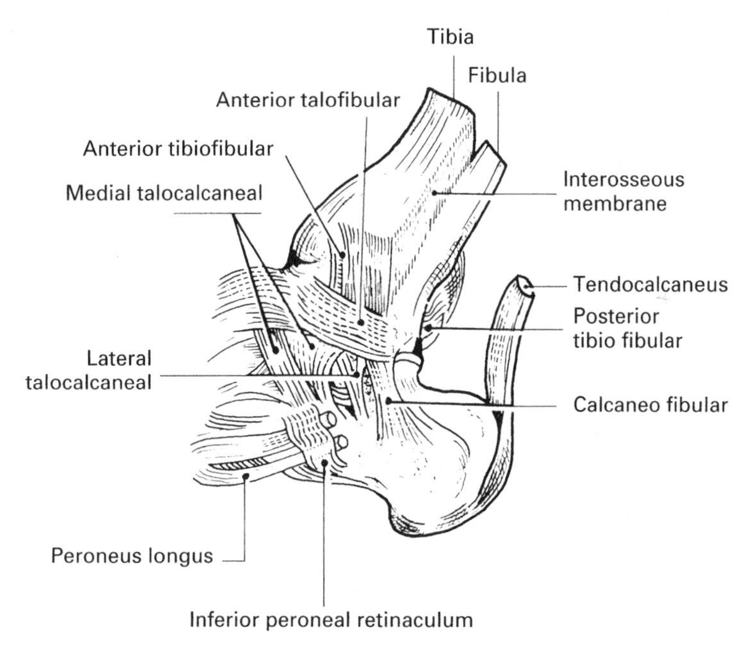

Fig. 8.30 Left ankle: lateral aspect, to show ligaments

and malleoli and below to margins of trochlear articular surface of talus, except anteriorly where attachment is to neck of talus.

Medial: Deltoid—a strong triangular ligament in 3 parts, attached above to medial malleolus and below to tuberosity of navicular (anterior), sustentaculum tali (middle) and medial surface of talus and medial tubercle (posterior).

Lateral: Three components:

(a) Anterior talofibular—attached above to anterior margin of lateral malleolus, passing medially to lateral side of neck of talus. (This is the ligament torn in a 'sprained ankle'.)

(b) *Posterior talofibular*—passes from lower part of malleolar fossa of fibula to posterior tubercle of talus.
(c) *Calcaneofibular*—rounded cord, passes downwards and posteriorly from apex of lateral malleolus to lateral side of calcaneum.

Blood supply: Malleolar branches of anterior tibial and peroneal arteries.

Nerve supply: Anterior and posterior tibial and deep peroneal nerves.

Movements and muscle:

Plantar flexion (flexion)—(this movement increases the angle between the dorsum of the foot and the anterior aspect of the leg)—gastrocnemius, soleus.
Dorsiflexion (extension)—the reverse of plantar flexion)—tibialis anterior, extensor digitorum longus, extensor hallucis longus, peroneus tertius.

Gastrocnemius: This muscle forms the rounded appearance of the calf. See under 'Hip Joint', p. 186

Soleus: (This is the other main muscle of the calf. It lies deep to gastrocnemius.)
origin—head and upper part of fibula, medial border of tibia below soleal line.
insertion—via tendon which joins that of gastrocnemius to form common tendon inserted into calcaneum (tendo calcaneus).
function—mainly concerned with the maintenance of the erect position.
nerve supply—tibial nerve.

Tibialis anterior:
origin—lateral condyle and upper half of tibia, interosseous membrane.
insertion—medial cuneiform and base of 1st metatarsal bone.
function—dorsiflexion of ankle, inversion of foot.
nerve supply—deep peroneal nerve.

Extensor digitorum longus:
origin—anterior crest of fibula, lateral condyle of tibia, interosseous membrane.
insertion—bases of middle and distal phalanges of lateral 4 toes.
function—dorsiflexion of ankle, extension of toes, eversion of foot.
nerve supply—deep peroneal nerve.

Extensor hallucis longus:
origin—middle third of anterior fibula, interosseous membrane.
insertion—base of distal phalanx of big toe.
function—extends ankle joints, extends big toe, eversion of foot.
nerve supply—deep peroneal nerve.

Peroneus tertius:
origin—lower third of medial fibula, interosseous membrane.
insertion—base of 5th metatarsal bone.
function—dorsiflexion of ankle, eversion of foot.
nerve supply—deep peroneal nerve.

Radiographic appearances of the ankle joint (Figs 8.31 to 8.33)

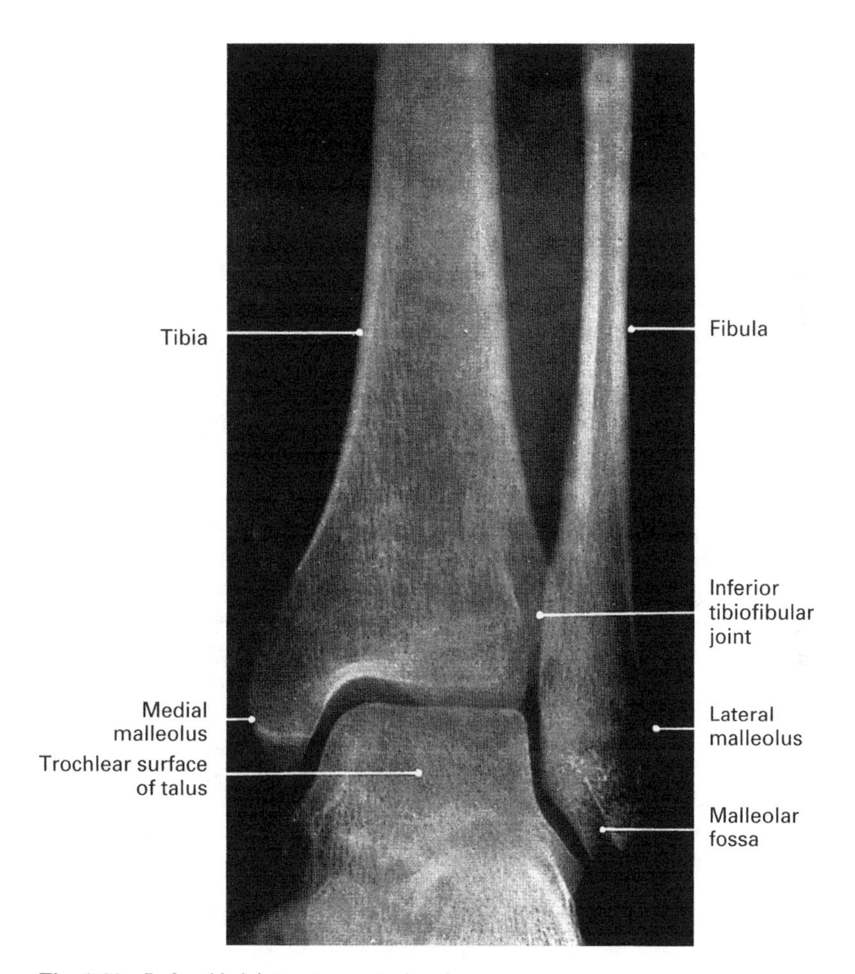

Fig. 8.31 Left ankle joint: anteroposterior view

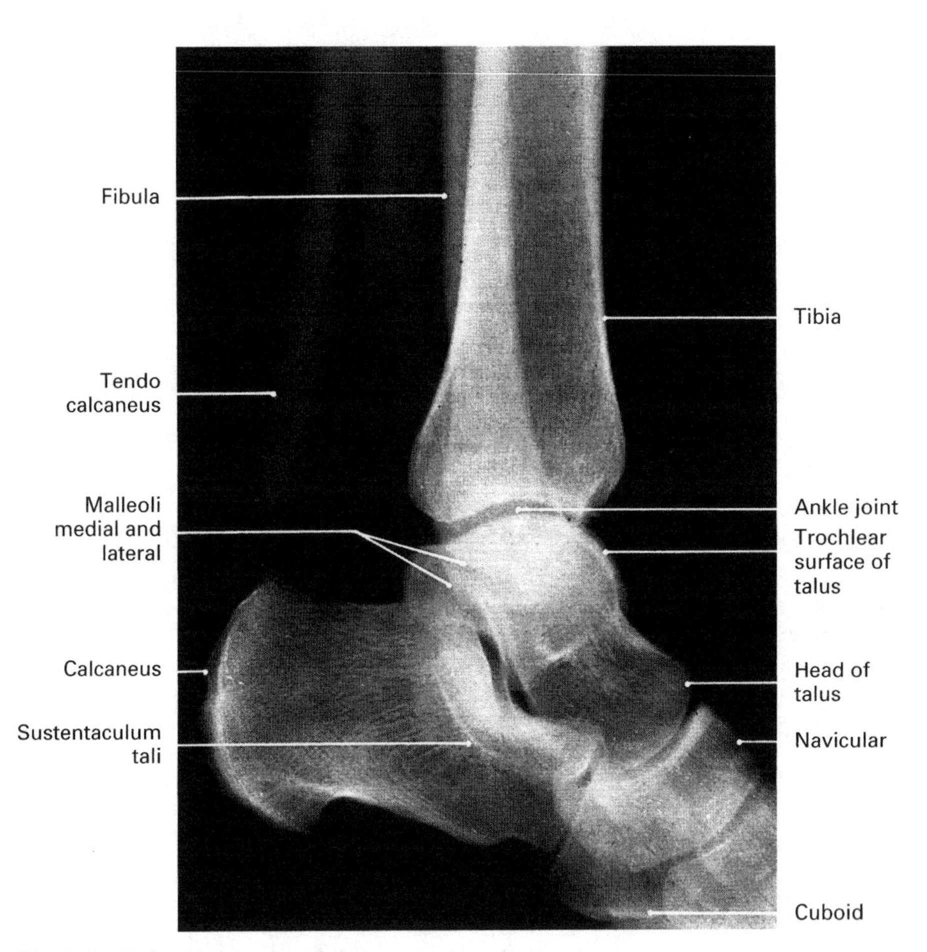

Fig. 8.32 Left ankle joint: lateral view

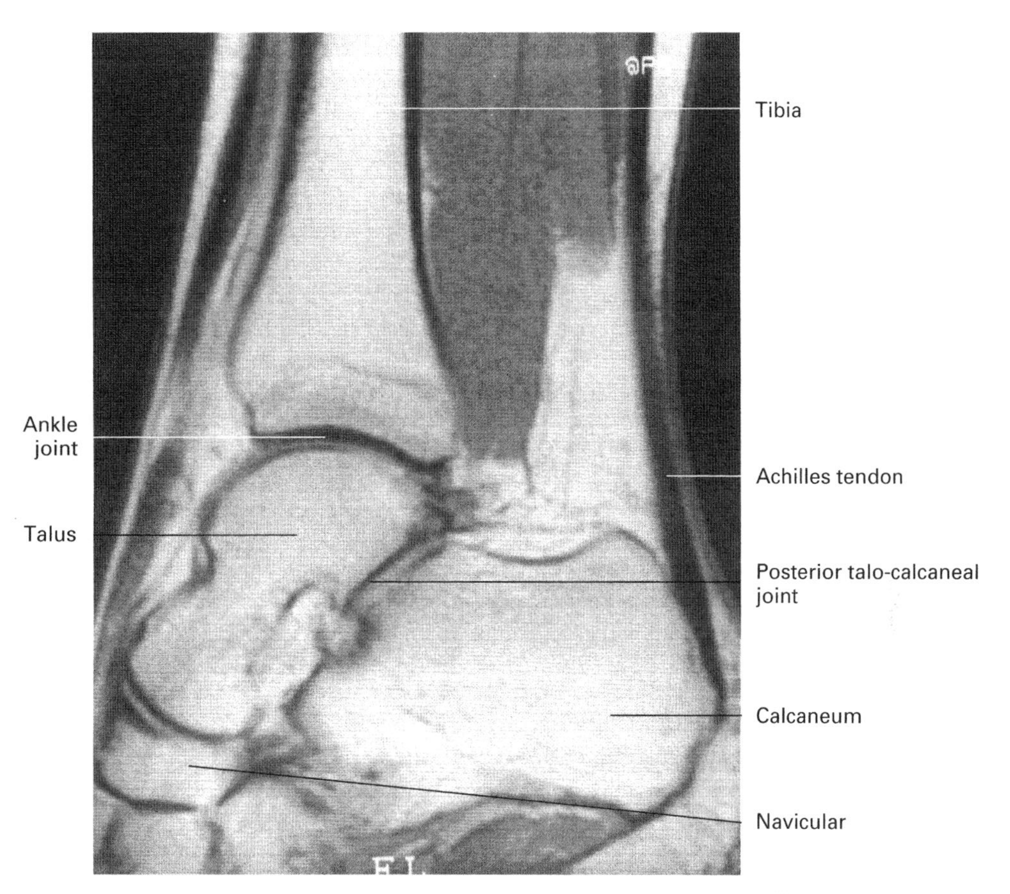

Fig. 8.33 Ankle joint: MR scan, sagittal section

RETINACULA AND SYNOVIAL SHEATHS OF THE ANKLE

Extensor retinaculum

Superior—a band attached anteriorly to the distal tibia and fibula above the ankle joint. It confines the extensor tendons of the ankle and foot.

Inferior—Y-shaped, lying anteriorly to the ankle joint. The stem of the 'Y' is attached laterally to the superior surface of the calcaneum and it runs medially, looping round the tendons of extensor digitorum longus and peroneus longus, then dividing into upper and lower bands. The upper band divides round the tendons of extensor hallucis longus and tibialis anterior and inserts on the medial malleolus. The lower band passes medially, merging with the thick plantar aponeurosis and thus enclosing the tendons of extensor hallucis longus and tibialis anterior.

Flexor retinaculum

This is a broad band which passes between the medial malleolus and the medial process of the calcaneum. It confines the tendons, vessels and nerves that pass around the medial malleolus into the sole of the foot.

Peroneal retinacula

These pass from the

- lateral malleolus to the lateral border of the calcaneum (superior retinaculum)
- inferior portion of the extensor retinaculum to the lateral malleolus (inferior retinaculum).

They confine the tendons of peroneus longus and peroneus brevis.

Synovial tendon sheaths

Synovial sheaths surround the individual tendons as they pass beneath the retinacula. Those of tibialis anterior and posterior commence well above the malleolus, those of extensor digitorum longus and peroneus tertius commence between the superior and inferior extensor retinacula. The tendon sheaths of peroneus longus and brevis are common at their origins but divide under cover of the peroneal retinaculum.

FOOT (Figs 8.34 to 8.53)

The bony structure of the foot is similar in many respects to that of the hand. The differences that there are result mainly from the fact that the feet support the weight of the trunk and assist in locomotion, whereas the hands are prehensile (capable of grasping). The bony structure of the foot is therefore stronger, but less mobile than that of the hand.

It should be noted that when referring to the foot, it is usual to employ the terms 'dorsal' and 'plantar' in place of anterior and posterior, but 'proximal' and 'distal' are employed in the usual sense that the terminal phalanges are the distal bones of the feet.

There are three groups of bones in the foot:

- the tarsus
- the metatarsus
- the phalanges.

They are arranged in a pattern similar to the bones of the hand but are set at right angles to the leg.

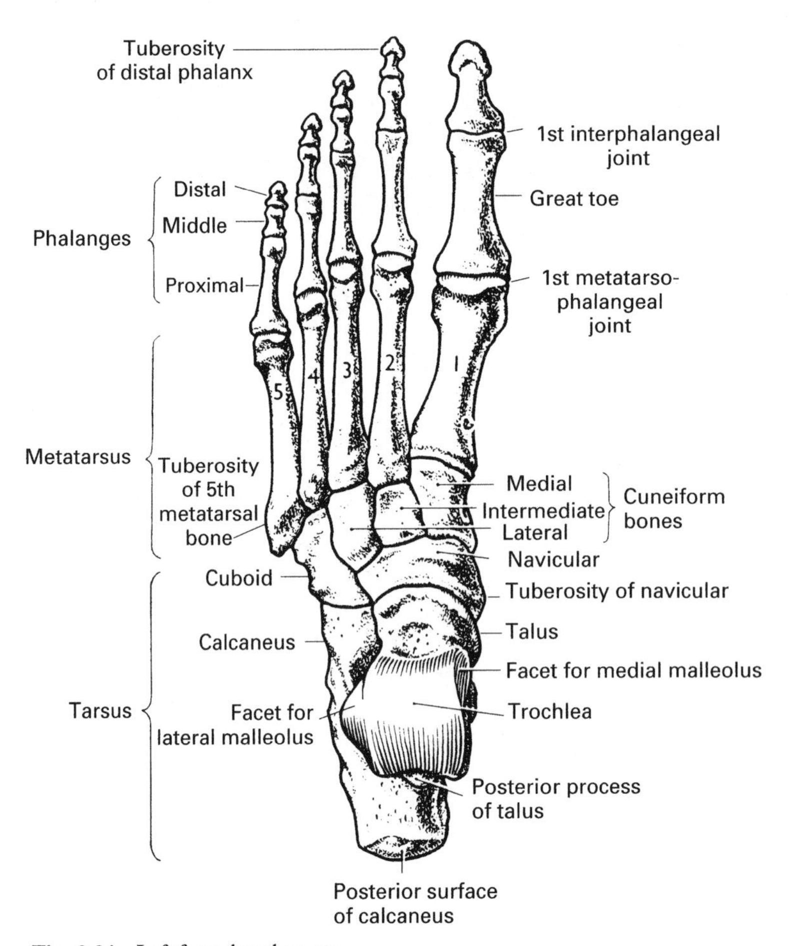

Fig. 8.34 Left foot: dorsal aspect

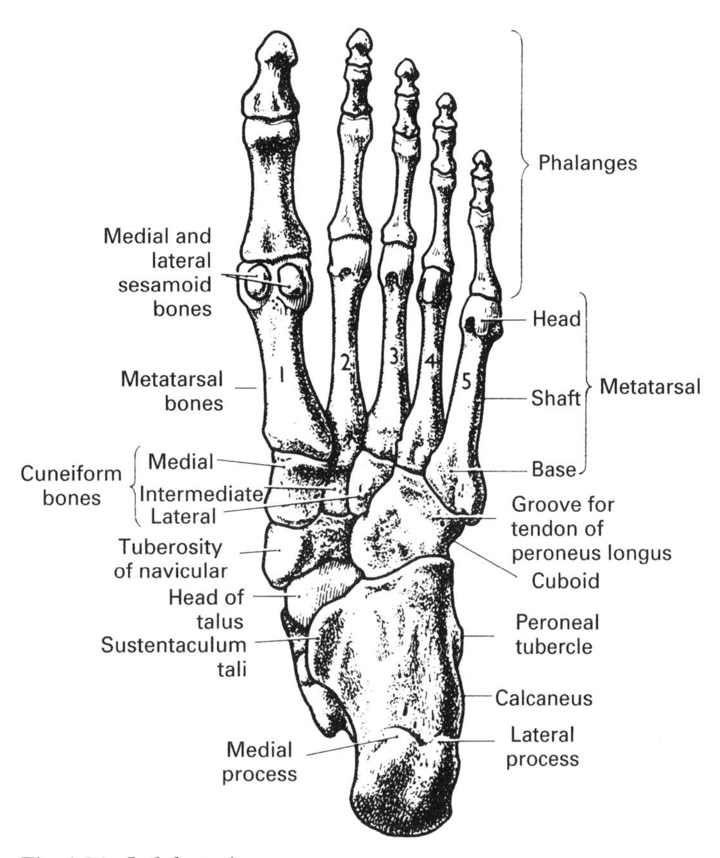

Fig. 8.35 Left foot: plantar aspect

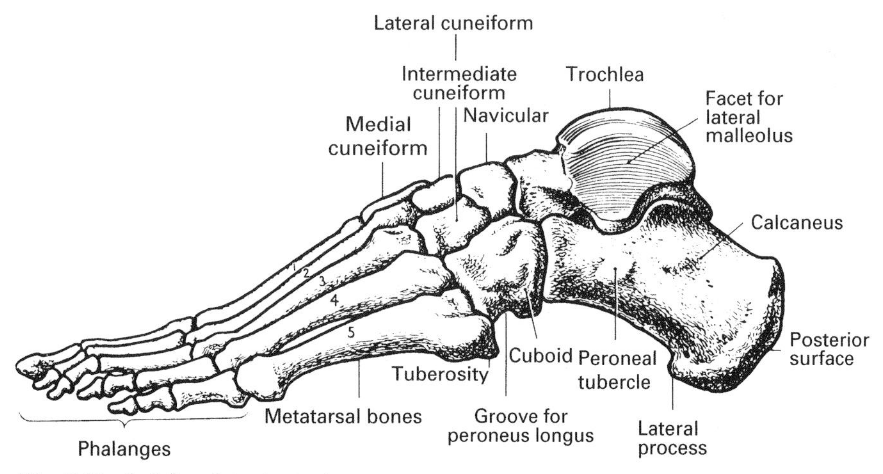

Fig. 8.36 Left foot: lateral aspect

TARSUS

There are seven bones in the tarsus. They are arranged in two rows but an additional bone is interposed between the two rows on the medial side.

The proximal row consists of two bones—the talus and the calcaneum. Unlike the carpus where the bones of the proximal row lie side by side, the talus is above the calcaneum with its long axis directed forwards, medially and downwards.

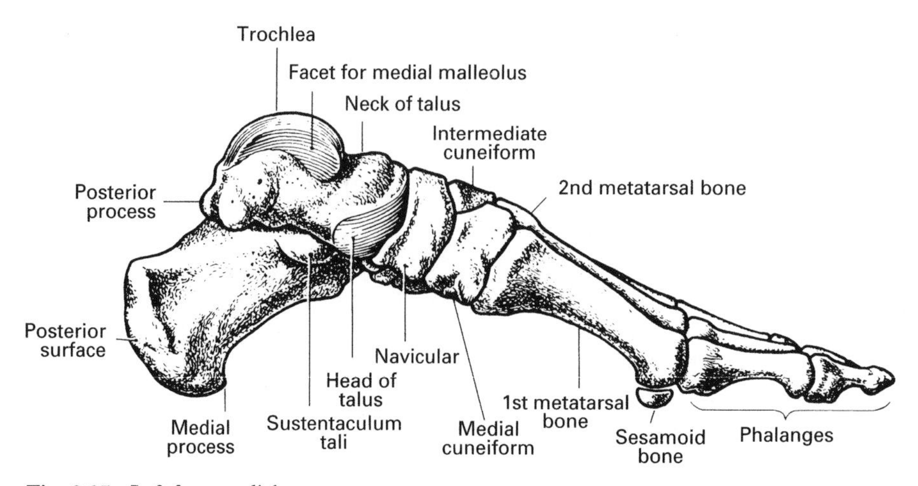

Fig. 8.37 Left foot: medial aspect

The distal row consists of the medial, intermediate and lateral cuneiforms and the cuboid. These four bones lie side by side and form part of the transverse arch of the foot (p. 241). On the medial side the navicular is interposed between the head of the talus and the cuneiform.

Talus (Figs 8.38 to 8.40)

The talus forms the connecting link between the leg and the foot. The talus consists of:

- the head
- the neck
- the body.

The head is directed forwards and slightly downwards and medially. The distal surface is convex and articulates with the navicular. The plantar surface articulates with the dorsal surface of the calcaneus by three articular facets—the anterior, middle and posterior calcanean articular surfaces. The posterior articular surface rests on the upper surface of the sustentaculum tali—a shelf-like projection on the medial side of the calcaneus. The anterior and middle articular surfaces are smaller than the posterior and are situated on the inferior surface of the head of the talus.

The neck is the short, slightly constricted part of the bone separating the head from the body. It inclines medially. On the medial part of the plantar surface is a deep groove, the sulcus tarsi. This groove forms the roof of a bony canal, the sinus tarsi, in which the interosseous talocalcanean and cervical ligaments lie.

The body is cuboidal in shape. The dorsal surface is convex and is covered by the trochlear surface which articulates with the lower end of the tibia to form the ankle joint (p. 217). The articular surface

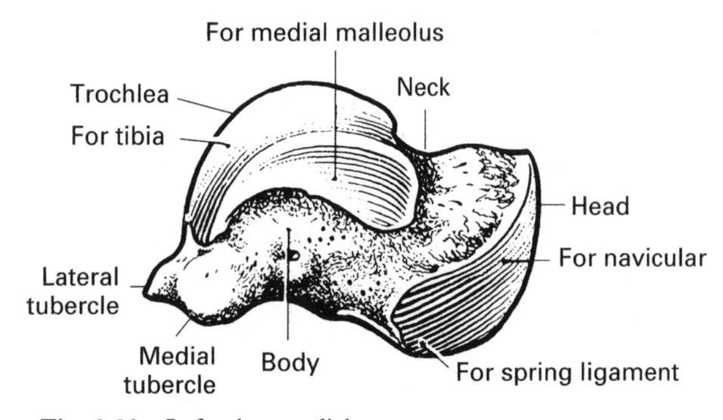

Fig. 8.38 Left talus: medial aspect

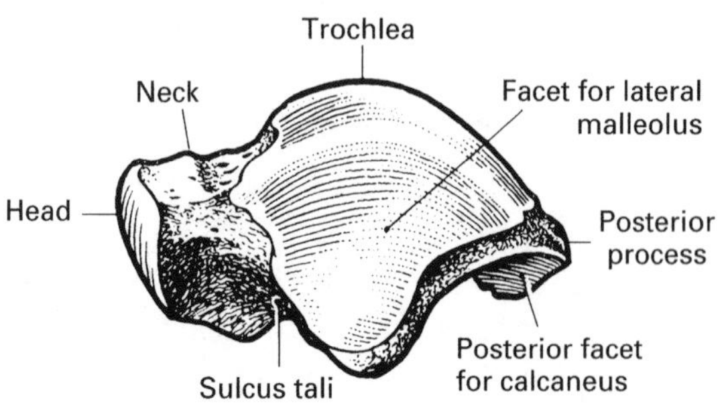

Fig. 8.39 Left talus: lateral aspect

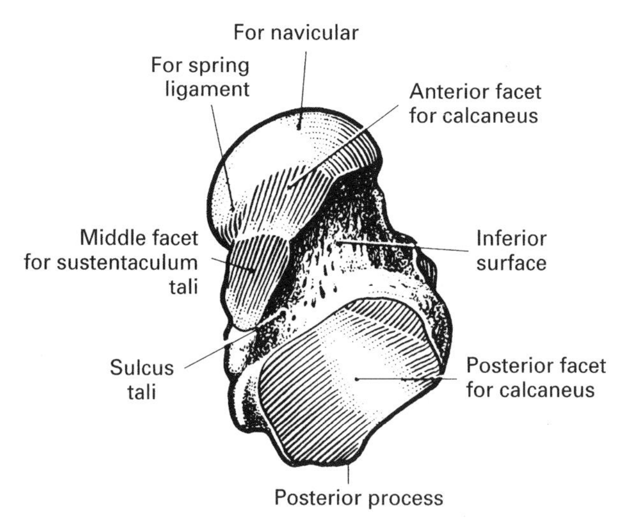

Fig. 8.40 Left talus: inferior aspect

extends on to the lateral and medial sides of the body where it articulates with the lateral malleolus of the fibula and the medial malleolus of the tibia.

Posteriorly the body narrows to a flat posterior process which is divided by a groove into a medial and a lateral tubercle. Occasionally the posterior process ossifies as a separate bone, when it is known as the os trigonum.

The talus, strongly held as it is on both sides by the malleoli, is capable of dorsiflexion and plantar flexion but lateral and medial movement is strongly resisted by the malleoli and the strong ligaments which surround the joint.

Calcaneum (Figs 8.41 to 8.43)

The calcaneum is the largest of the tarsal bones. It is roughly oblong and lies below and slightly on lateral side of the talus, with its long axis directed forwards. The greater and more massive part of the bone lies posterior to the plane of the ankle joint to form a lever for the muscles of the calf which are inserted into its posterior surface.

The posterior surface is large and convex; the tendo calcaneus (Achilles tendon) of the calf muscles is inserted into the roughened middle third of the posterior surface and is separated from the smooth upper third by a bursa.

The anterior surface is relatively small and articulates with the cuboid. The medial surface is slightly concave, but a shelf of bone—the sustentaculum tali—projects medially from the antero-superior part of this surface and supports the head of the talus. The lateral surface is flat except for the small peroneal tubercle which projects between the tendons of the peroneal muscles.

The upper surface has three articular facets towards the front of the bone, corresponding to the anterior, middle and posterior facets of the talus. The posterior facet is the largest; it is convex and articulates with the posterior facet on the under aspect of the talus, forming the talo-calcanean (subtalar) joint. The anterior and middle facets are small and are situated on the medial part of the upper surface of the bone; the middle facet lies on the upper surface of the sustentaculum tali, and a deep groove—the sulcus calcanei—separates it from the posterior facet. When the talus and the calcaneum are articulated, the sulcus calcanei and the corresponding sulcus tali form an oblique tunnel—the sinus tarsi—in which lies a strong interosseous ligament joining the bones.

The plantar surface of the calcaneum is slightly concave. At its posterior end is the calcanean tuberosity, divided by a shallow groove into a medial and a lateral process. The long plantar ligament is attached to the bone immediately in front of this tuberosity. Near the anterior end of the plantar surface is a rounded protuberance—the anterior tubercle—to which the short plantar ligament is attached.

Navicular

The navicular is roughly disc-shaped and is situated on the medial side of the foot. It lies between the proximal and distal rows of tarsal bones. The proximal surface is concave and articulates with the head of the talus. The distal surface is convex and is divided into three facets which articulate with the three cuneiform bones. A well-marked bony tuberosity projects downwards and backwards from the medial surface and receives the principal insertion of the tibialis posterior tendon.

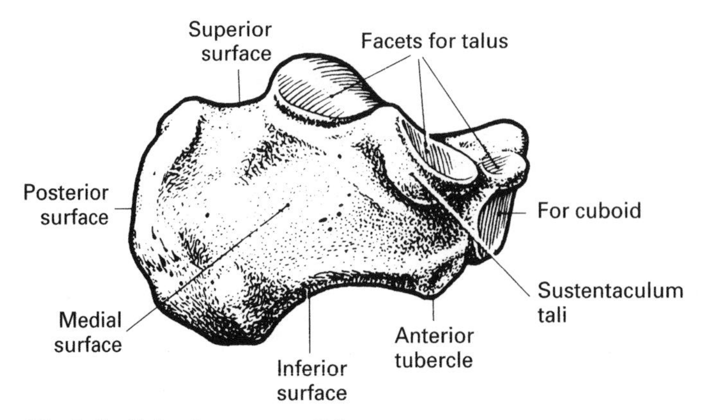

Fig. 8.41 Left calcaneum: medial aspect

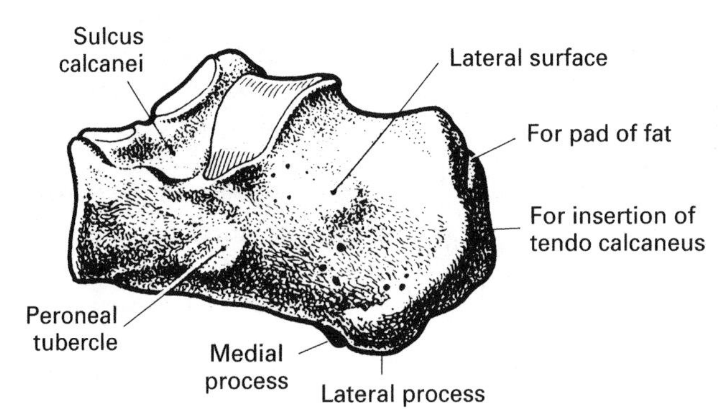

Fig. 8.42 Left calcaneum: lateral aspect

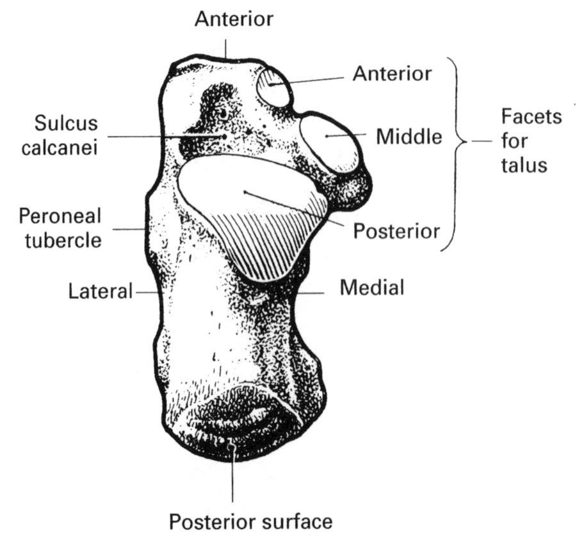

Fig. 8.43 Left calcaneum: superior aspect

Cuneiform bones

The medial, intermediate and lateral cuneiform bones lie side by side on the medial side of the distal row of the tarsus. The medial cuneiform is the largest and the intermediate the smallest. They articulate proximally with the navicular and distally with the medial three metatarsal bones. They are small, wedge-shaped and form part of the transverse arch of the foot (p. 241).

Cuboid

The cuboid lies on the lateral side of the distal row of the tarsus. It is a flattened six-sided bone which articulates proximally with the calcaneum, distally with the 4th and 5th metatarsal bones and medially with the lateral cuneiform and navicular. A deep oblique groove for the tendon of the peroneus longus muscle begins on the lateral surface of the bone and extends on to the plantar surface where it is bounded behind by a prominent oblique ridge. Frequently the tendon of the peroneus longus contains a sesamoid bone, which articulates with the lateral side of the cuboid.

METATARSUS

There are five metatarsal bones. They are miniature long bones and are similar in structure to the metacarpal bones. The *shafts*, with the exception of the first, are long and slender and are triangular in cross-section. The *bases* articulate with each other and with the distal row of the tarsus. The *heads* are rounded with slightly flattened sides. They articulate with the proximal phalanges of their own digits at the metatarsophalangeal joints (p. 239).

The first metatarsal bone is short and thick. The head is large and on its plantar aspect are two facets for articulation with two sesamoid bones. The base is large and kidney-shaped and it articulates with the medial cuneiform bone of the tarsus.

The second metatarsal bone is the longest bone of the metatarsus. Its base is wedge-shaped and extends a short distance proximal to the general line of the metatarsophalangeal joints. It therefore articulates not only with the intermediate cuneiform bone which directly opposes it, but also with the the medial and lateral cuneiform bones on either side.

The third metatarsal bone has a flat triangular base which articulates proximally with the lateral cuneiform bone.

The fourth metatarsal bone is smaller and articulates proximally with the cuboid bone and with the lateral cuneiform.

The fifth metatarsal bone articulates with the cuboid. On the lateral margin of the base there is a tuberosity—the styloid process.

PHALANGES

The phalanges of the foot resemble those of the hand (p. 152) in number and arrangement. There are two phalanges in the first (or big) toe and three in the other toes. The phalanges of the first toe are relatively large and strong but those of the other toes are small and short. The middle phalanges are very short. The terminal phalanges are similar to those of the fingers but are smaller and flattened, with a roughened tuberosity on the plantar aspect of the distal end for attachment of the pulp of the tip of the toe which provides a wide area to take pressure. The terminal phalanx of the first toe usually shows some degree of valgus deviation, i.e. lateral angulation.

Radiographic appearances of the foot (Figs 8.44 to 8.47)

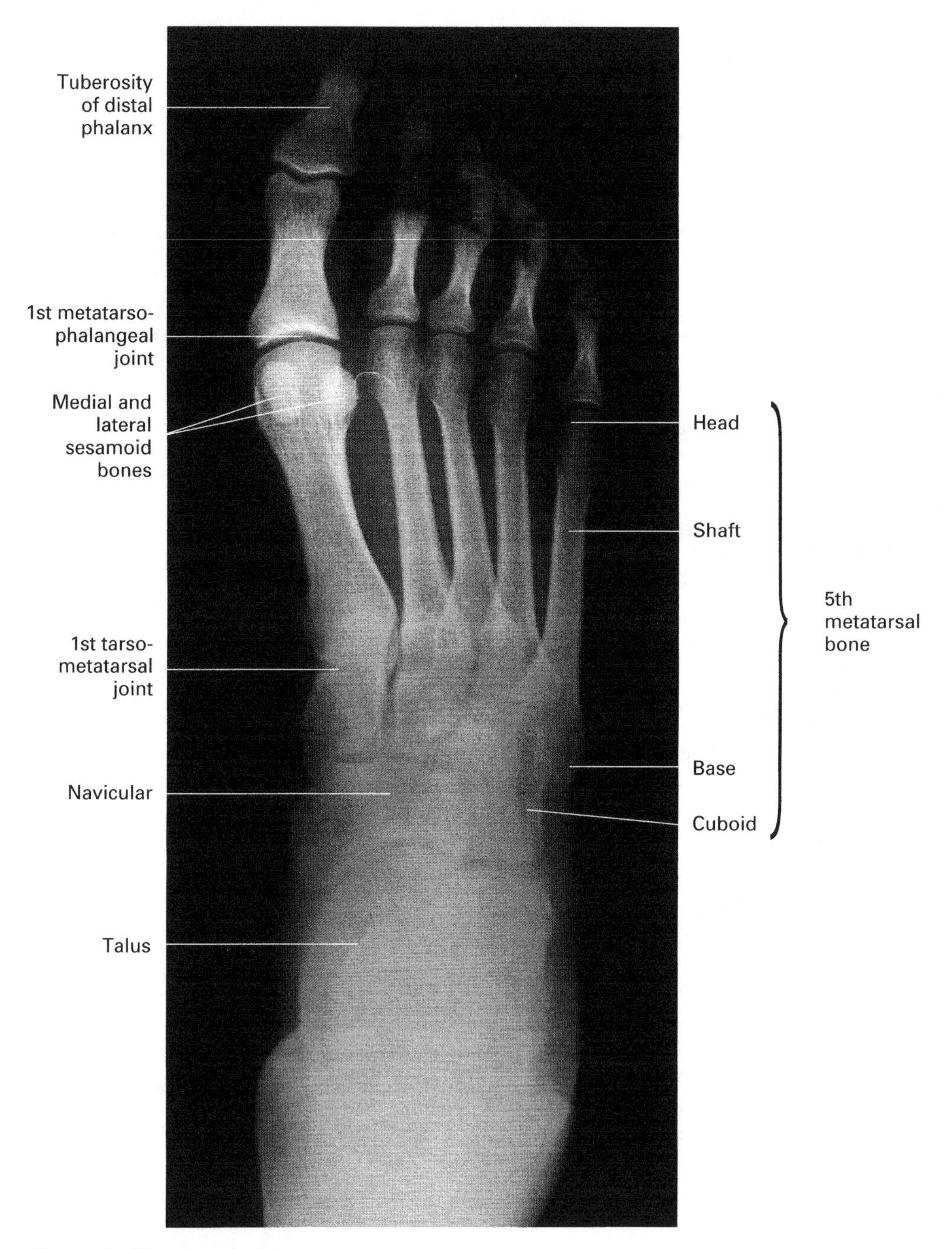

Fig. 8.44 Right foot: dorsiplantar view

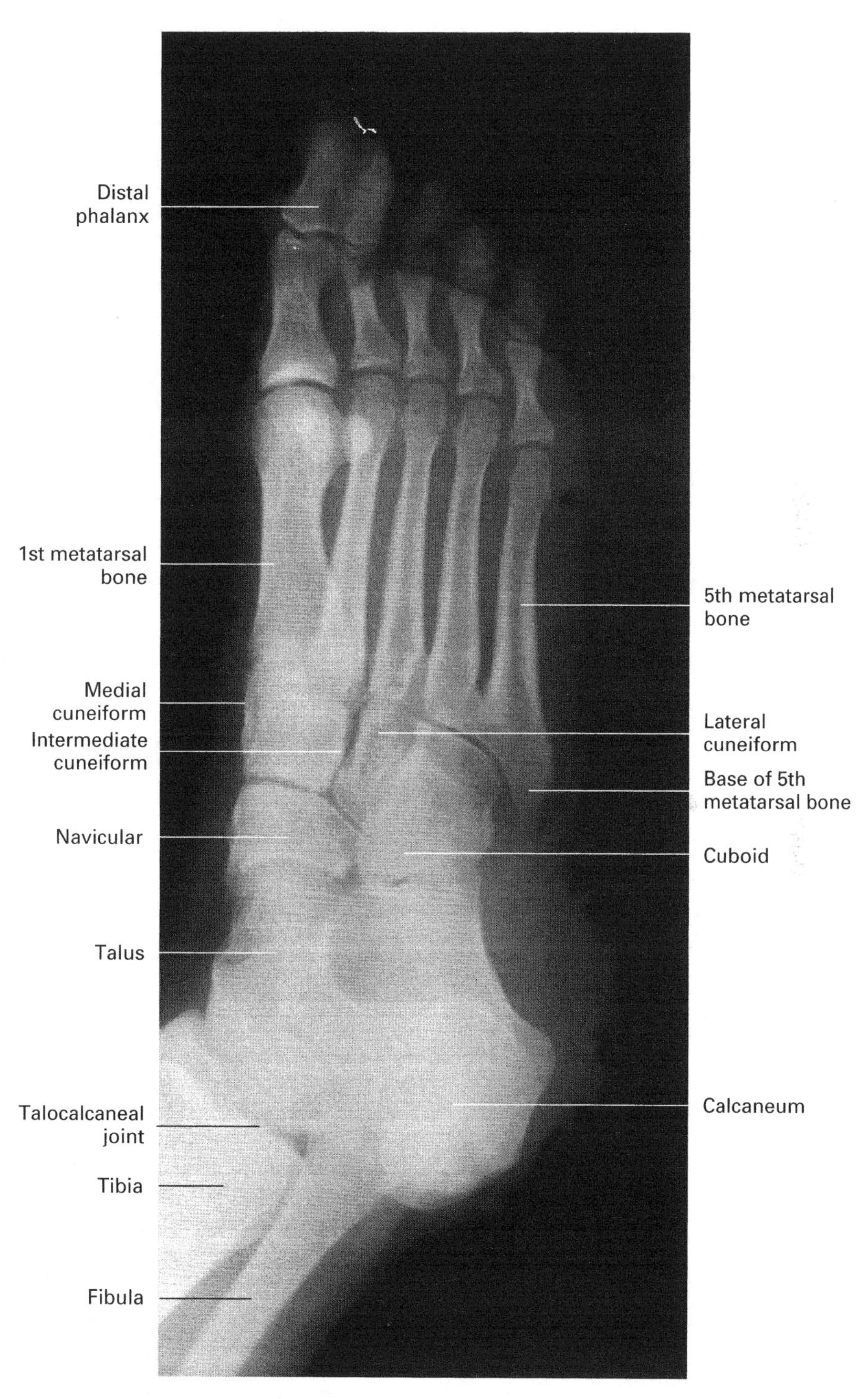

Fig. 8.45 Right foot: dorsiplantar oblique view

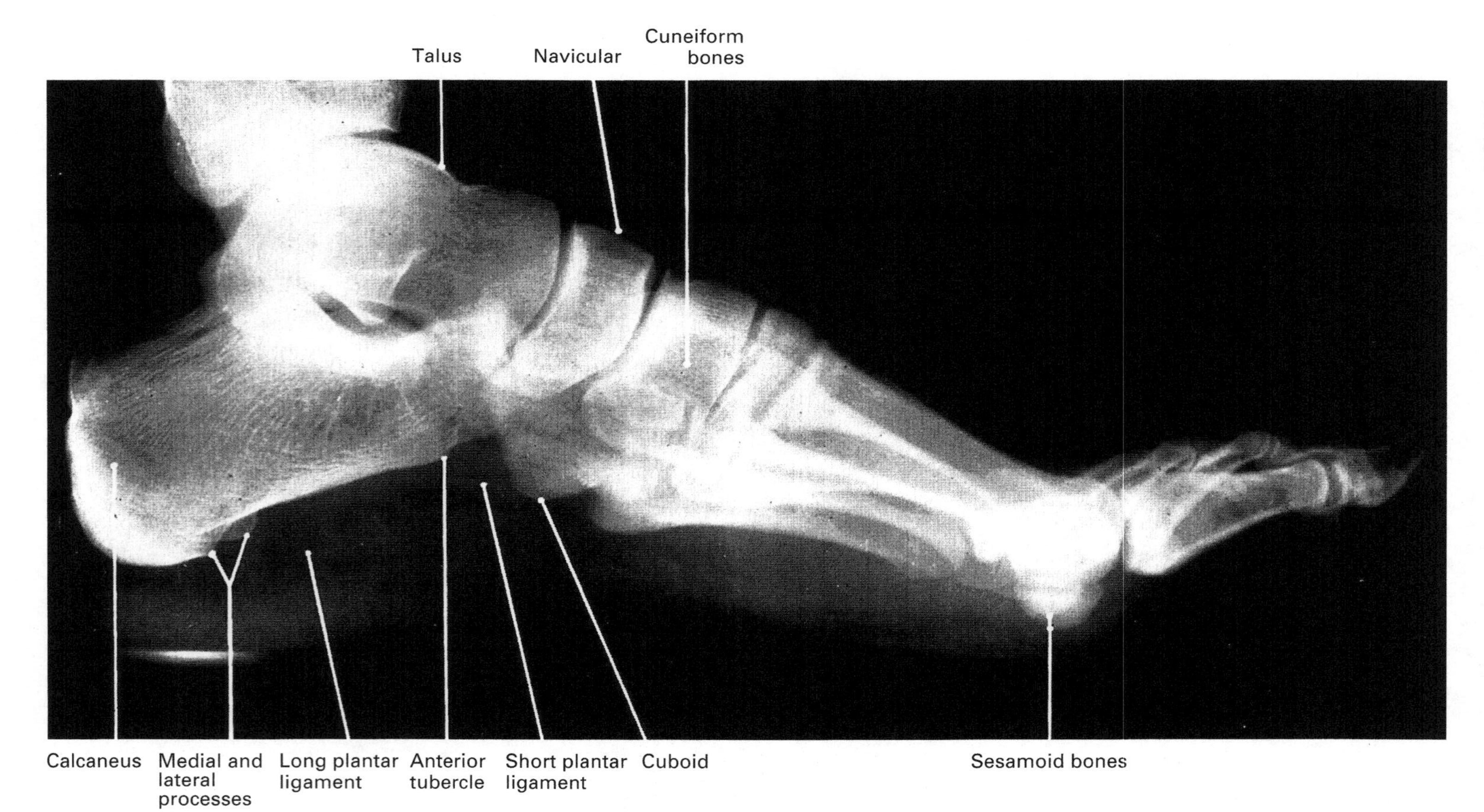

Fig. 8.46 Left foot: lateral view

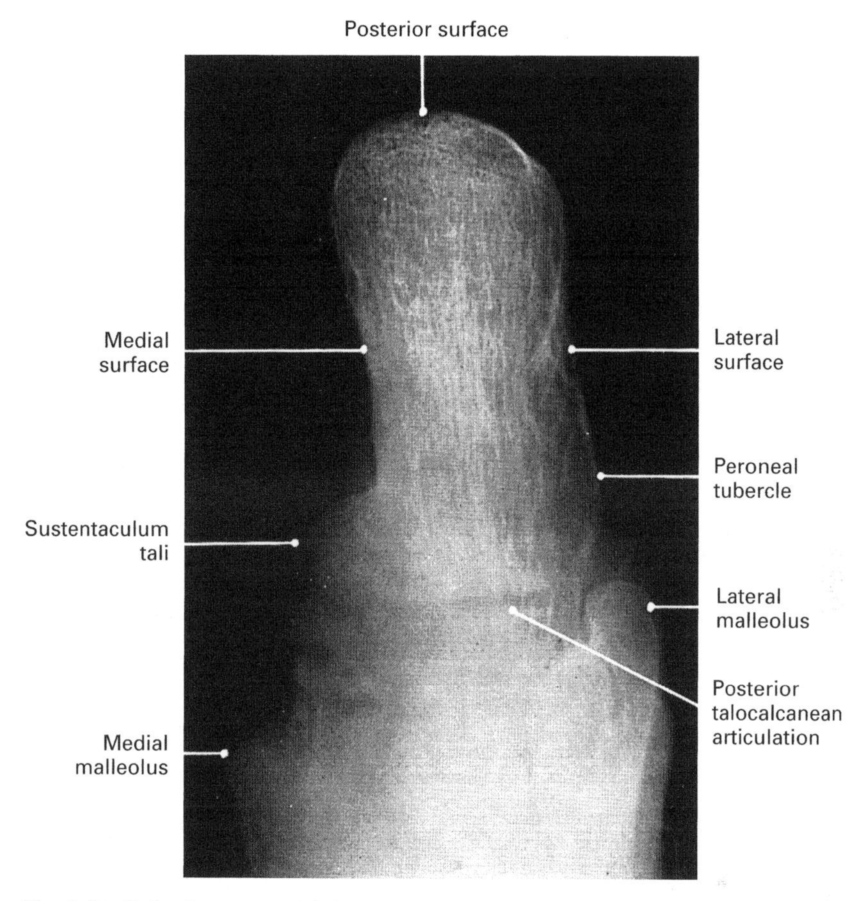

Fig. 8.47 Left calcaneum: axial view

OSSIFICATION OF THE BONES OF THE FOOT (Figs 8.48 to 8.51)

The general pattern of ossification in the foot is similar to that in the hand.

Primary centres

Tarsus—Each tarsal bone develops from one primary centre. (It should be noted that no ossification centres are present in the carpus at birth).

Calcaneum—appears about week 26 of intrauterine life
Talus—about week 28 of intrauterine life
Cuboid—at term, or just after
Navicular—4th year
Lateral cuneiform—1st year
Medial cuneiform—3rd year
Intermediate cuneiform—4th year

Metatarsus—Each metatarsal bone develops from one primary centre which appears in the shaft between the weeks 8–16 of intrauterine life.

Phalanges—Each phalanx develops from one primary centre which appears in the shaft between the weeks 8–16 of intrauterine life.

Secondary centres

Tarsus—Each tarsal bone ossifies from one primary centre only but a thin scale-like secondary centre appears for the extreme posterior part of the calcaneum. Occasionally the posterior process of the talus ossifies from a small additional centre. This centre may fail to unite with the main body of the talus and persist as a separate ossicle (os trigonum). The tuberosity of the navicular may do likewise.

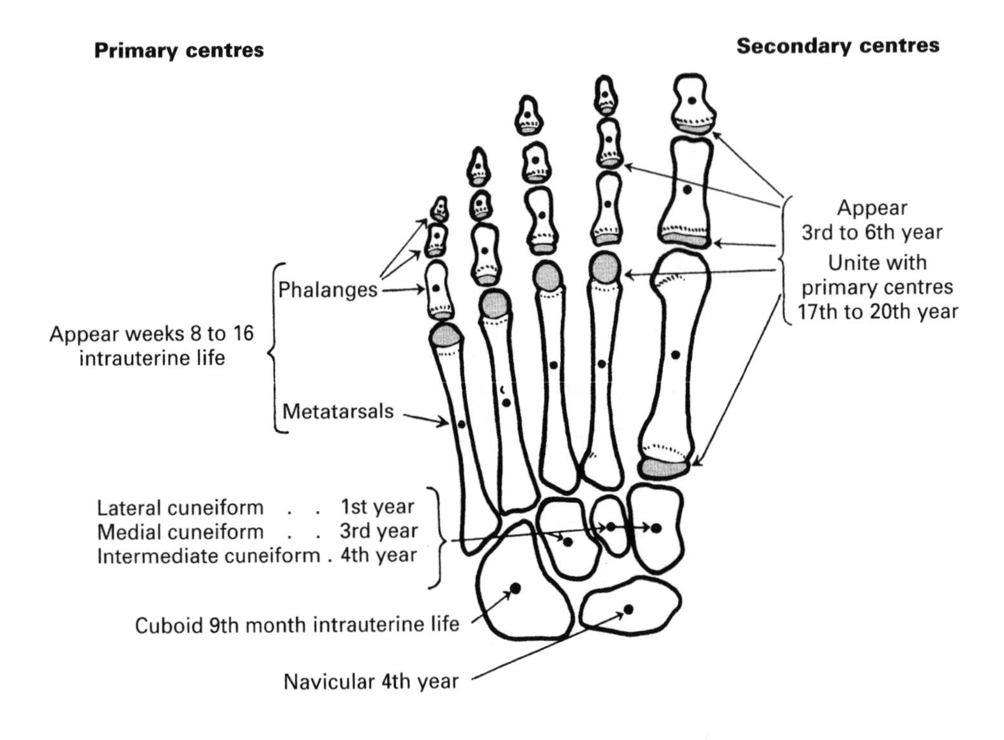

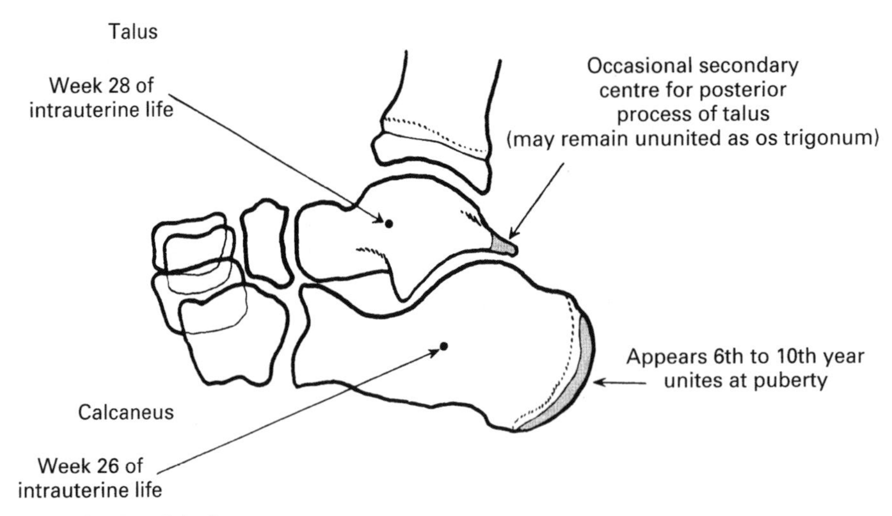

Fig. 8.48 Ossification of the foot

Metatarsus—Each metatarsal has one secondary centre. As in the hand, the secondary centre for the 1st metatarsal bone is present at the base of the bone whereas for the 2nd to 5th metatarsals it is at the head. Occasionally a small epiphysis is present for the tuberosity of the 5th metatarsal bone.

Phalanges—The secondary centres for the phalanges develop at the base of each bone between the 3rd and 6th years.

Fusion of the epiphyses and shafts of the metatarsal bones and phalanges takes place between the 17th and 20th years.

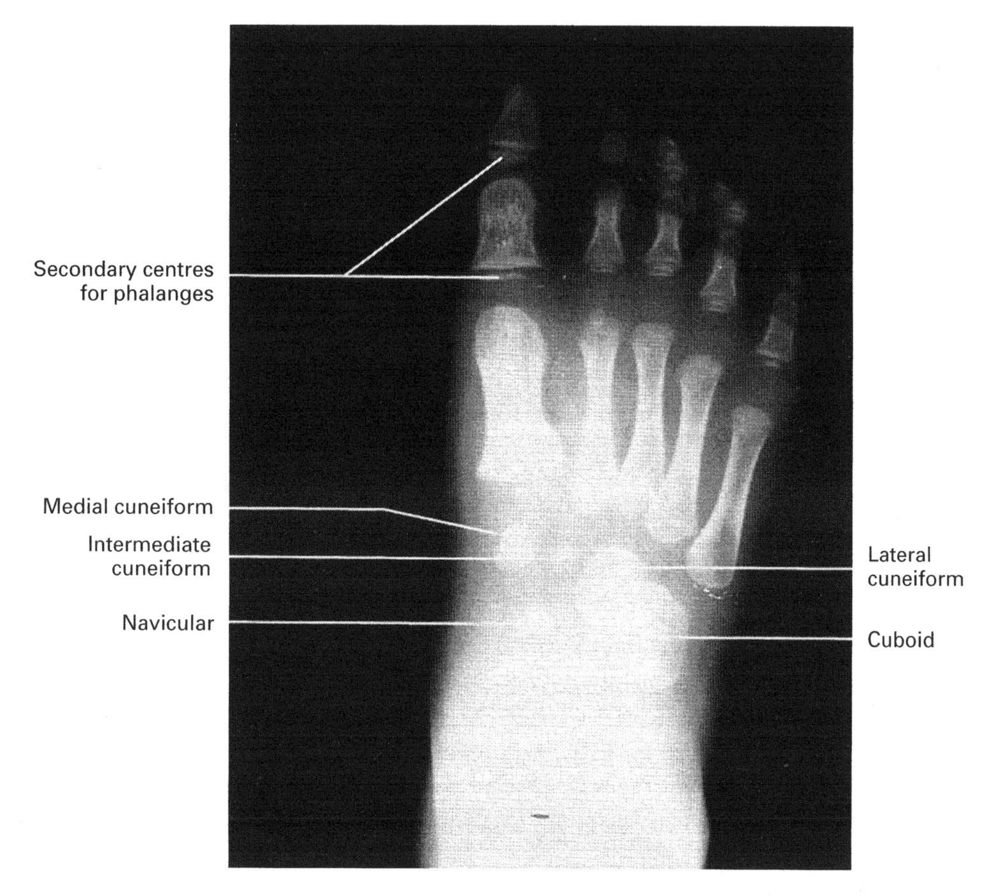

Fig. 8.49 Ossification of the foot: $2\frac{1}{2}$ years

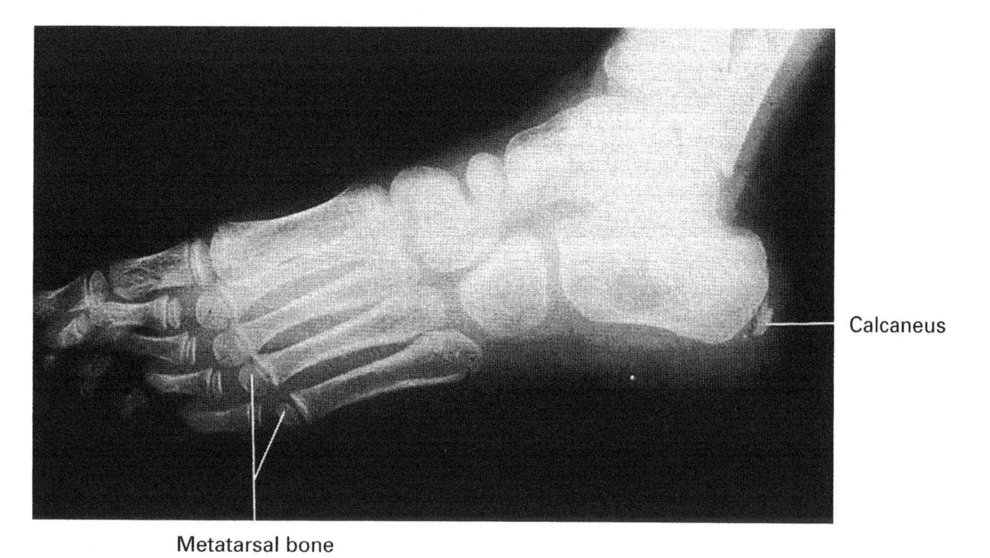

Fig. 8.50 Ossification of the foot: 6 years

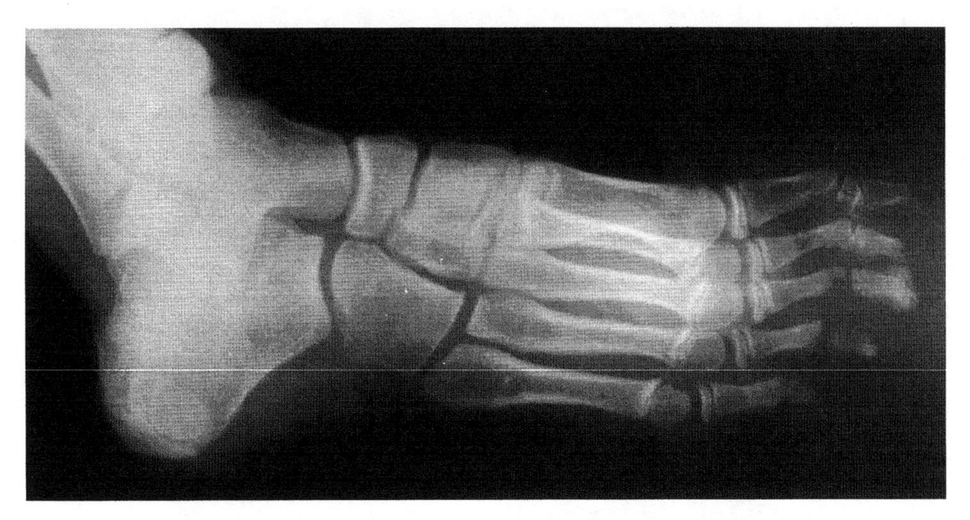

Fig. 8.51 Ossification of the foot: 12 years

SUBTALAR JOINT (Figs 8.52 and 8.53)

The subtalar joint is made up of two articulations—posterior and anterior. The anterior articulation lies within the talocalcaneonavicular joint.

Posterior. Talocalcanean joint

Type: Synovial, plane.

Articular surfaces: Concave posteroinferior facet on talus and convex facet on upper surface of calcaneum.

Ligaments:

Capsular—fibrous, lined with synovial membrane, enveloping the joint. There is no communication between this joint and the other tarsal joints.

Talocalcaneal—there are three parts:

(a) lateral—attached between lateral process of talus and lateral surface of calcaneum, anterior to calcaneofibular ligament.

(b) medial—attached to medial tubercle of talus, sustentaculum tali and medial surface of calcaneum. Fuses with deltoid ligament.

(c) interosseous—attached between talar sulcus and calcaneal sulcus, across tarsal sinus. It is the major union between these bones. It prevents excessive eversion (pronation).

Cervical—attached to upper surface of calcaneum and to tubercle on inferior part of neck of talus. It prevents excessive inversion of the foot (supination).

Anterior. Talocalcaneonavicular joint

Type: Synovial, ball and socket.

Articular surfaces: Head of the talus articulates with the proximal navicular surface, middle and anterior facets on upper surface of calcaneum and upper surface of the 'spring' (plantar calcaneonavicular) ligament.

Ligaments:

Capsular—fibrous, lined with synovial membrane.

Dorsal talonavicular—joins neck of talus with dorsal surface of navicular.

Plantar calcaneonavicular—'spring ligament'—strong, passes between sustentaculum tali to inferior surface of navicular, supporting head of talus by so doing. Head of talus articulates with part of this ligament.

Lateral calcaneonavicular—(this is the medial part of the bifurcate ligament. For lateral part, see p. 238)—joins calcaneum to lateral surface of navicular.

Movements and muscles: Both joints together.
Inversion—tibialis anterior, tibialis posterior
Eversion—peroneus longus, peroneus brevis

Tibialis anterior: see under 'Ankle Joint', p. 220.
Tibialis posterior:
origin—posterior tibial and fibular shafts, interosseous membrane.
insertion—tuberosity of navicular, sustentaculum tali, cuneiform bones, cuboid and bases of metatarsals 2, 3 and 4 (i.e. all bones except calcaneum).
function—plantar flexion of ankle, adduction and inversion of foot. Supports spring ligament and longitudinal arch of foot.
nerve supply—tibial nerve.
Peroneus longus:
origin—head and shaft of fibula
insertion—medial cuneiform inferiorly, base of 1st metatarsal lateral side.
function—plantar flexion of ankle, eversion of foot. Supports transverse arch of foot.

Peroneus brevis:
origin—lateral fibula, mid shaft.
insertion—base of 5th metatarsal (styloid process), dorsal surface.
function—plantar flexion of ankle, eversion of foot.

Radiographic appearances of the subtalar articulations (Figs 8.52 to 8.53)

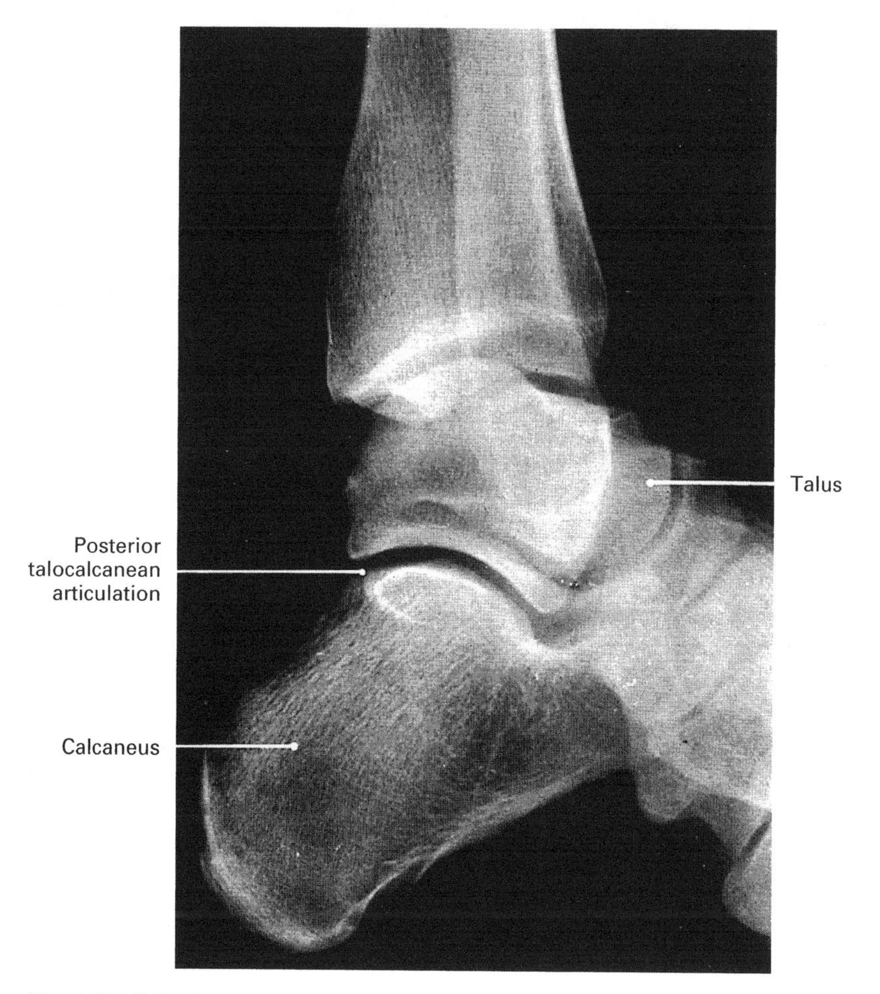

Fig. 8.52 Left talocalcanean joint: medial oblique view

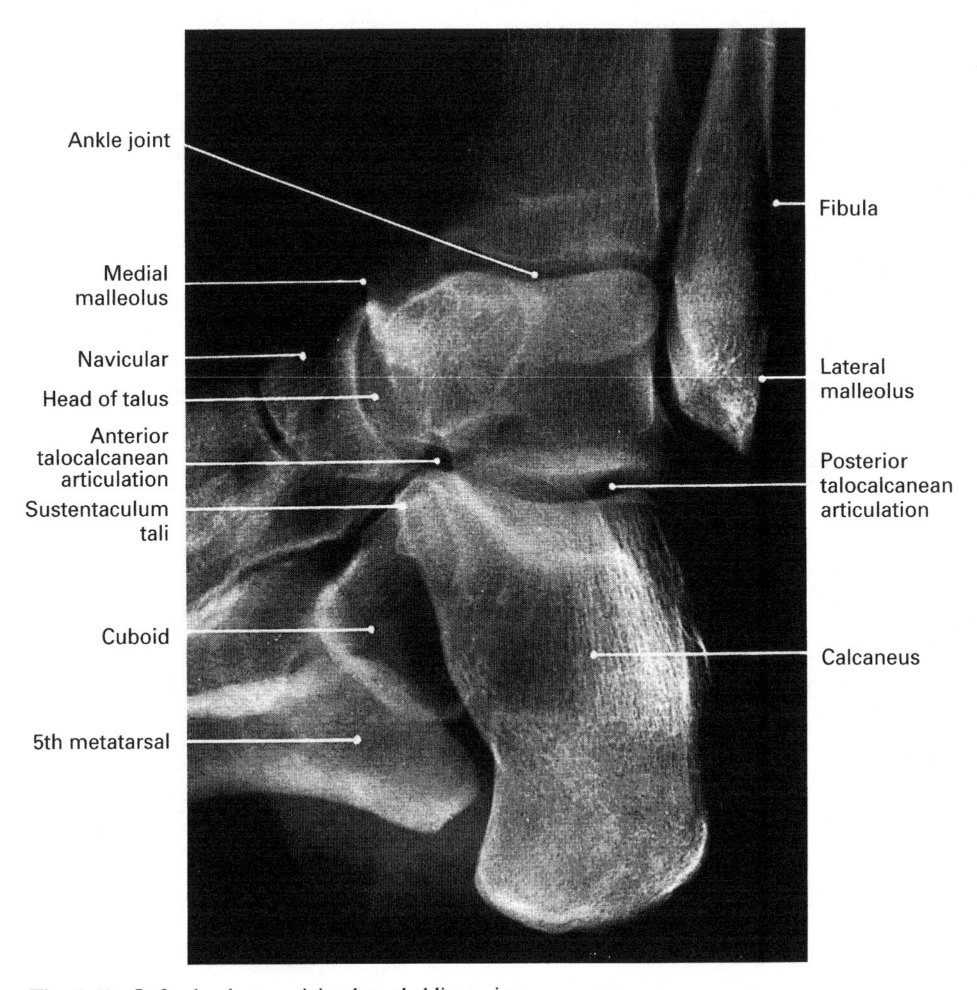

Fig. 8.53 Left talocalcanean joint: lateral oblique view

SYNOVIAL JOINTS OF THE FOOT

Separate joint cavities exist between the tarsal bones, except for the cuboidonavicular joint which is often a syndesmosis. This is in contrast with the carpus where several of the joints communicate with each other.

Calcaneocuboid joint

Type: Synovial, saddle.

Articular surfaces: Distal surface of calcaneum and proximal surface of cuboid.

Ligaments:

Capsular—attached to articular margins.

Calcaneocuboid (this is the lateral part of the bifurcate ligament. For medial part, see p. 236)—attached between medial surface of cuboid and dorsal surface of calcaneum.

Dorsal calcaneocuboid—joins dorsal surfaces of calcaneum and cuboid.

Short plantar—joins plantar surfaces of calcaneum and cuboid.

Long plantar—superficial to short ligament. Attached between posterior plantar surface of calcaneum, the cuboid and bases of metatarsals 2–4. Acts as a fibrous origin for short plantar muscles of foot and maintains longitudinal arch.

Movements: The calcaneocuboid and the talonavicular joints make up the transverse tarsal joint where the movements of inversion and eversion occur.

Intercuneiform and cuneocuboid joints

There are synovial plane joints between the cuneiform bones and between the lateral cuneiform bone and the cuboid.

Tarsometatarsal joints

Type: Synovial, plane.

Articular surfaces: Bases of the metatarsal bones with the three cuneiforms and the cuboid.

Ligaments:
Capsular—surrounds the articular surfaces. Capsule of the 1st joint is separate from the others. The capsule of the 2nd and 3rd joints is separated from that of the 4th and 5th joints.
Dorsal } these join the cuneiforms and cuboid
Plantar } to the metatarsal bones.
Interosseous cuneometatarsal—three, join the cuneiforms to the 2nd, 3rd and 4th metatarsal bones.

Movements: Slight gliding. Inversion and eversion in walking. (See also movement of subtalar joint, p. 237.)

Intermetatarsal joints

The bases of the second to fifth metatarsal bones form synovial joints with each other. The 1st metatarsal bone does not articulate with the 2nd. The bones are joined by interosseous, dorsal and plantar ligaments. The deep transverse metatarsal ligament, which joins the metatarsal heads, closely resembles the deep transverse metacarpal ligament (p. 166) except that all five metatarsal bones are joined.

Metatarsophalangeal joints

These are almost identical to the metacarpophalangeal joints of the hand (p. 166).

Type: Synovial, ellipsoid.

Articular surfaces: Convex heads of metatarsal bones with concave bases of proximal phalanges.

Ligaments:
Capsular—fibrous, attached to articular margins.
Plantar—flattened fibrous plate on plantar side of joints.
Collateral—strong, attached on each side of metatarsal heads to base of proximal phalanx.
Deep transverse metatarsal—wide bands which join the 5 metatarsal heads through their plantar ligaments. (It should be noted that the corresponding ligament in the hand excludes the thumb.)

Movements and muscles:
Flexion (less than in the hand)—flexores digitorum longus and brevis, lumbricals, interossei, flexores hallucis longus and brevis, flexor digiti minimi.
Extension (more than in the hand)—extensores digitorum longus and brevis, extensor hallucis longus.
Abduction (movement away from 2nd toe)—dorsal interossei, abductor hallucis (big toe), abductor digiti minimi (little toe).
Adduction (movement towards 2nd toe)—plantar interossei, adductor hallucis.

Flexor hallucis longus:
origin—posterior fibula and interosseous membrane.
insertion—base of distal phalanx of big toe.
function—flexion of toe, plantar flexion and inversion.
nerve supply—tibial nerve.
Flexor digitorum longus:
origin—posterior tibia, below soleal line.
insertion—base of distal phalanx of lateral 4 toes.
function—flexion of toes, plantar flexion, inversion.
nerve supply—tibial nerve.
Extensores digitorum brevis and longus: see below and p. 220.
Dorsal interossei: see below.

Interphalangeal joints

Type: Synovial, hinge.

Articular surfaces: Head of phalanx with base of next phalanx.

Ligaments:

Capsular: attached at articular margins.

Collateral: passes between corresponding heads and bases of articulating phalanges.

Movements and muscles:

Flexion—flexores digitorum longus, brevis and accessorius, flexor hallucis longus.

Extension—extensores digitorum longus and brevis, extensor hallucis longus.

MUSCLES OF THE FOOT

Dorsal surface

Extensor digitorum brevis (produces normal 'fullness' on lateral dorsum of foot).

origin—lateral and dorsal surfaces of calcaneum.

insertion—lateral side of long extensor tendons of 2nd, 3rd and 4th toes. (A separate tendon, extensor hallucis brevis, passes to base of proximal phalanx of big toe.)

function—extension of toes, except little toe.

nerve supply—deep peroneal nerve.

Dorsal interossei (four in number).

origin—between adjacent metatarsal bones.

insertion—bases of proximal phalanges. Interossei 1 and 2 insert on each side of proximal phalanx of 2nd toe; 3 and 4 on lateral side only of proximal phalanx of 3rd and 4th toes.

function—abduction of toes, except little toe which has its own abductor (digiti minimi) on the plantar surface.

nerve supply—lateral plantar nerve.

Plantar surface

Beneath the superficial fascia lies the plantar aponeurosis—a dense fibrous band which passes between the medial tubercle of the calcaneum and, dividing into five slips, joins the deep transverse metatarsal ligaments. Some fibres pass to the skin. The plantar aponeurosis helps maintain the longitudinal arch of the foot.

There are four layers, from superficial to deep.

Layer 1

- Abductor hallucis
- Flexor digitorum brevis
- Abductor digiti minimi

Layer 2

- Flexor hallucis longus (tendon)
- Flexor digitorum longus (tendon)
- Flexor accessorius
- Lumbricals 1–4.

Layer 3

- Flexor hallucis brevis
- Adductor hallucis
- Flexor digiti minimi brevis.

Layer 4

- Plantar interossei (three)
- Tibialis posterior (tendon)
- Peroneus longus (tendon).

Tendons of muscles of proximal origin pass through these layers. The muscles of layer 1 and flexor accessorius of layer 2 arise from the medial side of the tuberosity of the calcaneum. The lumbricals arise from the long flexor tendons, as in the hand.

In layer 3, adductor hallucis and flexor digiti minimi brevis arise from the peroneus longus tendon of the next layer down (i.e. layer 4). Flexor hallucis brevis arises from the lateral cuneiform and the cuboid.

In layer 4, the three interossei are adductors and they arise between the 3rd, 4th and 5th metatarsals. The big toe has its own adductor. The 2nd toe uses the dorsal interossei for abduction and adduction.

The nerve supply is from the medial and lateral plantar nerves—the terminal divisions of the posterior tibial nerve.

ARCHES OF THE FOOT

The bones of the foot form sprung arches which transfer the weight of the body to the ground without shock. This is achieved by:

- the shape of the bones
- the connecting ligaments

- the small muscles of the sole
- the sling-like action of some of the long tendons of the leg muscles.

These factors convert the foot into a resilient lever which propels the body in walking, running and jumping.

There are two arches—longitudinal and transverse. The longitudinal arch is composed of medial and lateral parts.

The medial longitudinal arch is formed by the calcaneum, talus, navicular, the medial, intermediate and lateral cuneiform bones and the medial three metatarsal bones.

The lateral longitudinal arch is formed by the calcaneum, cuboid, and the lateral two metatarsal bones.

The longitudinal arch is highest on the medial side of the foot where the talar head is the apex. The arch is maintained by the 'spring' (plantar calcaneonavicular) ligament, the plantar ligament, interosseous ligaments and the tendons of tibialis posterior and peroneus longus.

The transverse arch is formed by the bases of the metatarsal bones. It is maintained by the interosseous ligaments and the peroneus longus muscle.

Recommended further reading

Bull S 1985 Skeletal anatomy. Butterworth

Ellis H, Logan B, Dixon A 1991 Human cross-sectional anatomy. Butterworth-Heinemann

1973 Gray's Anatomy (35th edn) Longman

Middleton W D, Lawson T L (1989) Anatomy and MRI of the joints: a multiplanar atlas. Raven Press, New York

Pomeranz S J 1992 MRI total body atlas. MRI–EFI Publications

Ryan S, McNicholas M 1994 Anatomy for diagnostic imaging. Saunders

Index